Foundations of Biophilosophy

Springer
Berlin
Heidelberg
New York
Barcelona
Budapest
Hong Kong
London
Milan
Paris
Santa Clara
Singapore
Tokyo

Martin Mahner   Mario Bunge

Foundations of Biophilosophy

With 12 Figures

Springer

Martin Mahner
Institute of Zoology
Free University of Berlin
Königin-Luise-Strasse 1–3
D-14195 Berlin, Germany

Mario Bunge
Foundations & Philosophy of Science Unit
McGill University
3479 Peel Street
Montreal, QC
Canada H3A 1W7

Cataloging-in-Publication Data applied for

Die Deutsche Bibliothek - CIP-Einheitsaufnahme

Mahner, Martin:
Foundations of biophilosophy / M. Mahner ; M. Bunge. - Berlin ; Heidelberg ; New York ; Barcelona ; Budapest ; Hong Kong ; London ; Milan ; Paris ; Santa Clara ; Singapore ; Tokyo : Springer, 1997
ISBN 3-540-61838-4 Pp.

ISBN 3-540-61838-4 Springer-Verlag Berlin Heidelberg New York

Printed in Germany

Cover design: *design & production* GmbH, Heidelberg

Typesetting: Camera-ready by the authors

SPIN 10550057 31/3137 – 5 4 3 2 1 0 – Printed on acid-free paper

Preface

Over the past three decades, the philosophy of biology has emerged from the shadow of the philosophy of physics to become a respectable and thriving philosophical subdiscipline. While many areas in the philosophy of biology are still rife with controversy—which should come as no surprise in a philosophical discipline—one can clearly observe the formation of majority views in others. Sterelny (1995) even speaks of an emerging consensus with regard to certain topics, such as the species problem.

Obviously, though, where there is a majority or mainstream view, there are minority views. This book is likely to count as a specimen of the latter. For example, we shall defend the thesis that the concept of evolution only makes sense if species are immutable kinds rather than mutable individuals. In other words, we shall argue that species do not evolve, and that the most interesting, though not the sole, unit of evolution is the organism, not the population, let alone the species. (No promise for creationists here: our position is strictly naturalistic.) Furthermore, we shall argue that there is no such thing as a genuine teleological explanation in biology, and that there is no such thing as a genetic program steering development. We shall also reject both the neopositivist and the semantic views of scientific theories, and we shall reject the strong Artificial Life program, as well as the claims that thermodynamics and information theory are useful in evolutionary biology.

Some of these ideas may sound not only provocative but plainly wrong. After all, it is not uncommon to hear that those who disagree with a certain view just do not understand it: "You are an essentialist and just don't understand population thinking", or so the rejection schema goes. (Of course, in some cases such reproach is valid.) Therefore, if you choose to deviate from orthodoxy, you had better try hard and make a good case for your heresies. First, you should not be merely destructive but, whenever possible, constructive as well. In other words, you should not just criticize the prevailing view, but try to propose alternatives. Second, your alternatives should be viable, that is, consistent with solid biological knowledge. Third, your alternatives should be not stray but formulated in a systematic way. Last, but not least, it will be most helpful if you have a fully-fledged philosophical system at your disposal on which you can firmly anchor your philosophy of biology. In short, your views should cohere to form a systematic and comprehensive whole, enabling you to take a fresh look at both biology and its philosophy. This, then, is the ambitious goal of this book: to outline the foundations of a new philosophy of biology of interest to biologists as well as philosophers.

In order to make such a systematic view possible, as well as to disclose our *fundamentum argumentationis*, we had to write the rather extensive first part of the book dealing with general ontology, semantics, and epistemology. This part can be used as a general introduction to the philosophy of science. It is based on the senior author's philosophical œuvre, the content of which has been modified whenever necessary to better accommodate the specific and varied realm of the biological sciences. Although most of Part I is indispensable to follow the arguments in Part II, we shall not make explicit use of all the concepts introduced in Part I. Rather, some notions are mentioned or elucidated in order to provide a coherent picture of our philosophy. The point is that even those who disagree with our philosophy should be able to admit that it is at least a coherent view that can be discussed rationally in the light of contemporary biology. Other notions are explicated to justify why we do not (or cannot) make (as much) use of them as some fellow philosophers may expect. For example, the notion of probability is, in our opinion, often misused in the philosophy of science, hence in biophilosophy, because it involves illegitimate interpretations of the probability calculus. In particular, we have no use for the notions of probabilistic causality and probabilistic explanation. Nor shall we engage in any of the Bayesian probability games fashionable among many philosophers—though not among mathematicians or scientists.

In Part II, the philosophical system introduced in Part I is applied to some topical problems in the philosophy of biology. Needless to say, the issues dealt with are selective. The reasons for our selection are threefold. First, we have chosen issues, such as the species problem and the question of the unit(s) of evolution, about which our views are fundamentally different from those of most of our colleagues. Second, we have picked subjects, such as the problem of the nature of life and developmental biology, that have been somewhat neglected in the philosophy of biology. At the same time, we have largely disregarded more fashionable topics, such as population genetics, molecular biology, and sociobiology. However, it should be easy to extract our stance on these issues. Third, we have chosen some issues, such as the unit of selection question, that needed only some minor modifications rather than fundamental repairs. (The latter shows, incidentally, that not all of our views deviate from the mainstream.) Finally, personal interest can never be ruled out as a source of selectivity: research is always personal even if its results are not.

When we began writing this book we soon realized that there was no progress to be made below a certain level of analysis and without a (moderate) use of some simple formal tools such as elementary mathematical logic and naive set theory. However, the reader unfamiliar with such formal tools should still be able to follow the general course of argumentation. Moreover, in order to prevent this book from growing inordinately, we decided to presuppose an elementary knowledge of biological facts and concepts. For example, we will not explain what a zygote or an amniote is, for any good biological dictionary will do the job. Finally, we also presuppose some basic knowledge of the main topics in the philosophy of biology. For example, it will be helpful to have a minimal idea about what the problems concerning functional explanation, the units of selection, and biological species are. (For a quick and accessible review of the major topics in the philosophy of biology see Ruse 1988.) In sum, though possibly digestible for the industrious beginner, this book addresses the advanced reader, such as the philosopher with basic biological knowledge as well as the biologist with philosophical interests. It can therefore be used for senior level and graduate courses in the subject, as well as for independent study.

Since we have attempted to build a *systematic* biophilosophy, the chapters of this book must be read in the given sequence, even though one might not be interested in certain issues. (The less formally inclined reader may perhaps skip the chapter on semantics.) For one, before tackling Part II, it is necessary to take note of the philosophical fundamentals outlined in Part I. For another, when proposing certain postulates and definitions, we shall resort to postulates and definitions introduced in preceding chapters and sections. In other words, we adopt a (very moderate) axiomatic format, so that the chapters cannot be fully understood if read independently from one another.

Furthermore, the personality of the authors did not quite allow them to shut down or at least disconnect their limbic systems from their neocortices while philosophizing. In other words, though trying to do our best to hide the fact that we are passionate scientists and philosophers, we may on occasion have failed to philosophize completely *sine ira et studio*. May the critical reader forgive us, then, if there remain any traces of sarcasm or polemics in a few phrases or passages. But then again, other readers might enjoy some spice here and there.

Finally, a word about the authors. Both are scientists turned philosophers. The junior author (MM) has his doctorate in zoology, and the senior author in theoretical physics. The junior-senior order we have adopted reflects the distribution of workload.

Acknowledgments

Special thanks go to our friend Michael Kary, with whom we had many stimulating discussions. In addition, he was kind enough to prepare the figures. We also thank the following colleagues for discussions and suggestions: Peter Ax (Georg-August-Universität Göttingen), Graham Bell (McGill University), Lina Bettucci (Universidad de la República, Montevideo), Richard Lewontin (Harvard University), Luis Marone (Universidad de Cuyo), Norman Platnick (American Museum of Natural History), the late Osvaldo Reig (Universidad de Buenos Aires), Rolf Sattler (McGill University), Otto Solbrig (Harvard University), and Walter Sudhaus (Freie Universität Berlin). Needless to say, this does not imply that they agree with our views, and none of them is to be held responsible for our mistakes.

We gratefully acknowledge the permission of Kluwer Academic Publishers to use several short passages from the senior author's *Treatise on Basic Philosophy* (1974-1989).

The junior author thanks the *Deutsche Forschungsgemeinschaft* (Bonn, Germany) for funding his research while working with the senior author at the Foundations & Philosophy of Science Unit of McGill University during the 1993-96 period.

Last, but not least, we would like to thank Jean von dem Bussche and Ursula Gramm of Springer-Verlag (Heidelberg) for their constructive and pleasant cooperation.

Contents

Part I Philosophical Fundamentals

Part II Fundamental Issues in Biophilosophy

Special Symbols

$\neg p$	*not-p*
$p \,\&\, q$	*p and q* (conjunction)
$p \vee q$	*p or q* (or both; disjunction)
$p \Rightarrow q$	*if p, then q* (implication)
$p \Leftrightarrow q$ (p *iff* q)	*if p, then q, and conversely* (*p if, and only if, q*)
$A \therefore B$	premises A *entail* conclusion B (B follows from A)
Px	individual x has property P
Pxt	individual x has property P at time t
$(\forall x)Px$	*For all x*: x has property P (shorter: *All* x are P)
$(\exists x)Px$	*For some x*: x has property P (*At least one* x is a P)
$\{x \mid Px\}$	the set of objects x such that x possesses property P (shorter: the set of objects possessing property P)
$=_{df}$	identical by definition
$\emptyset$	the empty set
$x \in A$	individual x *is an element* (or *member*) of set A (object x *belongs to* set A, or x *is in* A)
$x \not\in A$	individual x *does not belong to* set A
$A \subseteq B$	set A is *included in*, or is *equal* to, set B
$A \subset B$	set A is *properly included* in set B (A is included in, but *does not equal*, B)
$A \cup B$	the *union* of the sets A and B (the set of objects in A *or* in B)

$\bigcup_{i=1}^{n} A_i$	the *union* of the sets A_i, where $i = 1, 2, ..., n$ (i.e., $A_1 \cup A_2 \cup ... \cup A_n$)
$A \cap B$	the *intersection* or *overlap* of the sets A and B (the set of individuals in A *and* B)
$\bigcap_{i=1}^{n} A_i$	the *intersection* or *overlap* of the sets A_i, where $i = 1, 2, ..., n$ (i.e., $A_1 \cap A_2 \cap ... \cap A_n$)
$A - B$	the *difference* between the sets A and B (the set of objects in A but *not in* B; the set A without the set B)
$\langle a, b \rangle$	the *ordered pair* of individuals a and b
$\langle a, b, ..., n \rangle$	the *ordered set* of n elements (an n-tuple)
$A \times B$	the *Cartesian product* of the sets A and B (the set of ordered pairs $\langle a, b \rangle$, where a is in A and b in B)
2^A	the *power set* (the set of all subsets) of set A
$\overline{A}$	the *complement set* of the set A (the objects *not* in A)
Rxy	individuals x and y are R-related
$x \sim_R y$	individual x is *equivalent* to individual y in respect R
$f: A \rightarrow B$	the *function* f mapping set A into set B
$f(x)$	the *value* of function f at x
$\mathbb{N}$	the set of *natural numbers*
$\mathbb{Z}$	the set of *integers*
$\mathbb{R}$	the set of *real numbers* (the *real line*)
$\mathbb{R}^+$	the set of *positive real numbers*
$a \sqsubset b$	thing a *is part of* thing b
$a \lhd b$	system a *is a subsystem* of system b

Part I

Philosophical Fundamentals

1 Ontological Fundamentals

1.1 Metaphysics and Science

Metaphysics (or ontology or philosophical cosmology) is a traditional branch of philosophy and as such it need not be justified in the eyes of the philosopher, unless he or she is a positivist. Some scientists, however, may still be somewhat suspicious about the relevance of metaphysics to their discipline. After all, it is still popular to equate metaphysics with either religion, wild speculation, or some unintelligible discourse about Being, Nothingness, *Dasein*, deconstruction, and the like. Thus, understandably, there are still antimetaphysicians among scientists, and even the odd philosopher expresses doubts as to whether ontology can be helpful for biology at all (e.g., van der Steen 1996, p. 121). Yet the fact that some ontologies are wrong or useless does not render all metaphysics objectionable: after all, every human belief and action involves some metaphysical presuppositions. For example, most of our actions presuppose that there is, in fact, a world external to the knowing or acting subject. Thus, as has been remarked many times, and rightly so, an antimetaphysician is just one who holds primitive and unexamined metaphysical beliefs.

However, some biologists interested in the philosophical underpinnings of their discipline have long known that there obviously are ontological problems in science (e.g., Woodger 1929; Beckner 1964). After all, the biophilosophical literature abounds with papers on, for instance, the "ontology of species", the "metaphysics of evolution", and the like. Still, what exactly is metaphysics or ontology?

As every philosopher is likely to answer this question differently, we shall briefly explain what we understand under "metaphysics". Along with Peirce (1892-93), Montagu (1925), Woodger (1929), and a few others, we take metaphysics to be *general science*. In other words, ontology is the science concerned with the whole of reality: that which studies the most general features of every mode of being and becoming. It attempts to answer such general questions as: What is matter? What is a process? What is spacetime? Are there emergent properties? Does every event fit some law(s)? Are there natural kinds? What makes an object real? Are there final causes? Is chance for real?

If ontology is general science, then the specific factual sciences, or sciences of reality, are *special metaphysics* or *regional ontologies*. In our view, both science and ontology inquire into the nature of things, but, whereas science does it in detail and thus produces theories open to empirical scrutiny, ontology is extremely general and can be checked solely by its coherence with science. Consequently, there is no gap, let alone an abyss, between science and ontology. Indeed, some of the most interesting scientific problems are at the same time metaphysical. Examples: What is life? What is a species? What is mind?

Depending on the metaphysical principles the scientist takes for granted, scientific research will be guided or misguided by them. For example, whereas a materialist ontology will rule out immaterial forces, such as the entelechy and the *élan vital*, an idealist metaphysics will condone them. It behooves the historian of science to dig up the ontological postulates of science, and the philosopher of science to formulate them clearly, to justify or criticize them, and eventually to systematize them. This, then, is the task of scientific ontology in general: to dig up, cleanse, generalize, and put together into a coherent whole (system) the metaphysical ideas actually used in scientific research.

Some of the ontological problems in the philosophy of biology may be exemplified by the following quotation:

> Genes, organisms, demes, species, and monophyletic taxa form one nested hierarchical system of individuals that is concerned with the development, retention, and modification of *information* ensconced, at base, in the genome. But there is at the same time a parallel hierarchy of nested ecological individuals – proteins, organisms, populations, communities, and regional biotal systems, that reflects the *economic* organization and integration of living systems. The processes within each of these two process hierarchies, plus the interactions between the two hierarchies, seems to me to produce the events and patterns that we call evolution. (Eldredge 1985a, p. 7, italics in the original).

This quotation is a rich mine of ontological problems, an incomplete sample of which reads thus: What is an individual? Are the entities referred to actually individuals? What is a system? What is a hierarchy? What is information? Are all the systems in the economic hierarchy alive? What constitutes the integration and cohesion of a system? As the two hierarchies are first called 'hierarchies of individuals', and later 'process hierarchies', what is the difference, if any, between an individual and a process? Are processes individuals, systems, or neither? Can hierarchies interact, and if so, how?

Although talk of the metaphysics of evolution is rampant in biophilosophy, in our opinion biophilosophers have not contributed much to elucidate ontological concepts such as those listed above. One reason for this claim is that the ontological analysis of such concepts in contemporary biophilosophy hardly goes beyond the application of two criteria of individuality or reality: spatiotemporal restrictedness and spatiotemporal unrestrictedness. Since we could not remain satisfied with this level of analysis, we had to write this chapter on ontology, which borrows freely from the senior author's earlier attempt to systematize the metaphysical presuppositions of science (Bunge 1977a, 1979a, 1981a).

1.2 Thing and Construct

Although this notion is much maligned by traditional and even contemporary philosophers (e.g., van Fraassen 1980; Feyerabend 1981; Putnam 1983), we begin with a basic assumption of *ontological* realism:

POSTULATE 1.1. The world (or universe) exists on its own (i.e., whether or not there are inquirers).

This axiom is not provable, and it tells us nothing about whether the world can be known and, if so, to what extent. (To affirm that it can, at least in part, is a thesis of *epistemological* realism.) It neither affirms nor denies that at least part of the world may be influenced or changed by inquirers. Yet the postulate implies a rejection of ontological constructivism, according to which the world is created or produced entirely by inquirers (individually or collectively). We submit that, like all scientists, biologists practice ontological realism, not constructivism. For example, they take it for granted that dinosaurs would have existed even if humans had never evolved. And the experimental biologist and the ethologist must note whether or how his or her observations or experiments exert any influence on the organisms to be studied—which again presupposes the inquirer-object distinction.

The next problem is: What kinds of objects do (really) exist in the world? We postulate that every fact involves some concrete (or material) thing: it is a state of a thing or a change of a thing. (There are neither states nor changes in themselves; nor are there abstract facts.) A concrete thing may be imperceptible like an electron or a biosphere, or tangible like a stone or a plant. The main characteristic of all material things is their changeability: they are all in flux. On the other hand, conceptual (or abstract) objects, such as numbers and theories, cannot be said to be changeable: only the brains that think of them are subject to change.

In this view, any collection of objects can be partitioned into two mutually disjoint subsets: a class of concrete or material objects (i.e., things) and its complement, a class of abstract, ideal, or conceptual objects (i.e., constructs). We compress this into:

POSTULATE 1.2. Every object is either a thing or a construct, i.e., no object is neither, and none is both.

Although according to this postulate there are no mixed objects, that is, objects composed of both things and ideas, there are concrete objects, such as human groups, within which certain ideas "rule" by virtue of being held (thought and believed) by the members of the group. Moreover, there are artificial things, such as written words, drawings, and graphs, which may stand for or represent ideas.

Postulate 1.2 is an axiom of *methodological* dualism, which should not be confused for an axiom of metaphysical dualism, because we are not claiming that there are real things of two kinds, i.e., (material) things proper and (immaterial) ideas. On the contrary, we adopt:

POSTULATE 1.3. The world is composed exclusively of things (i.e., concrete or material objects).

This is a central thesis of materialism. We take it that conceptual objects, whether useful or idle, scientific or mythical, are *fictions*, not real entities. That is, we *feign* that there are constructs, i.e., creations of the human mind to be distinguished not only from things (e.g., words) but also from individual brain processes and social circumstances. (More on this in Chaps. 3 and 6.) Since constructs are fictions, they are not part of the real world even when they take part in our representations of the latter. Consequently, we have to warn against the dual sins of reification and ideaefication.

Reification is the incorrect conception of properties, relations, or concepts as entities having autonomous existence. A classical example is the idea that sickness is an entity that the patient carries and may pass on to someone else. More recent examples are the structuralist idea that structures precede the corresponding structured things, or that processes are detachable from, or prior to, things—a well--known tenet of process metaphysics as well as of the fashionable idea that development consists in the embodiment of a genetic "program" or "instruction".

The dual of reification may be called *ideaefication*, which is the conception of material things or processes as self-existing ideas. Classical examples are Plato's idealist conception of ideas as detached from thinking brains (including Popper's World 3), or of science as a system of knowledge items, procedures, and behavior patterns. To be sure, it is permissible—and to do mathematics and philosophy it is indispensable—to *feign* that ideas are detached from brains, in order to be able to focus on certain features of concepts, such as their form and meaning, while disregarding everything else, in particular the thinker's circumstances. However, this is merely an instance of what may be called *methodological abstraction*: we should keep in mind that such conception of concepts in themselves is itself a fiction.

Another consequence of methodological dualism is that concrete objects (things) have no conceptual properties, in particular no logical and mathematical ones. What is true is that mathematics can deal with some of our ideas about the world, when detached from their factual referents. In other words, not the world, but some of our ideas about the world, are mathematical. Therefore, mathematics and logic are ontologically noncommittal or neutral (see Nagel 1956; Bunge 1974c, 1985a).

Any concept or statement that violates Postulate 1.2 will be declared *metaphysically ill-formed*. The attribution of conceptual properties to concrete things, and the attribution of substantial properties to constructs are in the category of metaphysically ill-formed concepts and statements. Examples of such metaphysical misfits are: "Nature is contradictory", "A biopopulation is a class of organisms", "Species are the units of evolution", "The Turtle Frog is a burrowing species that lives on termites", "Lineages evolve", "It is difficult to deduce human behavior from evolutionary theory", "Morphogenesis is guided by mathematical principles", "Homozygous genotypes produce only one kind of gamete", and "Selection is a vector with both direction and intensity".

We all use metaphysically ill-formed statements because they often are convenient *façons de parler*. For instance, we talk of a burrowing *species* instead of burrowing *animals* belonging to some species. Although such habits of speech, besides facilitating writing and reading, are harmless in most contexts, they are misleading in others. For one, if they are not recognized as what they actually are, they may be a source of conceptual confusion. Examples of such confusions with regard to the concept of species have been analyzed by Cracraft (1989), and we shall meet plenty of them in the course of our analyses. For another, if such metaphysical misfits occur in a work's central passages in which key concepts are supposed to be elucidated and defined, then the philosopher is entitled to suspect that they are indicators of either a flawed underlying ontology or a lack of theoretical or philosophical depth. Anyway, in both cases, is it the task of the philosopher of science to unearth, analyze, criticize, and, if possible, repair such flaws.

Let us now take a brief look at the notion of a thing. The most basic or general concept of a thing is that of a bare substantial individual, that is, an entity devoid of any peculiarities (properties) This may be defined as anything that can join another individual to form a third individual. (For a formalization of this concept see Bunge 1977a.) Of course, real things have many additional properties, such as energy. (Since some people maintain that energy is a substance on the same footing as matter, it seems important to emphasize that, according to physics, energy is a *property* of things. Moreover, it is a *universal* property in that *all* things, and only things, possess energy. The basic stuff the world consists of is not, as can often be read, "matter-energy"—a term suggesting that energy could be somehow converted into matter, or conversely. Views like these seem to rest on a misinterpretation of Einstein's famous formula "$E = mc^2$". The correct interpretation of this formula is: the amount of energy *of a thing* equals the amount of *its mass* multiplied by the square of the speed of light in void. That is, the formula applies only to things endowed with a mass. Yet there are massless things such as photons and electric fields, which possess energy. Consequently, we warn against conflating matter or materiality with mass. Moreover, as we shall see in a while, matter is strictly speaking not material—only concrete things are.)

Returning from physics to metaphysics, a real thing, then, is a substantial individual endowed with all its properties. To emphasize that properties do not exist apart from things, we formulate:

> DEFINITION 1.1. Let x represent a bare substantial individual and call $P(x)$ the collection of all the (known and unknown) properties of x. Then the individual together with its properties is called the *thing* (or *concrete* or *material* or *real object* or *entity*) X; i.e., $X =_{df} \langle x, P(x) \rangle$.

Note that, although usually a small number of a thing's properties will suffice to *distinguish* it from other entities, nothing short of the totality of its properties will *individuate* it, i.e., render it ontically distinct from every other entity.

1.3 Properties

1.3.1 Properties Proper

Although properties cannot be (physically) detached from the things that possess them, we can distinguish them (conceptually). In particular, it will be useful to distinguish several kinds of properties. A first important distinction is that between intrinsic and relational properties. An *intrinsic* property of a thing is one that the thing possesses regardless of other things, even if acquired under the action of other things. For example, composition and mass are intrinsic properties; so is the property of being alive. By contrast, a *relational* property of a thing is one that the thing possesses by virtue of its relation to other things. Examples are weight (in a given gravitational field), parenthood, adaptedness, being a host of a parasite (or conversely), or being an alpha male (or female).

With Galileo and Locke, we further distinguish *primary* from *secondary* (or *phenomenal*) properties. Whereas primary properties are objective, or subject-independent, and either intrinsic or relational, all secondary properties, such as color and loudness, are relational properties. More precisely, color is wavelength *as perceived by some subject*, and loudness is *perceived* sound intensity. In short: no sentient organism, no phenomenal property. Note that, in our view, secondary properties are neither purely objective nor purely subjective, for they are possessed by the subject-object system rather than by either component separately. (Only hallucinations are entirely subjective, even though their content depends on former experiences.) Note also that the intrinsic/relational distinction occurs in the very definition of "objectivity", as well as in the principle that physics, chemistry, and biology study only primary properties, leaving the secondary ones to psychology.

Another partition of properties is that between essential and accidental properties. An *essential* property of a thing is one that the thing loses if it is transmuted into a thing of a different kind or species. An *accidental* property, on the other hand, is one that makes little, if any, difference to any essential properties. For example, possessing a (functional) brain is essential for being a human being, whereas wearing a green shirt or even being a biologist is not. An accidental property of a thing is not necessarily connected to any other property of the thing, whereas every essential property is lawfully related to at least one other property. Hence there are no stray essential properties: they all come in natural clusters or property systems (see Sect. 1.3.4).

Another partition of properties is that between *qualitative* and *quantitative* ones. The former admit of no degrees, whereas the latter do. Examples of qualitative properties are pregnancy, parenthood, lichenization, and being alive. Examples of quantitative properties are mass, weight, length, temperature, age, fitness, and population density.

Finally, it is useful to distinguish manifest properties from dispositions (or propensities or potentialities). A *manifest* property is one possessed by a thing under all circumstances as long as the thing exists and remains in the same kind. Dispositions may be partitioned into causal dispositions and chance propensities. A *causal disposition*, such as solubility, electric conductivity, viability, or reproducibility, is the propensity to acquire certain manifest properties under certain circumstances. A *chance propensity* is the disposition to acquire a certain manifest property with a certain probability, depending (or not) on the circumstances. That is, whereas in the case of causal dispositions actualization requires that the thing in question join another real entity, in the case of chance propensities actualization may occur independently of any external circumstances, that is, it may be uncaused, as in radioactive decay or in spontaneous (stimulus-independent) neuron firing.

1.3.2 Properties and Predicates

Whereas mathematicians, idealists, naive realists, and antirealists need not distinguish between properties and predicates, realists must do so, because a property of a thing cannot be detached from the latter: there are neither substantial properties without things, nor things without properties (see Definition 1.1) By contrast, a *predicate* (or *attribute*), if attributed (truly or falsely) to a concrete thing, is a conceptual representation of a thing's property (Bunge 1977a). However, with a few exceptions (e.g., Sober 1982; Sober and Lewontin 1982; Suppe 1989, p. 214), most philosophers do not care for this distinction.

How are properties conceptualized? Intrinsic properties are represented by unary predicates, while relational properties are conceptualized as binary, ternary, or, in general, n-ary predicates. (Note that the n-arity of a predicate is not a property of a property but a second-order predicate.) The latter holds in particular for secondary properties, so that for instance "is red" is not a "unary property"—*pace* Sober (1982). Rather, it must be construed as an (at least) binary predicate: "x is red for perceiving animal y".

A unary predicate may be analyzed as a function mapping a set of things into a set of propositions involving the predicate in question. For example, the predicate "metabolizes", or M for short, is a function from the collection Ω of all organisms to the collection P of propositions of the form "x metabolizes", where x is in Ω. More precisely, $M: \Omega \rightarrow P$. That is, if b is an organism (i.e., $b \in \Omega$), then the value of M at b is $M(b)$, and is to be read as 'b is attributed Mness'. Thus, the proposition "b metabolizes" can be abbreviated as "Mb". All intrinsic qualitative properties are representable as such unary predicates (For details see Bunge 1974a, 1977a; more on semantical notions in Chap. 2.)

A binary predicate, such as "is homologous with" is a function from ordered couples of organs (or any subsystem of an organism) to the set of propositions of the form "x is homologous with y". (In obvious symbols, $H: O \times O \rightarrow P$, where $\times$ symbolizes the Cartesian product. Thus, if b and c are in O, then $H(b, c)$ is read 'H is predicated of the pair $\langle b, c \rangle$'.) A somewhat finer analysis may reveal that "is homologous with" is a ternary predicate, relating the homologous things x and y to the respect r in which x and y are homologous. H is now predicated of the triple

$\langle b, c, r\rangle$. A well-known example is the wings of bats and the wings of birds, which are homologous *as forelimbs*, not as wings. This example confirms the need to distinguish properties from the concepts representing them. While the former are objective, the latter depend on the student and the level of analysis he or she adopts. In other words, one and the same property of a concrete thing may now be conceptualized as a certain attribute and, later, in the light of fresh information or deeper analysis, as a different one.

Another important reason for drawing the property-predicate distinction—and one that necessarily escapes the naive realist as well as the antirealist—is that not all predicates represent properties of real things. Whereas for every attribute there is another attribute equal to the negation of the first, things have only "positive" properties. A thing either possesses *P* or does not possess *P*, but it cannot "possess" not-*P*: negation is *de dicto*, not *de re*. For example, tapeworms do not think, but this is not to say that they exert the function of not-thinking. Negation affects the proposition "tapeworms think", not the property of thinking. In short, *pace* Russell (1918) and others, there are no negative properties.

A consequence of this distinction for biological classification is that organisms do not possess "negative" characters, although in traditional systematics there are groups such as Apterygota (primitively wingless insects) and Invertebrata that are characterized by negative attributes, such as "wings absent" or "vertebrae absent". Yet since such attributes do not refer to any properties of the organisms in question, the resulting groups are not natural ones. Nevertheless, negative attributes such as "abiotic" or "anaerobic" are undoubtedly indispensable, if not in classification, then in our discourse for comparative reasons.

What holds for negation also holds for disjunction: there are no disjunctive properties although there are disjunctive predicates. For instance, there is no such thing as the property of being alive *or* dead, although the predicate "is alive or dead" is perfectly respectable. In sum, negative and disjunctive predicates occur in our discourse about things but they do not represent real properties of things.

A further rationale for distinguishing predicates from properties is that, whereas the former satisfy a theory, namely predicate logic, properties "satisfy" objective laws. (More precisely, while the set of predicates has the Boolean algebra structure, the set of substantial properties has the structure of an inf-semilattice, which is a much poorer mathematical structure than Boolean algebra; see Bunge 1977a.) Thus, "For all x and all P: if x is a P, then x is a P or x is a Q" [in symbols: "$\forall x\ (Px \Rightarrow Px \vee Qx)$"], where P and Q are predicates, is a formula of ordinary logic, which states something about predicates and logical implication, not about the world. On the other hand, the statement "An increase in the concentration of hormone x elicits behavior y" asserts a factual (not a logical) relation between two properties of an animal, namely between hormone concentration and behavior. The logical law does not refer to anything in particular, hence it cannot be tested empirically. The physiological generalization, in contrast, refers to actual individuals and it can be confirmed or disconfirmed by observation.

1.3.3 Generic and Individual Properties

We have just distinguished a property of a thing from the attribute, predicate, or function that represents the property. Now, even though there are no two exactly identical things, all things share some properties. For example, at a given time every organism has some age or other, which can be represented by a function as explicated above. We say that the age of an organism is a *generic* property, while the particular age of that organism is an *individual* property. An individual quantitative property is represented by a particular number (or list of numbers)—in our case the value of the age function. In obvious symbols, $F(a, t) = n$, where a is the name of the individual in question, t the time at which the property is measured or calculated, and n the value of F for a at t. On the other hand, an individual qualitative property, such as being alive, may be represented by a dichotomous variable, i.e., one that can take only two values, such as 1 (alive) and 0 (dead).

The totality of individual properties of a thing (at a certain time) constitutes its individuality or uniqueness (at a certain time). Indeed, all things are unique in the sense that they do not and cannot possess exactly the same individual properties, although they may possess the same generic properties. (If two things had exactly the same individual properties, i.e., if they were strictly identical, they would be one.) Thus, two organisms may possess exactly the same generic properties while the individual values of the latter vary. For example, although the *particular* fingerprint pattern of each and every human is unique, all (or, rather, most) humans share the generic property of having *a* fingerprint pattern. This spells trouble for the methodological vitalists, who argue for the autonomy and methodological uniqueness of biology, as well as the antiessentialists, who argue that organisms cannot be grouped into equivalence classes. Indeed, neither of them can refer to the uniqueness of organisms in order to make their case, unless they explicitly refer to uniqueness with respect to the possession of generic properties.

Occasionally, philosophers and scientists speak of properties of properties. For example, one could say that the weight of an organism (a generic quantitative property) has the property that it varies over time. This is, however, just a way of saying that organisms have a *variable weight*, which is a property of theirs, not a second-order property. We conclude that, while every predicate has some (second--order) predicates, there is no such thing as a second (or n-th)-order substantial property. (Logic, on the other hand, studies second- and higher-order predicates.)

Another reason for speaking of properties of properties in the philosophy of biology may arise from the ambiguity of the terms 'character', 'feature', and 'trait', which are used in the senses of both "property" and "part" (or "component" or "subsystem"). (This ambiguity has been addressed by various authors, e.g., Woodger 1929; Ghiselin 1984; Colless 1985; Fristrup 1992.) But a part of a thing is a thing, not a property. What is a property (of a whole) is the possession of a certain part. For example, each individual hair on a mammal's body is a part of it. The corresponding property, however, is "having hair" or "hairiness". Still, do those properties not have second-order properties, such as selective value (Sober

1981)? No, only concrete parts of an organism can have selective values. That is, in our ontology the selective value of a trait is a (relational) property of the trait *qua* concrete part of the given organism, not a property of the trait *qua* property of the given organism. In short, *pace* Sober, selective value is not a property of a property, although it may be treated as a second-order predicate in biological discourse.

This confusion seems also to be at the root of the distinction between characters and character states in systematics. For example, if eye color is referred to as a character, then this character is said to come in certain "states" such as brown or blue. However, as we shall see in Section 1.4, only things can be in certain states, not properties. Indeed, the concept of state of a thing is defined by means of the concept of property. That is, being in a certain state amounts to having certain (individual) properties at a given moment. Thus, what can be in a certain state is the eye as a concrete subsystem of a given organism; and the state of this eye comprises of course a certain pigmentation and thus color. Therefore, the expression 'character state' is misleading.

Moreover, it is superfluous because it does not coincide with the distinction between generic and individual properties, although it is at first sight similar to it. Indeed, both characters and character states can involve generic properties. For example, brown-eyed is a generic property, as being shared by many animals. Since systematists are interested only in generic properties, not individual ones—they are not concerned with what makes an individual an individual but with what makes an individual a member of a class of equivalent individuals, i.e., a taxon—there is no need for the notion of a character state. The systematist has use only for (organismal) characters *simpliciter*, which are represented as predicates referring to generic properties of organisms. (For further criticisms concerning the notion of character state see Platnick 1979, as well as Mayr and Ashlock 1991.)

1.3.4 Laws

We assume that every essential property is lawfully related to some other essential property or, in other words, that all entities "satisfy" some laws. However, before we can introduce the concept of a law, we need the concept of the scope of a property, which is introduced in:

> DEFINITION 1.2. The *scope* S of a property is the collection of entities possessing it.

(Note that "scope" is an ontological notion defined on substantial properties, whereas "extension" is a semantical concept defined on predicates. Thus, the latter is defined for negation and disjunction, while the former is not; see also Sects. 2.2 and 7.2, as well as Bunge 1974b, 1977a.)

DEFINITION 1.3. If P and Q are (essential) properties of things, then P and Q are said to be *lawfully related* if, and only if, $\mathcal{S}(P) \subseteq \mathcal{S}(Q)$ or $\mathcal{S}(Q) \subseteq \mathcal{S}(P)$.

In other words, if the scopes of two properties are coextensive, or if the scope of a property is included in that of another, the two properties are said to be *lawfully related.* A property not so related to any other property can be said to be *stray* or *lawless.*

Laws, then, are constant relations between two or more properties (Bunge 1967a, 1977a; Dretske 1977). They are *represented* by *law statements*, that is, propositions such as "$\mathcal{S}(P) \subseteq \mathcal{S}(Q)$" or its converse or, equivalently, "$(\forall x)Lx$ ", with $L =$ "x possesses $P \Rightarrow x$ possesses Q" or its converse. The term 'law' usually refers to both laws and law statements. (To our knowledge, the first to distinguish laws explicitly from law statements was Ampère 1843.)

We can now formulate the hypothesis that there are no stray or lawless essential properties:

POSTULATE 1.4. Every essential property is lawfully related to some other essential property. That is, for any two essential properties, P and Q, either $\mathcal{S}(P) \subseteq \mathcal{S}(Q)$ or $\mathcal{S}(Q) \subseteq \mathcal{S}(P)$.

We submit that this ontological *principle of lawfulness* underlies all science and technology. This principle must not be mistaken for the principle of *uniformity,* which states that the laws are the same across the cosmos at all times. In an even stronger version, the latter states that the same events occur everywhere and at all times. If this thesis were true there would be no evolution.

Since the principle of lawfulness is itself a lawlike proposition concerning either objective patterns (constantly related properties) or law statements, it is also subsumed under the term 'law'. Furthermore, the word 'law' is also used to designate law-based rules or procedures, such as "If you want to achieve B, do A". Because of this ambiguity we must distinguish four different concepts designated by the word 'law', which we identify by a subscript each:

Law_1 = Objective pattern of being and becoming.
Law_2 = Law statement = Proposition(s) representing a law_1.
Law_3 = Law-based rule or nomopragmatic statement.
Law_4 = Metanomological statement = Proposition about some law_1 or law_2.

We shall have more to say about law statements, nomopragmatic and metanomological statements in Section 3.5.8. Here we focus on the ontological concept of a law.

Since laws_1 interrelate properties of things, *laws_1 themselves are (complex) properties of things*. So much so, that they can be represented by formulas of the basic form "$(\forall x)Lx$", i.e., laws_2. (To be precise, laws_1 are properties only in a broader sense than conceived here: for details see Bunge 1977a.) Since objective laws are in the nature of things, i.e., essential properties of theirs, laws_1 cannot be broken the way a legal norm can, nor can they be bent as a result of human action;

that is, $laws_1$ hold independently of human knowledge or will. Further, since $laws_1$ are *in rebus*, not *ante res*, it makes no sense to say that things "obey" $laws_1$, or that $laws_1$ "govern" the behavior of things—except as a convenient, if misleading, *façon de parler*. Neither does it make sense to say that the laws of nature are "eternal and immutable". When a thing undergoes a qualitative change for the first time in the history of the cosmos, new $laws_1$ emerge and old $laws_1$ may submerge. What is immutable, on the other hand, are $laws_2$, for law statements are constructs. However, not even constructs are eternal because they are fictions which do not exist apart from brains that may think of them or give them up. In short, no brains, no $laws_2$.

A law_1 may be said to be "spatially and temporally boundless" when it is possessed by (or "holds" for) all things, as is the case with the basic physical laws. Or it may be said to be "bounded in space and time", as is the case with the known biological $laws_1$, which came into existence on our planet only about three to four billion years ago together with the first organisms. In other words, not all $laws_1$ are universal, in the sense that they "hold" everywhere and at all times. If one demands—as many people do (see, e.g., Smart 1963; Rosenberg 1994)—that all laws be universal, then it is clear that biology will never discover any. Yet physics, chemistry, and the social sciences are in the same boat. For example, the $laws_1$ of liquids and solids did not emerge before planets or at least asteroids were formed. Likewise, the $laws_1$ of chemical reactions do not exist wherever the temperature is either too low or too high for such processes to occur. The case of biological $laws_1$ is parallel.

However, in biology, the problem is that biological $laws_1$ are as varied as the organisms possessing them. Thus, the range of biological $laws_2$ may be rather small. For instance, although there are laws holding for all organisms *qua* biological systems, there will also be laws holding only for some subspecies, race, or variety. Worse, at the extreme, there might be $laws_1$ possessed by only a single individual, as is the case with the last member of a species on the verge of extinction. Biology is, however, not unique in this respect: think of geology, where many statements refer solely to Earth. (Note that, in cases like this, the corresponding law statement is still general in that it refers to *all* individuals in its reference class, and perhaps to all times. Only the reference class is a singleton.) Furthermore, lawfully related properties of organisms may emerge and submerge in the course of evolution. If we finally consider the immense variety of habitats on this planet, which accounts for the multitude of different circumstances to which organisms are subjected, it is not surprising that biologists have a hard time coming up with law statements.

In sum, $laws_1$ are objective patterns of being and becoming, which can be represented (truly or falsely) by law statements. Yet the notion of becoming still awaits elucidation.

1.4 State

1.4.1 State Function

As we saw above, every concrete thing, no matter how simple, has a number of properties. The totality of properties of a thing at a certain time determines the *state* of the thing at the given time. Of course, we are unlikely to get to know all of these properties. Hence our knowledge of a given thing at a certain time reduces to a list of its known individual properties at that time. This list represents the state of the thing at the time as known to the investigator.

If we know n properties of a thing, we can represent each of them as a function F. For example, the *generic* property of having a mass can be represented by the numerical function $M: B \times T \times U_M \rightarrow \mathbb{R}^+$, where B designates the set of bodies, T the set of time instants, U_M the set of mass units, and $\mathbb{R}^+$ the set of non-negative real numbers. A particular value $M(b, t, u) = r$ of this function represents an *individual* property of organism b, for instance, that of having a mass of 100 g at time t.

As with mass, so with the remaining properties of the thing of interest: every one of these can be formalized as a function. The list of n such functions is called a *state function* of things of the kind concerned. (Often they are also called *state variables*.) In other words, if we have n functions F_i, the state function F of the given thing is the list or n-tuple $F = \langle F_1, F_2, ..., F_n \rangle$. The value of F at time t, i.e., $F(t) = \langle F_1(t), F_2(t), ..., F_n(t) \rangle$, represents the *state* of the thing at time t. (Actually, the state of any thing at any time is also frame-dependent but we need not deal with this complication here.) $F(t)$ may be visualized as the tip of an arrow in an n-dimensional abstract space; more on this next.

1.4.2 State Space

The set of possible states of a thing can be represented in a *state space* or *possibility space* for the thing. This is the abstract space spanned by the corresponding state function $F = \langle F_1, F_2, ..., F_n \rangle$. If only two properties of the thing are either known or taken into account, the corresponding state space is a region of the plane determined by the axes F_1 and F_2 : see Fig. 1.1. A state space for a thing with n known properties is n-dimensional.

Every state of a thing of a given kind can be represented as a point in the corresponding state space. "As time goes by", the values of some of the properties of the thing are bound to change, and so the representative point "moves" along some trajectory. The stretch of such trajectory over a time period τ is called the *history* of the thing concerned during τ: see Fig. 1.2. (The total history of an organism is usually called its 'life history' or 'life cycle', but the latter term is a misnomer because the word 'cycle' suggests that the organism returns to some prior or initial state, which is, of course, not true. After all, death is neither a return to birth nor to fertilization.) However similar to one another, two different things are bound to have more or less different histories, because of differences in either their composition or their environment, or both.

Note that, although the state of a thing is definite and objective, it may be conceptualized in different ways, depending on our knowledge of the thing. This is why there is no such thing as *the* unique state function for things of any given kind, and why we cautiously spoke of *a* state function and *a* state space. Indeed, there are as many state functions, hence spaces, as representations or models of the thing that we can conceive. However, the choice of state function is neither arbitrary nor a matter of taste, because whichever state functions are chosen, they are supposed to satisfy the law statements included in the theory—and this is far from being a matter of convention.

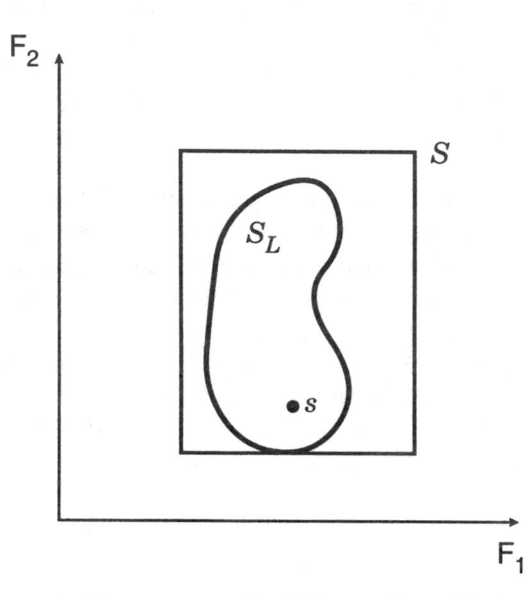

Fig.1.1. The conceivable state space S of things of some kind and the corresponding lawful (or nomological) state space S_L, i.e., the set of all really possible states of the things in question. Only two state functions, F_1 and F_2, are involved in a two-dimensional state space. (Actually the axes are the ranges or sets of values of the functions, not the functions themselves.) The point s represents the *state* of a thing of the given kind at some time

1.4.3 Nomological State Space

The principle of lawfulness (Postulate 1.4) states that all essential properties of a thing are lawfully related. Accordingly, any logically or conceptually possible state function of a thing will not take values in its entire codomain but will be restricted to a subset of the latter, namely to those values which are compatible with the laws of the thing in question. Hence, only a certain subset of the conceivable states of the thing in question will be really possible for it. We call the subset of really possible states of the logically possible state space of the thing in the given representation its *lawful* or *nomological* state space: see Fig. 1.1. For example, every organism is caged inside a space of (possible) states—an abstract space of course—peculiar to its taxon, in particular its species (see Sect. 7.2.1.8). Among the laws restricting the (logically) possible range of forms of organisms, i.e., its so-called morphospace (Alberch 1982; Lauder 1982; Gould 1989; Goodwin 1994), are those properties which are discussed as developmental and phylogenetic constraints. Yet what functions as a constraint in evolution originates in a law in development, for this is the period in an organism's life history when qualitative novelties emerge. (See also Sects. 8.2.4.3 and 9.3.6, as well as Levins and Lewontin 1982.)

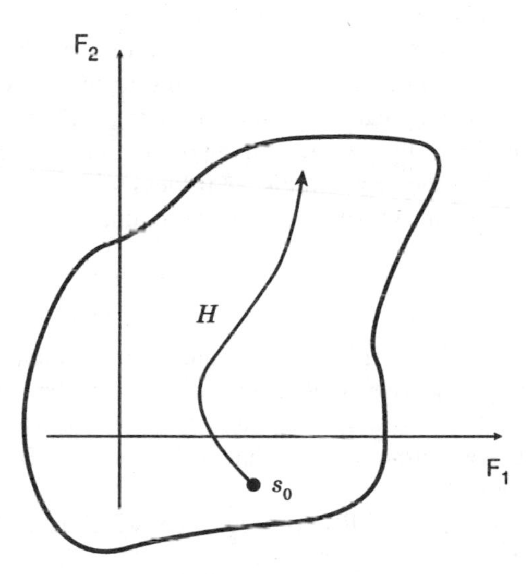

Fig.1.2. The nomological state space of things of some kind, spanned by the two axes F_1 and F_2. The trajectory H represents the *history* of a particular thing of the given kind over some period of time, assuming that it has acquired no new properties. If new properties do appear during the given period, either the corresponding new axes must be added, or a higher-dimensional state space must be constructed from scratch

1.5 Event, Process, and History

A philosophical principle shared by most philosophical schools is that all things are in flux. (To be sure, although this principle has never been refuted, it is hard to prove: this is why we must postulate it.) Thus we introduce:

POSTULATE 1.5. Every thing changes.

That is, "as time goes by", some of the properties of every thing change. In other words, the state of any concrete thing changes in the course of time. More precisely, an *event* can be represented as an ordered pair of states $\langle F(t), F(t')\rangle$, where $t < t'$. An alternative construal is as follows. If S_K designates a state space for things of kind K and $s, s' \in S_K$ denote two states of a thing x of kind K, then the net change or event in x when passing from state s to state s' is representable by the ordered couple of these states, i.e., by $\langle s, s'\rangle \in S_K \times S_K$. (Note that in this construal we do not need the notion of time. In fact, time is not prior or external to things but is constituted by the change of things. Thus the above expression 'as time goes by' is actually metaphysically ill-formed.) Such a transition from one state to another can be represented by the motion of the representative point in a state space (Fig. 1.2).

As with the logically possible states of a thing, we can collect all the logically possible events in (or changes of state of) a thing to form the conceivable *event space* of the thing in question. Since, according to the principle of lawfulness (Postulate 1.4), every thing can only be in nomologically possible states, it can only undergo nomologically possible changes of state, i.e., events. Thus, from Postulate 1.4, which is the principle of lawfulness with regard to being, we can infer a principle of lawfulness with regard to becoming. This principle is:

THEOREM 1.1. Every thing can undergo only lawful changes (i.e., events or transformations).

Note that the assumed lawfulness of all change does not imply that all change is strictly deterministic or causal. In fact, some events may be random, which, however, is not the same as lawless: since there are stochastic or probabilistic laws, randomness, just like causation, is a kind of lawfulness (Bunge 1959a, 1977a; Levins and Lewontin 1982, 1985).

Theorem 1.1 entails an immediate consequence, which is held tacitly by the vast majority of scientists:

COROLLARY 1.1. There is no total disorder, and there are no miracles.

Due to the assumed lawfulness of all change, the set of really possible events is restricted to the *nomological event space* of the thing in question. The notion of an event space will come in handy for the elucidation of the concept of causation (see Sect. 1.9.3).

The study of most events, such as fertilization and birth, reveals that they are not point events or "quantum jumps" but processes. (Even the quantum jumps are processes, though they occur very fast.) By definition, a *process* is a sequence of states or, if preferred, a sequence of events. As the notion of process figures prominently in biophilosophical discourse, we had better spell this out in:

DEFINITION 1.4. A complex event, i.e., one formed by the composition of two or more events, is called a *process*.

Thus, whereas a point event is describable by the ordered couple ⟨initial state, final state⟩, a complex event or process is described by a sequence of more than two, perhaps infinitely many states, i.e., a curve or trajectory in some space. For example, a process $\langle s, s'' \rangle$ can be either analyzed as the sequence of states $\langle s, s', s'' \rangle$ or, alternatively, as the sequence of events $\langle <s, s'>, <s', s''> \rangle$. We emphasize that not every old set of states or events constitutes a process. For example, an arbitrary collection of events occurring in different things that are comparatively isolated from each other does not constitute a process, even if it is ordered in time. For a set of events to constitute a process it must satisfy two conditions: (a) the events must involve or concern just one thing, however complex, and (b) the events must be ordered intrinsically, i.e., they must be representable as a curve or trajectory in some state or event space. Accordingly, to speak of evolution as an (individual) process makes sense only if there is just a single evolving thing, e.g., the biosphere. However, if evolution takes place in biopopulations, as is commonly believed, what we actually have are different evolutionary processes, so that 'evolution' can only designate the *collection* of all evolutionary processes in biopopulations. (More on this in Sect. 9.1.) In short, the word 'process' denotes both individual processes and process classes, and is thus a possible source of conceptual confusion.

Whereas a process (or partial history) of a thing is any (ordered) sequence of states of (or, alternatively, events in) a thing, the (total) *history* of a thing is the ordered set of *all* its successive states (or events). More precisely, if s_1 designates the initial and s_n the final state of a thing, then the (total) *history* H of the given thing can be represented either by (a) the ordered set of successive values of the state function F of the given thing, e.g., $H = \langle F(t_1), F(t_2), ..., F(t_n) \rangle$ or $H = \langle s_1, s_2, ..., s_n \rangle$, or (b) by the sequence of events in the thing, e.g., $H = \langle <s_1, s_2>, <s_2, s_3>, ..., <s_{n-1}, s_n> \rangle$.

Note that we did not need the notion of cause in our conceptions of event, process, and history. For example, what we call simply 'process' is often called a 'causal process' by philosophers (e.g., Salmon 1989). In our view, a state cannot be the cause of a later state. For example, the state of motion of a body does not cause any later state of motion—if only because states of motion constitute a continuum, and in a continuum there is no such thing as the point that comes next to a given point. Thus, the movement of a body along some trajectory is not a causal process. (See also Woodward 1989.) We shall speak of an (external) cause only when a change of state, i.e., an event, in a given thing generates the change of state, i.e., an event, in some other thing (see Sect. 1.9). Internal causes are parallel: an event in one part of a thing may cause an event in another part.

With regard to the fashionable ontology of "historical entities" in the philosophy of biology, we emphasize that these are not entities but reifications of the histories of things. The history of a thing is not an entity (or concrete individual) because, being a succession of states of a thing, it is in no state whatsoever itself and, thus, the category of changeability cannot be attributed to it (see Sect. 1.6). In other words, a life history is a timeless, hence an ahistorical, object. Moreover, a sequence of states of a thing is not a concrete system (see Sect. 1.7). However, 'historical entity' can also refer to a thing at a certain time. Thus Hull (1989, p.187) asserts that the "time-slices" of historical entities would be historical entities, too. Yet since all things change and thus have a history, 'historical entity' in this sense is just as redundant as 'historical process'. In sum, in Hull's metaphysics, apparently both changeable and timeless (i.e., ahistorical) objects are said to be historical. Thus, the notion of historical entity is either trivial or contradictory. (For the idea that taxa are historical entities see Sect. 7.3.)

Although we do not need the notion of a historical entity, knowledge of the history of many things is required to understand their present state. That is, the current state of certain complex things depends either on their whole past or at least on part of it, so that, in describing the former, we must take its history into account. This is the case with all "hereditary" systems, or systems with "memory", such as certain alloys (elastic hysteresis), all ferromagnets (magnetic hysteresis), all organisms (DNA), many animals (learning), and all social systems (tradition): they all retain traces of their past.

A particular feature of most organisms (as well as of many nonliving things) is that sometimes they also undergo qualitative transformations, such as in development or in evolution. That is, they acquire or lose properties. Whereas quantitative change is represented by a trajectory in some state space with fixed axes, qualitative change amounts to either the sprouting or the pruning of axes in the original state space. Figuratively speaking, we could say that during a qualitative change the history graph of the thing in question jumps to a different state space, possibly one of either higher or lower dimensionality: see Fig. 1.3. This saltational feature of qualitative change is both scientifically and philosophically important: Leibniz notwithstanding, *natura facit saltus*, however small these jumps may be. (See also Sects. 1.7.3, 4.1, and 9.1.)

This might be the place for a few remarks on the occasional sympathy for Whitehead's own variety of process metaphysics expressed by some biologists and biophilosophers (e.g., Woodger 1929; Løvtrup 1974; Ho and Fox 1988; Goodwin 1990; G.C. Williams 1992a.) Undoubtedly, we often study and describe processes rather than things, such as metabolic cycles, gastrulation, or selection. And in scientific explanation we explain the current state of things in terms of mechanisms, i.e., processes (see Sect. 3.6.3). In so doing, we may forget that there are no events or processes in themselves, i.e., apart from changing things, and speak of processes as if they were self-existing entities. This is admissible as long as it is seen as an instance of methodological abstraction, i.e., the formation of process classes, not as an ontological thesis about the primacy of events and processes over things. Yet such a metaphysical thesis was advanced by Whitehead (1929), who thought that, because the concept of an event is so important, it deserves to be taken as primitive or undefined. So he proposed to define a thing as a bundle of events. However, this strategy is not logically viable because it is impossible to define the concept of property of a thing in terms of that of an event. Indeed, since an event is by definition a change in some properties of a thing, the concepts of thing and property are logically prior to that of an event. Consequently, a precise description of any event or process ought to mention the changing thing(s) in question. If we describe a process without the changing thing(s) in question, we are led to ask later what the "units" of that process are or, more precisely, what the members of the given process class are. Witness the units of selection controversy (see Sect. 9.2).

A final remark concerns the popular statement that (concrete) individuals "participate" or "function" in various natural processes. Phrases like these make no sense in our ontology, according to which things change and may undergo processes, but since the latter are not prior to things, they cannot "impinge" *ab extrinsico* on things. What is true, however, is that whereas things are changeable, constructs are not. More precisely, the latter cannot even be said to be so.

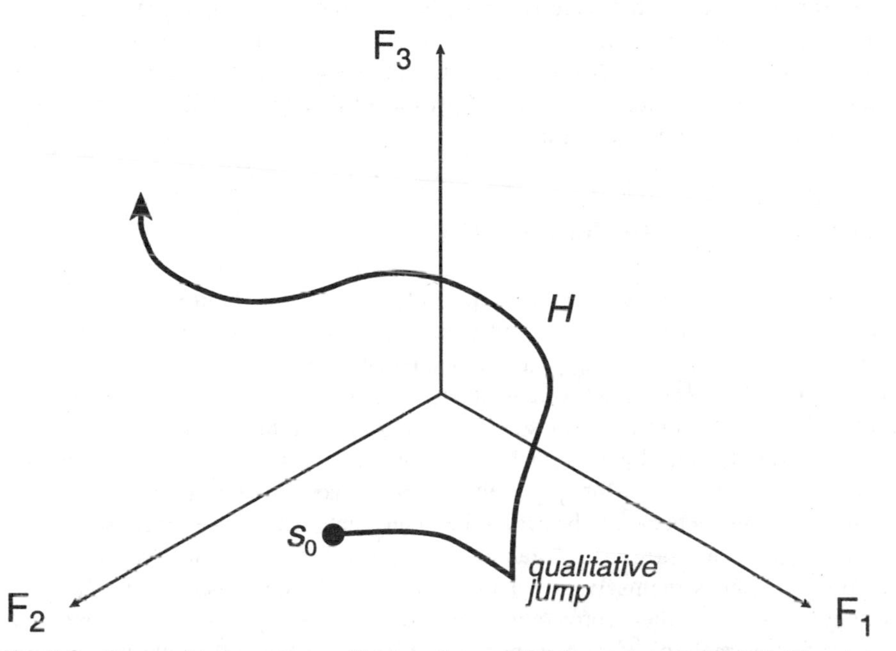

Fig. 1.3. Representation of a qualitative change of things of some kind K. The state space is given by three functions, F_1, F_2, and F_3, and the trajectory represents a partial history of a particular thing of kind K. In the beginning, the thing has only two properties, and the change it undergoes is quantitative, i.e., the history graph moves along the plane spanned by F_1 and F_2. Then the thing acquires a new property represented by F_3, so that its state space is now three-dimensional, and the trajectory moves within this three-dimensional space. The qualitative jump happens at the junction of the two partial trajectories

1.6 More on Things and Constructs

1.6.1 Materiality and Changeability

The concept of change is so central to factual science that it can be used to define that of a concrete, material, or real thing (Bunge 1981a). Indeed, as an alternative to Definition 1.1, we propose:

DEFINITION 1.5. For any x: x is a *concrete* (or *material*, or *real*) thing (or *entity*) $=_{df} x$ is changeable.

Note that this definition of a concrete or material thing is strictly objective, for it does not include the notion of an inquiring subject. Hence, a subjectivist has no use for it, because he or she is interested only in what is real for him or her: what I feel or perceive is what there is. Thus, Feyerabend (1981, p. xiii) tells us: "We decide to regard those things as real which play an important role in the kind of life we prefer". This reality criterion is acceptable neither to the scientist nor the technologist, for in science and technology the attribution of reality is not a matter of life-style but of objective tests.

1.6.2 Ideality and Unchangeability

Let us now turn to ideal (or conceptual) objects, such as numbers and theories when studied in themselves, i.e., independently of the people who think of them. *Prima facie*, an ideal object, such as a mathematical figure or a biological theory, is unchanging. If a construct were to change it would become a different construct. By contrast, a concrete object such as an organism must change to stay alive. Now, to say that an object is unchanging amounts to saying that it remains forever in the same state. (More precisely, its state space would have a single point.) However, constructs cannot be said to be in any state whatsoever because conceptual objects do not possess substantial properties: their properties are as fictional as the individuals themselves. For instance, it makes no sense to ask 'How is 7 doing today?'. (In other words, constructs have no state spaces.) Thus, constructs are neither eternal objects (Plato) nor changeable ones (Hegel). What does change from person to person are the brain processes when constructs are thought. Accordingly, we propose the following convention:

DEFINITION 1.6. For all x: x is an *ideal* (or *abstract* or *conceptual*) object (or *construct*) $=_{df} x$ is neither unchanging nor changeable.

Note that Definition 1.6 does not imply the autonomous existence of ideas—far from it. It is just a convention concerning the meaning that we shall attribute to the concept "ideal object"—or, equivalently, the signification to be attributed to the expressions 'ideal object', 'conceptual object', and 'abstract object' as well as to the word 'construct'.

If Definition 1.5 is admitted, then *reality* or *matter*, i.e., the collection of all things (or real or material objects), turns out to be identical with the collection of all changeable objects. Yet collections are conceptual (or ideal) objects, not real ones: only their members are real. Hence, matter or reality, though consisting of (the collection) of all (real) things, is not real itself, i.e., it is not a thing. (Similarly, motion does not move, and evolution does not evolve.) What is real is the world or universe, that is, the concrete system composed of all concrete things.

1.6.3 Spatiotemporality and Individuality

It is usually assumed that, unlike ideas, material things "exist in space and time". That is, "existence in space and time" is taken to define the very concept of a material object in contrast to that of a nonmaterial one. In turn, what is meant by "x exists in space and time" is that x is located in a region of space and endures throughout a time interval.

Statements such as these presuppose the autonomous existence of the spatiotemporal framework, which, in turn, would be a nonphysical object. This idea may have been appropriate in Newton's time, but it is no longer tenable in the aftermath of general relativity, which suggests adopting a relational, not an absolute, theory of spacetime (Bunge 1977a). Accordingly, things do not float in spacetime but hold spatiotemporal relations among each other. That is, space and time are not self-existing objects (or containers of things). Instead, spacetime is a network of relations among distinct changing things. In short, no distinct and changing things, no spacetime. This is why we can, in principle, elucidate the notions of thing and event without recourse to the notions of space and time, though we refrained from doing so above for the sake of simplicity. Thus, in our ontology it makes as little sense to say that a thing *exists in* spacetime as to say that spacetime *exists in* a thing.

Another notion that is popular in contemporary biophilosophical discourse is that of an individual. Usually, an individual is defined as a particular in contrast to a class or a universal (see Chap. 2). Regrettably, this logico-semantical individual-class distinction does not answer the ontological question of whether the object in question is material or conceptual. For example, both the planet Jupiter and Zeus are (logico-semantical) individuals but, whereas the former is a material object, the latter is a construct. On the other hand, all sets and classes are conceptual objects. The members of a set may be concrete things or constructs, or both (moreover, a set may consist of either individuals or other sets), but a class of real objects is itself no more real than a class of constructs. We shall see in Chapter 7, however, that whereas some classes are arbitrary collections, others are realistic (not real) in that they are collections of things with objective commonalities, such as common descent.

We finally emphasize that the spatiotemporal location of things has no bearing whatsoever on whether the things in question are parts of an individual (or system) or members of a class. The fact that all known biological systems are "spatiotemporally restricted", i.e., that they occur only on our planet and that they are, moreover, related by common descent, does not suffice to establish that these organisms are parts of an individual. And the fact that a free electron is "spatiotemporally unrestricted" in the sense that it is spread all over the universe does not suffice to say that an electron is a class or kind rather than an individual. What is at issue is (a) whether or not the things in question are coupled together, forming a system (i.e., a "higher-order individual"), or (b) whether we put certain things together conceptually, i.e., whether we form collections or classes. We shall elucidate the concept of a system in the following section. The analysis of the conceptual operation of classification is left for Chapter 7.

1.7 Wholes

1.7.1 Aggregate and System

Most entities are not simple things, such as quarks and leptons, but are composed of other things. Things that are composed of other things may be called *complex things* or *wholes*. There are two ways in which a whole may come into being: by *aggregation* and by *combination*. The accretion of dust particles or of sand grains, as well as garbage dumps, water pools, and clouds, exemplify aggregation. What characterizes all of these wholes is the lack of a specific structure composed by strong bonds: that is, such wholes are not cohesive. Consequently, once formed, they may break up rather easily under the action of external forces. For example, the addition of a single grain to a sandpile may cause its collapse.

When two or more things get together by interacting strongly in a specific way, they constitute a *system*, i.e., a complex thing possessing a definite structure. Hence, systems, unlike aggregates, are more or less cohesive. Atomic nuclei, atoms, molecules, crystals, organelles, cells, organisms, biopopulations, ecosystems, families, business enterprises, and societies are systems. It is obvious from these examples that the objects of scientific study are mostly systems. Only elementary particle physicists study simple things.

Note that we do not adopt the standard definition of a system as a *set* of interrelated elements, because sets are concepts, whereas we are dealing here with concrete systems. Besides, a set has a fixed membership: once a member always a member. By contrast, the composition of a concrete system may be variable over time: think of a cell or an ecosystem.

An assumption of our ontology is that there are no stray things: that every thing, whether simple, aggregate, or system, interacts with other things, so that all things cohere, forming systems (Bunge 1979a). We therefore propose:

POSTULATE 1.6. Every concrete thing is either a system or a component of one.

POSTULATE 1.7. Every system, except the universe, is a subsystem of some other system.

POSTULATE 1.8. The universe is a system, namely the system such that every other thing is a component of it.

Postulate 1.8 should not be mistaken for a holistic thesis. The hypothesis that the world is a system is weaker than the holistic cosmology of Plato, the Stoics, or Hegel, according to whom every thing is bonded to every other thing, so that the world as a whole is an "organic whole" or "holon". If this holistic hypothesis were true, then it would be impossible to study any particular thing without taking cognizance of every other thing, i.e., the whole universe, and it would be impossible to act on any particular thing without disturbing the entire universe.

Systemism ("Every thing hangs together with *some* other thing or things") steers a middle course between holism ("*All* things hang together") and atomism ("Everything goes its own way"). No part of the universe is completely isolated but every thing is isolated *in some respects* from some other things. This partial interconnectedness of the parts of the world renders their study possible, for the study of every thing is partial and rests on the possibility of making contact (however indirect) with the thing. Witness any scientific experiment, which presupposes that the system under investigation can be sufficiently (though never completely) isolated from its environment (see Sect. 3.7.4.2.).

1.7.2 The CES Analysis of Systems

As the notion of a system is pervasive in science and technology, it will be useful to introduce a general model of concrete systems of any kind, living or nonliving. Obviously, a concrete system s consists of (at least two) parts or components. Let us call the collection of all its parts (i.e., the set $\{x \mid x \sqsubset s\}$, where $\sqsubset$ designates the part-whole relation) the *composition* (or *constituent class*) of s: $C(s)$.

The complement of a system's composition is, of course, the collection of things other than the given system, i.e., its environment. Yet in modeling a system s we usually do not need to take the rest of the universe into account but only those things that can be influenced by s or that may act upon s. Only the collection of those things other than s that are thus related to s will be called the *environment* of system s: $E(s)$. Thus, we use the term 'environment' always in the sense of 'immediate or proximate environment'. We could also say that the immediate environment of a system is the composition of its next supersystem. Incidentally, a similar distinction is made by some ecologists, who speak of the "general environment" and the "operational environment" of an organism (e.g., MacMahon et al. 1981).

Note that the environment of a system is defined as the *collection* of things external to a given system. Hence the environment is always relative to a given system. That is, neither are there empty environments to be populated, nor is the environment a system itself. For this reason, the environment cannot act as a whole on the given system. In particular, we cannot speak of a system-environment interaction. Being defined as a collection, only the members of the given system's environment can act upon the system or upon some of its parts. Of course, certain items in the environment of a given system may be connected to form a system, which then may act as a whole upon the system in question. Still, the environment of a system as conceived here is not an entity. If we need the concept of a self-existing environment, as we do when we ask questions such as 'What will happen when we move fish *x* from lake *y* into lake *z*?', we suggest using the term 'habitat' (or something else) rather than 'environment'. The habitat is "out there", but the environment of the system in question comprises only those items interacting with it during a given time interval. (For a similar concept of environment see Lewontin 1983b.)

In order to constitute a system, rather than a heap, the parts of a system must act on, or be acted upon by, other components of the system. That is, there must be links, ties, or bonds among the components of *s*. With the help of the concept of state introduced above, we can be more precise and formulate:

> DEFINITION 1.7. A relation between a thing *x* and a thing *y* is a *bonding* relation if, and only if, the states of *y* alter when the relation to *x* holds.

It is precisely because of the existence of such bonds among the parts of a whole that we call a system a 'cohesive or integrated whole'. Note that the cohesiveness of a system does not necessarily require that the composing parts be spatially contiguous. For example, a business firm with several branches dispersed all over the world is bonded together by, say, telecommunication, mail, shipments, traveling employees, and so on. Note further the difference between the *integration* and the *coordination* of a system. We take the former to consist in structural and the latter in functional integration. The distinction between these two concepts is important because there can be integration without coordination, but not conversely. For example, a complex machine out of kilter, such as a crashed computer, is integrated but not coordinated. On the other hand, organisms are coordinated and *a fortiori* integrated as long as they are alive.

Not all relations among entities are bonding relations: there are also spatial and temporal relations, which do not produce a change of state in the relata. For example, the fact that a thing *x* is located 7 meters apart from (or close to) a thing *y* with respect to a certain reference frame *z*, does not change their states by itself. It may allow them, however, to act upon one another, but this is a different relation. The same holds for temporal relations: a thing that, at a given time, does no longer or not yet exist cannot act upon a thing existing at that time. Only simultaneously existing things can do so, though they need not do so, of course. Simultaneity is not a bonding relation either.

What about the ancestor-descendant relation? Is there such a bond as a "genealogical nexus" (Hull 1987, 1989)? Does reproduction "glue" organisms together into cohesive individuals such as species and monophyletic taxa (Eldredge 1985a)? Our answer is a straightforward *no*. Ancestors, if they no longer exist, cannot act upon their descendants. If the histories of ancestors and descendants overlap, then the ancestors do not act on their descendants *qua* ancestors but, if at all, *qua* fellow organisms. The fact that the nomological state space of the descendant can be accounted for only by reference to its ancestry must not be confused with the existence of a bonding relation between ancestor(s) and descendant(s). To qualify as a bonding relation, a relation must alter the state of the relata here and now. Since we shall meet this problem again in later chapters (Chaps. 4 and 7), we now return to the structure of systems.

To sum up, the collection of relations among the parts of a system is split into two nonoverlapping subcollections, namely the collection of bonding relations (or the bondage) B and the collection of nonbonding relations $\overline{B}$. Let us call the union of the collection of the bonding and the nonbonding relations among the parts of a system s the *internal structure* (or *endostructure*) of s: $S_I(s)$. It goes without saying that we are usually interested in the bonding endostructure of a system.

Now all systems, or at least some of their parts, also interact with some environmental items. Let us call the collection of all bonding and nonbonding relations among s, or its components, and environmental things the *external structure* (or *exostructure*) of s: $S_E(s)$. Again, scientists are mostly interested in the bonding exostructure of things, such as the ecological relations between an organism and the things in its environment.

Finally, the union of $S_I(s)$ and $S_E(s)$ may be called the (total) *structure* of the system s: $S(s)$.

We now have all we need to build the simplest qualitative model of an arbitrary material system s (except the universe), namely the ordered triple

$$m(s) = \langle C(s), E(s), S(s) \rangle.$$

We call any such triple a *CES model* of a given system. However, such a CES model is only a minimal model of a concrete system, because it says nothing about the history and the laws of the system. Evidently, the history of a system is of particular interest in the case of biosystems and sociosystems.

An alternative and much finer representation of a concrete system can be attained by means of the concept of state space. The state space of an aggregate or conglomerate of things is uniquely determined by the state spaces of its parts. Moreover, since the contributions of the latter are all on the same footing, we may take the total state space of an aggregate to equal the union of the partial state spaces. In the case of a system, however, the state of every component is determined, at least partly, by the states other components are in, so that the total state space is no longer the union of the partial state spaces.

Before proceeding with material systems, it should be noted that the CES model can also be applied to conceptual systems. A *conceptual system* is a system composed of constructs linked together by logical or mathematical relations. The environment of a conceptual system is the body of knowledge to which it belongs, e.g., algebra, paleontology, or economics. Classifications and theories, for example, are conceptual systems. Although conceptual systems are individuals (particulars), the attribution of change and evolution to such individuals, e.g., theories, is in the category of metaphysically ill-formed statements. (Of course, people may change their mind about theories and may alter or even drop them. For example, biologists have produced different theories of evolution, but there has been no such process as the evolution of evolutionary theory. This is not to deny the relevance of any "history of ideas" but a warning against mistaking fiction and analogy for fact.) It will further be obvious that, in line with our conceptual/material (or formal/concrete) dichotomy (see Postulate 1.2), we deny the existence of mixed, i.e., half-conceptual and half-material, systems.

Most concrete systems are composed not only of simple things but of subsystems, which, in turn, may be composed of further subsystems. In other words, many systems consist of nested systems. With the help of the CES analysis of systems we can define the notion of subsystem as follows:

DEFINITION 1.8. A thing x is a *subsystem* of a system s, or $x \lhd s$, iff
(i) x is a system;
(ii) $C(x) \subseteq C(s)$, $E(x) \supseteq E(s)$; and $S(x) \subseteq S(s)$.

Note that the subsystem relation $\lhd$ —like the part-whole relation—is an order relation, i.e., it is reflexive, asymmetric, and transitive.

The complexity of the real systems studied in most sciences forces us to analyze the concepts of composition, environment, and structure of a system into as many levels as needed. For example, the molecular biologist may be interested in the molecular level of a cell, the cytologist in the cellular level of a (multicellular) organism, the histologist in its tissue level, and the gross anatomist in its organ level. In other words, it is often useful to relate the CES model of a system to a certain level L of individuality, in which the scientist is interested in. (More on levels, especially biolevels, in Sect. 5.3.) In doing so, we speak of the *L-composition*, the *L-environment*, and the *L-structure* of a system s. The CES model of s now reads thus: $m_L(s) = \langle C_L(s), E_L(s), S_L(s) \rangle$, where $C_L(s) = C(s) \cap L$, and analogously for $E_L(s)$ and $S_L(s)$.

Furthermore, it may be useful to restrict the analysis of a system not only to a certain level of interest but also to a certain time t or to some time interval τ. In the former case, a CES model of any system s will be symbolized thus: $m_L(s, t) = \langle C_L(s, t), E_L(s, t), S_L(s, t) \rangle$. The notation for a time interval τ is analogous. We shall use any of these notations in the following whenever appropriate.

A systemic analysis of systems should always be carried out in the form of a CES model. Neglecting one or more of the three coordinates leads to a nonsystemic, in particular reductionist, approach. (See also Wimsatt 1982a.) Atomists focus on composition and ignore environment and structure; environmentalists and holists focus on environment or supersystem (or even "holon") and thus ignore composition and structure; and structuralists neglect composition and environment. Only the scientist who pays attention to the multiplicity of levels will tend to cross some of the artificial frontiers between fields of inquiry: he or she will tend to adopt an *interdisciplinary* approach.

With regard to the CES analysis of a system, we must finally warn against the careless usage of the term 'structure' in biology, especially in morphology. When talking about subsystems (organs) or morphological features of organisms, biologists usually speak of "structures". Given the concept of structure introduced above, it is obvious that this is an instance of reification. A subsystem of an organism *is* not a structure but *has* a structure. If both the system itself and its structure are termed 'structure', we are led to speak of the structure of a structure. To avoid such problems, we shall use the term 'structure' only in its proper sense, which is the one elucidated in mathematics. (See also Young 1993.)

1.7.3 Emergence

Evolutionary biology has spawned at least three major ontological concepts: those of evolution, level of organization (or, better, level of system or individuality), and emergence of qualitative novelty (or of things possessing radically new properties). It also suggested that the three concepts are mutually related in a straightforward manner, constituting the hypothesis that *new levels emerge in the course of evolution*. (Strictly speaking, this is a metaphysically ill-formed statement: see Sect. 5.3.) This hypothesis has spilled over from biology to all the factual sciences. Indeed, it is found, either explicitly or tacitly, in such diverse fields as the studies of molecular evolution, astrophysics, developmental and comparative psychology, and social history (or historical sociology).

Despite its importance to evolutionary biology, the ontological concept of emergence is sometimes resisted because it is misunderstood (e.g., Ayala 1983, as well as in Dobzhansky et al. 1977, p. 489; Rosenberg 1994) For example, emergence is sometimes equated with ignorance of the mechanism resulting in the assembly of new things from their constituents or their precursors. It is argued that, if we only knew the exact composition of a thing and the bonds among its components, emergence would be seen for what it is, namely an illusion. An obvious rejoinder is that explained novelty is no less novel than unexplained novelty, and predicted novelty is no less novel than unpredicted (or perhaps even unpredictable) novelty: the concept of emergence is ontological, not epistemological.

The concept of emergence was introduced by G.H. Lewes (1879), who distinguished between resultant and emergent properties. (See Blitz 1992 for a history of the concept.) A property of a whole which is also possessed by some of its parts is said to be *resultant* . If, on the other hand, a property of a whole is not possessed by any of its components, it is called *emergent*. For example, the property of being alive is an emergent property of cells, but a resultant property of multicellular organisms.

To repeat, these definitions have to do with things and their properties, not with our knowledge of them. That is, they belong in ontology, not in epistemology. However, there are also emergent "properties" (attributes) of conceptual wholes, in particular of geometrical objects. For example, connecting three straight lines at their ends yields a (conceptual) whole with emergent attributes, namely a triangle. A more important example for biophilosophy are the statistical artifacts such as averages, modes, and variances, which also depend upon the knowing subject. Such statistical parameters may be called *aggregate* or *collective* attributes, for they result from the aggregation of properties of the individuals constituting the population in question (see Lazarsfeld and Menzel 1961). A collection, or aggregate, or population *in the statistical sense* is characterized by such aggregate attributes. However, these statistical attributes do not correspond to any substantial (emergent) property of a material whole. For example, when saying that a certain population of snakes has an average body length of 95 cm, we refer to the statistical attributes of a collection of individuals. A concrete (physical) aggregate or population of snakes may have a certain height and length when we pile the snakes on top of each other; but this population as a material whole has neither a body nor a body length, and, *a fortiori*, it has no average body length. In other words, the so-called aggregate or collective "properties" have no ontological status: they are conceptual, not substantial or material, emergents. (See also Horan 1994.) We shall deal only with material emergents in the following.

The thing-construct and property-predicate distinctions are especially important with regard to the ambiguous notion of biological population, which can mean either a statistical aggregate of organisms, or a concrete aggregate, or perhaps a system composed of organisms, such as a reproductive community (more on this in Sect. 4.5). Similarly, in the group selection debate, the answer to the question of what genuine group properties are depends on whether groups are conceived of as statistical or concrete entities. For example, the usage of average fitness values in models of group selection does not render the groups in question material wholes, in particular units of group selection. On the other hand, emergent properties of groups as material wholes would, for instance, be the density and age "structure" of a population, as well as the cohesion of a social group. (More on this in Sect. 9.2.)

We are now ready for a somewhat more formal elucidation, namely:

DEFINITION 1.9. Let P represent a property of a thing b. P is an *emergent* property of b if, and only if, either

(i) b is a complex thing (a system), no component of which possesses P;

or

(ii) b is a thing that has acquired P by virtue of becoming a component of a system (i.e., b would not possess P if it were an independent or isolated thing).

The dual notion of *submergence*, or loss of properties, is definable in a similar way. Relating the notions of emergence and submergence to the state space approach introduced above, we can say that the emergence of a new property is represented by the budding of a new axis in the state space, and the submergence of a property by the pruning of an axis (Figs. 1.2 and 1.3). The two concepts in question occur in:

POSTULATE 1.9. All processes of development and evolution are accompanied by the emergence or the submergence of (generic) properties.

Our Definition 1.9 is more inclusive than Lewes's. Indeed, it encompasses two types of emergence: *intrinsic* (or global) and *relational* (or structural, or contextual). The first refers to a system as a whole. The second refers to a property a thing acquires when becoming part of a system. (See also Levins and Lewontin 1985, p. 3.) Examples of (i) are the properties of having a structure (a complex, not a simple thing), being alive (a cell, not a molecule), or being conscious (a highly evolved brain, not a neuron). Examples of (ii) are those of being a gene (not an isolated string of DNA), an offspring, a mating partner, a predator, a prey, or an alpha (fe)male.

We emphasize again that material emergence is an ontological category, not an epistemological one. In other words, we assume emergence to be a feature of the real world. However, as mentioned previously, emergence is often identified with unexplained (or even unexplainable) or unforeseen (or even unpredictable) novelty (e.g., Salt 1979). Yet to a realist, qualitative novelty is independent of our ability to explain or predict it. Only the introduction of radically new ideas depends upon the knowing subject. Such ideas may be conjectured to consist in the emergence, through spontaneous self-assembly, of new plastic neuronal systems (see Sect. 3.1.1, as well as Bunge 1993).

Now, material emergence is emergence *from* precursors or emergence *in* a process of incorporation into a system. Thus the occurrence of a new property or properties E depends upon the previous existence of some basic or precursor property or properties B. The two cases distinguished in Definition 1.9 above can then be described by propositions of the types:

1. A thing, possessing properties B at time t, develops into a thing with new properties E at time t', with $t' > t$.
2. A thing, possessing properties B at time t, acquires new properties E upon becoming a component of a system at time t', with $t' > t$.

The concepts of gain and loss of properties of a thing over a period $[t, t']$ can be elucidated as follows. Let $P(t)$ and $P(t')$ designate the set of all properties of a thing at times t and t', respectively. The gain (γ) and the loss (λ) of properties over this period are $\gamma(t, t') = P(t') - P(t) = E$, and $\lambda(t, t') = P(t) - P(t')$, respectively.

When the emergent properties are either important or numerous, we are justified in referring to one thing at time t, and a different thing—i.e., either one in a different phase or one of a different kind—at time t'. Examples: change in state of condensation (e.g., liquid → vapor), metamorphosis, and speciation. However, in the case of development, the emergence and submergence of qualities is included in the very characterization of the (biological) species to which the individual belongs. Thus, for instance, the butterfly belongs to the same species as the caterpillar, from which it emerges through metamorphosis. Only in the case of evolution do we speak of speciation.

How do these processes occur: what are the emergence mechanisms? In other words, how does E depend upon B? Again: what is the function that maps the set B into the set E? Only a pure mathematician or a Hegelian metaphysician could come up with such a general question. A biologist, or any other factual scientist, knows that there is no general answer: the answer depends on the nature of the case. In other words, there are uncounted emergence mechanisms, and they have hardly anything in common except for the occurrence of new properties. Thus, the process whereby two germ cells combine to form a zygote is totally different from the process whereby two atoms join to constitute a molecule. Likewise, the process whereby an organism acquires a mating partner is utterly different from the process whereby an ion captures an electron. Moreover, neither of the four processes just mentioned is ever described as a mapping from B to E detached from the things that possess B or E. Instead, emergence processes are described, for instance, as changes occurring in things through self-assembly and self-organization, through internal rearrangement, through interactions with the environment, or any combination thereof.

The preceding casts severe doubts on the ability of the concept of *supervenience* to cope with qualitative novelty and to succeed as an alternative to the concept of emergence. Indeed, talk of supervenience is talk about properties in themselves (i.e., detached from the things possessing them), some of which—the negative and disjunctive ones—cannot be possessed by anything concrete (i.e., the property-predicate distinction is missed), which do not arise in the course of a process occurring over time, and which do not involve any systems. To show that this is indeed the case, let us examine briefly the standard logical (rather than ontological) treatment of supervenience proposed by Kim (1978) and adopted by Rosenberg (1978, 1985) and others.

Let M be a set of unary properties (actually predicates) and M^* the superset formed by adding to M the complements of all the members of M as well as the disjunctions and conjunctions of any two members of M. Take now a second set N of unary properties of entities in some domain D of objects and form the superset N^* formed in the same way as M^*, i.e., adding the complements, disjunctions, and conjunctions. (One may think of M and N as basic mental and neurophysiological, or else moral and physical, predicates respectively.) M is said to be *supervenient on* N just in case, necessarily, objects in D which share all properties in N^* also share all properties in M^*. (Kim's original construction is more complicated: it involves two further sets, one derived from M^* and the other from N^*. But he makes no use of them in defining "supervenience". Furthermore, Kim himself admits that the notion of necessity occurring in his definition is unclear. It could be logical necessity, as in deduction, or natural necessity, as in lawfulness.)

Kim claims that his concept elucidates Donald Davidson's fuzzy notion that "mental characteristics are in some sense dependent, or supervenient, on physical characteristics". However, this is not so, for the N above is not constructed out of M. Indeed, according to Kim's definition, the relation of supervenience is an atemporal relation between two distinct sets that, after being defined independently from one another, are found to mirror one another. In other words, all we know upon being told that M is supervenient on N, is that N^* mirrors M^* and conversely (shorter: supervenience is bijective, not injective). Since the supervenience relation is symmetric, one might as well say that N is supervenient on M rather than the other way round. In other words, no one-sided dependence is involved in supervenience, much less any idea of process or time. Consequently, Kim's concept of supervenience might be used to exactify the idea of psychophysical parallelism, but not that of emergence of mental functions as a concomitant of the organization or reorganization of neural systems.

Rosenberg (1978, 1985) believes that Kim's notion of supervenience would help account for properties such as fitness (or adaptedness), which may result from very different underlying properties. That is, two organisms belonging to different taxa, hence possessing different morphological and physiological properties, may happen to have the same fitness value. Rosenberg claims this to be a case where the fitness of an organism *supervenes* on its "physical" (actually: biological) properties. However, there is no need to invoke a (pseudo)technical term in this case, any more than there is to account for cases such as 1 kg of bread and 1 kg of butter, or the nearly equal longevity of humans, parrots, and tortoises. All three are cases of coincidence. Moreover, they do not involve the occurrence of qualitative novelty. (The fact that the individual values of properties like weight or adaptedness may be coincidental in the various individuals that happen to possess these properties does not preclude that we use the notion of weight in a physical theory or that we use the notion of adaptedness in a biological theory, particularly in the theory of selection. After all, weight is a generic property of all things endowed with a mass that happen to be in a gravitational field, and adaptedness is a generic relational property shared by all living systems *qua* living systems occurring in a given environment: see Sect. 4.2.)

To conclude: the notion of supervenience—a static and symmetric relation—is irrelevant to qualitative novelty, which always occurs in the course of some process or other, particularly in developmental and evolutionary processes. Moreover, involving the confusion of properties with predicates, we see no use for it in a scientific metaphysics.

1.7.4 Assembly and Self-Organization

Two important notions in any naturalistic ontology are the concepts of self-assembly and self-organization. Any process whereby a system is formed out of its components is called *assembly*. An assembly process can occur either in a single step or, more likely, in a number of steps, and it can happen either spontaneously (naturally) or artificially (i.e., man-guided as, for example, in a technological lab or industrial plant). Since in biology we are mainly interested in natural processes, we suggest the following:

DEFINITION 1.10. Let x denote a thing that, at time t, is composed of uncoupled parts—a mere heap or aggregate of things without bondage—i.e., $B(x, t) = \emptyset$. Then

(i) x *assembles* into y at time $t' > t$ iff y is a system with the same composition as x but a nonempty bondage, i.e., $C(y, t') = C(x, t) \neq \emptyset$ & $B(y, t') \neq \emptyset$;

(ii) the assembly process is one of *self-assembly* iff the aggregate x turns spontaneously (i.e., naturally rather than artificially) into the system y;

(iii) the self-assembly process is one of *self-organization* iff the resulting system is composed of subsystems that were not in existence prior to the onset of the process.

Examples of self-assembly are the formation of crystals out of solutions, microspheres out of thermal proteins, and microtubules out of globular protein molecules. An example of self-organization is the formation of an embryo's organs, which constitute subsystems that did not exist prior to its development. As self-assembly and self-organization occur at all levels, the corresponding concepts are not exclusive to biology but genuine ontological concepts. (For a voluminous treatise on self-organization in biology see Kauffman 1993.)

1.8 Fact

1.8.1 Objective Fact

Having elucidated the concepts of state and event, we can now say that a *fact* is either the being of a thing in a given state, or an event occurring in a thing. Constructs do not qualify as facts since they are not objects that can be in a certain state, let alone undergo changes of state. Therefore, we should not call a true factual proposition a 'fact'. (The view that facts are theory-dependent or else empirical data rather than things "out there" is rampant in the philosophy of biology: see, e.g., Sattler 1986; Shrader-Frechette and McCoy 1993; van der Steen 1993.) In other words, a well-confirmed hypothesis, such as the hypothesis of descent with modification, is not a fact: it *refers* to a fact, i.e., a process or, more precisely, a number of processes. Similarly, there are no "scientific facts": only a procedure to attain knowledge can be scientific (or not), not the object of our investigation. Accordingly, scientists neither "collect" facts nor do they come up with or, worse, "construct" facts, but advance hypotheses and theories referring to or representing facts. Of course, some of those hypotheses may turn out to be false, either for referring to purely imaginary objects, or for describing incorrectly real facts.

Furthermore, the nonoccupancy of a state and the nonoccurrence of an event are not facts. Thus, not having been inoculated against polio is not an event, hence it does not count as a cause of polio. In short, there are no negative facts (neither are there disjunctive facts). Finally, although facts are objective, they do not—*pace*

Wittgenstein (1922)—constitute the world. The world is not the totality of facts but the totality of things, i.e., of material or changeable objects.

Finally, we must distinguish microfacts from macrofacts. Consider a system and its components at some level—e.g., a multicellular organism and its organs. A *macrofact* is a fact occurring in the system as a whole. A *microfact* is a fact occurring in, or to, some or all of the parts of the system at the given level. For instance, a shark's roaming the sea in search of food is a macrofact, whereas its digesting the devoured prey is a microfact, i.e., a process in one of its subsystems. We will need the notions of microfact and macrofact when dealing with micro- and macroexplanation (see Sect. 3.6).

1.8.2 Phenomenon

A *phenomenon* is a perceptual appearance to someone: this is the original meaning of the word. Yet in ordinary language, in the scientific literature, and even in philosophy 'phenomenon' is now often used as a synonym of 'fact'; so much so, that some authors even speak of observed and observable as well as of unobserved and unobservable phenomena. (For some such misfits see, e.g., van Fraassen 1980.) However, although there are imperceptible (or transphenomenal) facts (or *noumena*), there are no phenomena without sentient organisms. Appearance, then, is an evolutionary gain, which emerged together with the first animals equipped with sufficiently complex sensors and nervous systems. Accordingly, 'observed phenomenon' is a pleonasm, and 'unobservable phenomenon' is a *contradictio in adjecto*.

Since phenomena are events occurring in some nervous system or other, they are facts. (More precisely, they are subject-object relations or semisubjective facts.) So, phenomena are real, but the set of phenomena is a smallish subset of the collection of facts. And since different animals are never in the same state and can never adopt the same standpoint, an objective fact is bound to appear differently, or not at all, to different animals in different circumstances. In sum, there is no one-to-one correspondence between facts and appearances. As this thesis is part and parcel of scientific realism, it deserves a postulate of its own (Bunge 1980):

POSTULATE 1.10. Let Φ designate the totality of possible facts occurring in an animal b and its (immediate) environment during the lifetime of b, and call Ψ the totality of possible percepts of b (or the phenomenal world of b) during the same period. Then Ψ is properly included in Φ, i.e., $\Psi \subset \Phi$.

This assumption suggests two rules:

RULE 1.1. All sciences should investigate possible real facts and should explain phenomena (appearances) in terms of them rather than the other way round.

RULE 1.2. A science-oriented ontology and epistemology should focus on reality, not appearance.

Thus, whereas the phenomenalist formulates subject-centered sentences such as 'Phenomenon x appears to subject y under circumstance z', the realist will say 'Phenomenon x, appearing to subject y under circumstance z, is caused (or is indicative of) noumenon w'—a sentence involving subjective as well as objective items. Eventually, the realist will attempt to construct strictly objective sentences such as 'Fact x occurs under circumstance z'. (For phenomenalism see Sect. 3.2.2.)

1.9 Causality

1.9.1 Broad (or Inflationary) Use of the Term 'Cause'

The term 'cause' is used in a very broad sense in philosophy as well as in biology, where almost anything that "makes a difference" is regarded as a cause. On the one hand, this is due to the fact that the term 'cause' is part of ordinary language which makes it seemingly comprehensible to everyone without further analysis. On the other hand, it is due to Aristotle's still influential theory of causation, which distinguished four types of cause: the *causa materialis* (or stuff), the *causa formalis* (or shape), the *causa efficiens* (or force), and the *causa finalis* (or goal). This doctrine of causation still holds strong among biologists (see, e.g., Riedl 1980; Rieppel 1990). Final causes survive in the concept of teleonomy and of genetic program (Mayr 1982, 1988); formal causes are referred to in what is called 'downward causation' (Campbell 1974; Popper and Eccles 1977; Riedl 1980; Petersen 1983); and material causes underlie the dual of 'downward causation', namely 'upward causation', as it is invoked, for instance, in the talk of selection *for properties* (Sober 1984). Furthermore, material causes are involved in the notion of ultimate causes as opposed to proximate ones, which may be material or efficient causes (Mayr 1982, 1988).

Aristotle's doctrine was strongly criticized at the beginning of modern times. Especially, Francis Bacon dismissed final causes as being both barren and dedicated to religion, and Galileo accepted only efficient causes, which he equated with motive forces. One century later, David Hume rejected even efficient causes, claiming that there are only conjunctions and successions of events. This became the dominant view on causation because it matched the most popular philosophy of science, namely positivism. Of course, many scientists never actually followed the Humean analysis of causation, because scientists are seldom satisfied with the mere description of successive events, but are seeking explanation, which amounts to the search for the mechanisms underlying the observable conjunctions and successions of events.

1.9.2 Causation as Energy Transfer

We shall adopt here a restricted concept of causality, which boils down to the thesis that causation is a mode of event generation by energy transfer from one entity to another (Bunge 1959a). More precisely, we submit that:

1. The causal relation relates events. That is, only changes can be causally related. When we say that thing *x* caused thing *y* to do *z*, we mean that a certain event in *x* generated a change of state *z* in *y*. In other words, causation is a mode of becoming, not of being. Consequently, neither things, nor properties, nor states (particularly antecedent and succedent states of the same thing—*pace* Mill 1875 and others), let alone ideas, are causally related.

2. Unlike other relations among events, such as that of simultaneity, the causal relation is not external to them: every effect is somehow produced by its cause or causes. In other words, causation is a mode of event generation.

3. There are at least two different causation mechanisms, which we call *strong energy transfer* (or [complete] *event generation*) and *weak energy transfer* (or *triggering signal*, or *event triggering*). The proverbial bat hitting a baseball exemplifies the first kind of causal mechanism; a zebra fleeing at the mere sight of a lion is an instance of the second kind of causal mechanism, so is a flower blooming in sunshine. In the first case, all the energy needed for the change of state of the *patiens* is provided by the causative *agens*. In the second case, the energy transfer is too small to produce the entire effect in the *patiens*. Still, the small amount of energy transferred by some signal triggers some macroevent in the *patiens*.

4. The causal generation of events is governed by laws. That is, there are causal laws or, at least, laws with a causal range. (Yet not all laws are causal.)

5. Causes can modify propensities (e.g., to flee or fight) but they are not propensities, and therefore causation cannot be elucidated in probabilistic terms. (See also Bunge 1973c.) In the expression 'Event *c* causes event *e* with probability *p*' (or 'The probability that *c* causes *e* equals *p*') the terms 'causes' and 'probability' are not interdefinable. As a matter of fact, the concept of causation is logically prior to that of conditional probability. Moreover, strict causality is nonstochastic.

6. Although the world is determinate, it is not strictly causal. That is to say, not all interconnected events are causally related, and not all regularities are causal. So causation is just one of the modes of determination—albeit a rather pervasive one. Hence, determinism should not be conceived narrowly as causal determinism. Science is deterministic in a broad sense: it requires "submission" to laws (of any kind) without magic. In other words, scientific determinism in the broad sense assumes that there are no miracles, and that things do not come out of nothing or go into nothingness (Bunge 1959a).

7. Properties and states do somehow "make a difference" to future states of things or events in things. However, since they do not generate events, we have to regard them as conditions or determinants (in the broad sense), not as causes. In other words, an elucidation of causality in terms of, say, "anything that makes a difference" is too coarse to distinguish determination in general from causation or causal determination.

Our (ontological) view of causality can be rendered more precise with the help of the previously introduced state space approach. We proceed to provide such an elucidation. (The less formally inclined reader may skip this section.)

1.9.3 A State Space Approach to Causation

Consider two different things, or parts of a thing, of some kind(s) or other(s). Call them x and y, and call $H(x)$ and $H(y)$ their respective histories over a certain time interval. Moreover, call $H(y \mid x)$ the history of y when x acts on y. Then we can say that x *acts* on y if, and only if, $H(y) \neq H(y \mid x)$, that is, if x induces changes in the states of y. The *total action* A (or *effect*) of x on y is defined as the difference between the forced trajectory of y, that is, $H(y \mid x)$, and its free trajectory $H(y)$ in the total state space of x and y. In symbols, $A(x, y) = H(y \mid x) - H(y)$; likewise for the reaction of y upon x. The *interaction* between things x and y is the union of $A(x, y)$ and $A(y, x)$.

Finally, consider a change c (event or process) in a thing x over a period τ_1, and another change e (event or process) in a thing y over another period τ_2. (One of the things may be a part of the other, and the changes and periods are taken relative to a common reference frame.) Then we say that c is a *cause* of e if, and only if, (a) e begins later than c, and (b) the history $H(y \mid x)$ of y over τ_2 is included in the total action $A(x, y)$ of x on y over the period $\tau_1 \cup \tau_2$. In this case e is called an *effect* of c.

Having defined the notions of cause and effect, we may now state the *principle of strict causality*, namely "Every event is caused by some other event". To be more precise: "Let x be a thing with nomological event space $E(x)$ (relative to some reference frame). Then, for every $e \in E(x)$, there is another thing $y \neq x$, with nomological event space $E(y)$ relative to the same reference frame, such that $e' \in E(y)$ causes e." Having stated the principle of strict causality, we hasten to note that it holds only for a proper (and perhaps small) subset of all events, because it neglects two ever-present features of becoming, namely spontaneity (self-movement) and chance.

Even if they agree with our analysis of causality, many philosophers will not be happy with it, because it restricts the usage of the concepts of cause and causality. In particular, many instances of what are now regarded as causal laws and causal explanations would not be properly causal on our construal. For example, Einstein's law_2 "$E = mc^2$"is not causal, because neither the energy nor the mass of a body are causes of each other: they are just constantly related; and the explanation of development in terms of genes is not a causal explanation, because genes do not cause anything: they are just passive templates (see Sect. 8.2.3.2). Still, such laws and explanations may be called 'deterministic' in the broad sense. Thus, we must distinguish deterministic events (in the broad sense) from causal events proper.

1.9.4 Causes and Reasons

The conflation of causes and reasons has haunted ordinary language up until today. This may be due to radical rationalists, such as Descartes, Spinoza, and Leibniz, who equated causes with reasons and demanded that a reason be advanced for whatever exists and happens. Thus, one of Leibniz's favorites was his principle of sufficient reason: "Nothing happens without a sufficient reason". He conflated the ontological principle of causality with the rule or procedure which demands that we give a reason for, or justify, our beliefs and actions. Yet, in accordance with our thing-construct distinction, we maintain that reasons must be distinguished from causes, because the former are constructs, while the latter are (real) events. For example, reasons, but not causes, can be either logically correct or invalid; and causes, but not reasons, can change the world. However, thinking of and giving a reason, i.e., reasoning, is a brain process that may trigger an action in the animal in question. Hence, reasons regarded as thoughts (not as constructs) can be causally efficacious.

1.9.5 Causation in Biology

Having introduced our view of causality as a mode of event generation, we proceed to examine its consequences for biology, as well as some uses and misuses of the term 'cause' in biology.

A first important consequence is to realize that neither properties of things nor laws are causally efficacious. (For the view that properties are causally efficacious see, e.g., Sober 1982, 1984; Sober and Lewontin 1982; Walton 1991. We presume that, what, e.g., Sober and Lewontin have in mind, is that certain properties are lawfully related.) Another consequence is that prior states of a thing are also not causes of its posterior states. For instance, today is not the cause of tomorrow, youth is not the cause of old age, and the caterpillar is not the cause of the butterfly. More precisely, an ontogenetic stage is not a cause of another, although the former certainly codetermines (or "makes a difference" to) the latter. Since things are not causes either, it follows that, for example, ancestors are not causally related to descendants via reproduction—*pace* Wilson (1995), so that the ancestor-descendant relation is not a bonding relation. Neither can DNA sequences or genes be causes of anything, such as phenotypic characters, behavior patterns, or what have you. Genes contribute to determining phenotypic traits but they do not act upon them. This consequence is of particular interest in the age of molecular biology, when genes are usually regarded as the *prima causa* or *primum movens* of living beings (see Sect. 8.2.3.2). In sum, there are no material causes, but only efficient ones.

With the rise of systemic thinking in biology and, particularly, with the notion of levels of organization or, better, levels of systems (see Sect. 5.3), it became obvious that a multilevel system, such as an organism, can be viewed either "bottom-up" or "top-down". Thus the upward (or Democritean) view takes the higher-level properties and laws of a system to be determined by (hence reducible to) the properties and laws of its components (e.g., by its genes); shorter: the parts determine the whole. The rival (or Aristotelian) view is the downward one, according to which the higher-level properties and laws of a system determine its components. In other words, the whole determines its parts. We submit that the truth lies in a synthesis of the upward and downward views, and that neither of these should be formulated in terms of causation, as is done, e.g., by Campbell (1974), Popper and Eccles (1977), Riedl (1980), Petersen (1983), Vrba and Gould (1986), and Rieppel (1990). What we do have here is not *causal* relations but *functional* relations among properties and laws at different levels. Again, there are neither material nor formal causes.

The distinction between *proximate* and *ultimate* causes has become commonplace in biology (Mayr 1982, 1988). In order to explain any morphological or behavioral feature, such as the behavior of migrating birds, we would have to take two levels of (alleged) causation into account. The proximate level consists in the physiological mechanism that produces or triggers the behavior, such as the effect of diminishing daylight and temperatures on the physiology of the bird, or the developmental pathways in the case of a morphological character. The ultimate cause, by contrast, would be the evolutionary history of the organ or behavior as contained in the "genetic program", which thus has to be regarded as a material as well as final cause.

As will be obvious from the previous considerations, what is called 'proximate causes' may indeed be such, but there are no such events as ultimate causes. Undoubtedly, the history of an individual and, particularly, the history of its genetic material are determinants of the developmental processes leading to its current morphology and behavior, but they do not cause it. What is true is that the history of a system provides some of the *conditions* or *circumstances* of the system's possible changes. Therefore, the expression 'ultimate cause' should be replaced by the expression *historical condition*, or *distal cause* in the case of a genuine past cause.

1.10 Chance and Probability

1.10.1 Chance or Randomness

The word *chance* designates at least the seven following different concepts:

1. *Chance as a manifestation of fate or divine will.* Most primitive and archaic world views make no room for either accident or randomness. According to them, everything happens either by natural necessity or by supernatural design. Hence, chance would be only apparent: it would be a name for our ignorance of either necessity, fate, or divine will. Needless to say, science has no use for this concept of chance.

2. *Chance as ignorance of necessity*. The world unfolds according to deterministic (usually called 'causal') laws, but we have only partial knowledge of such laws as well as of circumstances, so our forecasting powers are limited. In other words, chance is in the eye of the beholder. Example: games of chance. (God, who is supposed to be omniscient, could not gamble in good faith, for he would be able to foresee the outcome of every game of chance.)
3. *Chance as intersection of initially independent "causal" lines*. This is an application of the foregoing to two or more processes. Examples: the chance encounter of two friends, Brownian motion, and the (causal) propensity to survive and reproduce in a particular habitat.
4. *Chance as a feature of either extreme instability or complexity*. Example 1: a lever in equilibrium can tilt now to the right, now to the left, under the action of a minute perturbation. Example 2: a tiny alteration in the initial position or velocity of the ball in a roulette game is likely to have a momentous effect.
5. *Chance as an outcome of random (arbitrary) sampling*. Even if the objects of study are not random, we introduce chance into them every time we pick arbitrarily (blindly) only some of them.
6. *Chance as a result of the congregation of mutually independent items of a kind*, such as students in a course coming from different families and social backgrounds.
7. *Chance as a natural (nonartificial) and basic (irreducible) disposition*. So far, only quantum physics and genetics have used this concept. Indeed, such quantum events as radioactive decay and certain gene mutations, are assumed to be irreducibly random, as being the outcome of quantum processes at the nuclear and molecular level, respectively. While some random processes are endogenous, others are triggered by collisions with the surrounding material or with cosmic rays. (Some mutations are random not because they are not designed or functional—an obsolete meaning of 'chance'—but because they are results of collision or scattering processes satisfying the laws of quantum mechanics, which is a radically probabilistic theory. Thus, an individual gamma ray photon has a definite propensity of ionizing an atom or dissociating a molecule, which event, in turn, has a definite propensity of triggering a chemical reaction constituting a genic change.)

We conclude that modern science, in particular since the advent of quantum physics, takes it for granted that there are chance events out there and that there are things possessing chance propensities. That is, chance (or randomness) is objective, and it is an ontological category. In other words, the world is no longer seen as strictly deterministic in the way described classically by Laplace. Indeed, if the universe is strictly causal, randomness is resorted to only because of our ignorance of details and of the ultimate causes. Accordingly, the philosopher who believes in strict determinism must take chance and probability to be epistemological categories. For him or her, chance is a substitute for ignorance, and probability only measures his uncertainty about facts. (This is the tacit ontological thesis of subjective or Bayesian probability: see below, as well as Rosenberg 1994)

However, the objectivity of randomness does not imply that the world is completely indeterminate, chaotic, or lawless. Randomness is a type of order: there are stochastic or probabilistic $laws_1$, and even random events often depend on circumstances. The fact that some random facts may be dependent on the occurrence of some other facts shows that randomness and nonrandomness are not dichotomic: randomness comes in degrees, i.e., probability values range between 0 and 1. Accordingly, we may speak of fully random and partially random events and processes. Thus, to assert, for instance, that natural selection is not a random process (Sober 1993) can only mean that it is not a *fully* random process, though it may still be *partially* random—which, of course, is a matter of debate (see Sect. 9.2).

In sum, the universe is not strictly deterministic, but rather deterministic in the broad sense of lawful.

1.10.2 The Mathematical Theory of Probability and Its Interpretations

If chance propensities are objective properties of things, the concept of probability can be regarded as providing a *measure* of the tendency (or propensity or possibility) of a thing to undergo a certain event. Yet philosophers have used (or misused) the concept of probability to elucidate notions as diverse as those of uncertainty, plausibility, credibility, information, truth, confirmation, simplicity, causality, and others. Since we do not regard the concept of probability as an all-purpose concept, we must briefly examine it in more detail (see Bunge 1977a, 1981b, 1985a, 1988.)

1.10.2.1 The Propensity Interpretation

Probability theory is a branch of pure mathematics. In fact, a probability measure is a real valued and bounded function P (often also abbreviated Pr) defined on a family S of abstract sets. The only conditions that S must satisfy, i.e., the ones that guarantee the (formal) existence of P, are the following purely formal ones: (a) the complement of every member of S is in S; (b) the intersection of any two members of S is in S, and (c) the countable union of any elements of S is in S. The function P is specified by two or more axioms, depending on the theory. (One of them is that, if $a, b \in S$ and $a \cap b = \emptyset$, then $P(a \cup b) = P(a) + P(b) = 1$. Another is that the probability of the complement $\bar{a}$ of a set a equals $1 - P(a)$.) Nothing is said in these theories about events (except in the Pickwickian sense of being members of the family S) or their frequencies, let alone about methods for measuring probabilities. These other notions occur in applications.

As long as the probability space S is left uninterpreted, probability has nothing to do with anything extramathematical: $P(x)$, where x belongs to S, is just a number between 0 and 1. Therefore, if we want to apply the notion of probability to anything real, we must *interpret* this purely formal probability concept, i.e., endow it with some extramathematical referents. (Incidentally, this semantical interpretation has nothing to do with definition or explanation.)

In doing so, we submit that there is only one correct interpretation in science and technology: the *realistic* or *propensity* interpretation. More precisely, we assume that the probability space S has to be interpreted as a collection of *random factual items*, and $P(x)$, for every x in S, as the quantitation of the objective possibility of x. This assumption presupposes the hypothesis that chance is objective and a property of every one of the individual facts forming the basic probability space S: in short, *no randomness, no probability*. (This interpretation was introduced by Poisson in 1837, has been widely used by physicists since then, and popularized by Popper under the name 'propensity interpretation': see, e.g., Popper 1957b.)

Using the notion of a state introduced above we can, for instance, assume that every member of the probability space S is a bunch of states. Then $P(x)$ can be interpreted as the strength of the tendency (or propensity) the thing has to dwell in the state or states x. Similarly, if x and y are states (or sets of states) of a thing, the conditional probability of y given x, i.e., $P(y \mid x)$, is interpreted as the strength of the propensity for the thing to go from state(s) x over to state(s) y.

1.10.2.2 Objections Against the Propensity Interpretation

The propensity interpretation of probability has been criticized by some philosophers for sundry reasons; suffice it to mention only two of them.

The Propensity Interpretation is Allegedly Circular. The propensity interpretation of probability has been suspected of being circular. That is, it has been claimed that the notion of propensity is used to define or explain that of probability, and the concept of probability, in turn, is used to define or explain that of propensity—an objection recently repeated by Sober (1993). This objection dissolves upon correctly distinguishing the notions of interpretation, definition, explanation, and exactification (or formalization). We submit that what is involved here are interpretation and exactification. More precisely, we must distinguish the *propensity interpretation* of probability from the *probability exactification* of the presystematic notion of propensity. An interpretation attaches factual referents to a mathematical concept, e.g., things possessing chance propensities to a probability space S. By contrast, an exactification consists in endowing an intuitive factual concept with a precise mathematical structure, e.g., in mathematizing or formalizing the concept of propensity in terms of probability theory (Bunge 1974b, 1977a). Thus, no circularity is involved.

The Propensity Interpretation Would Not Match Bayes's Theorem. Earman and Salmon (1992) have recently claimed that the propensity interpretation is inadmissible because Bayes's theorem—a formula in the mathematical theory of probability—cannot always be interpreted in terms of propensities. This is true, but it does not affect the propensity interpretation. To see why this is so, let us begin by reviewing the theorem in question, which, in turn, requires us to recall the notion of conditional probability.

The probability of a fact may or may not depend on that of another fact. If it does, one speaks of the *conditional* probability $P(B \mid A)$ of fact B with respect to fact A. If, on the other hand, A makes no difference to B, the two facts are said to be stochastically independent: $P(B \mid A) = P(B)$. For example, the probability of drawing a black ball from an urn containing black and white balls depends on whether the previous ball was replaced. If it was, the two facts are independent; otherwise, the second depends on the first, but of course not conversely: there is no symmetry.

These notions are exactified as follows. Let A and B be two "events" (in the technical sense of the probability calculus), and call $P(A)$ and $P(B)$ their respective probabilities. (The said "events" are just subsets of a given set S, not necessarily events in the ontological sense of the word, and the corresponding probabilities are measures of such subsets. It is useful to think of S as the possibility set, and of the measures as areas.) The conditional probability of B given A is defined as $P(B \mid A) = P(A \cap B)/P(A)$. (Roughly speaking, if A and B are chance events, the chance that B will occur if A has happened equals the chance of the joint occurrence of A and B divided by the chance of A. If A and B are stochastically independent, $P(B \mid A) = P(B)$, i.e., the occurrence of A makes no difference to that of B.) The probability of A given B, or $P(A \mid B)$, is derived from $P(B \mid A)$ upon exchanging A and B in the preceding formula: $P(A \mid B) = P(A \cap B)/P(B)$. Finally, dividing this formula by the former results in $P(A \mid B) = P(B \mid A) \cdot P(A)/P(B)$. This is *Bayes's theorem*, the axis around which the subjectivistic (or Bayesian) interpretation of probability revolves (see Sect. 1.10.2.4). Note the symmetry of the preceding formulas with respect to the independent variables A and B, but beware: this mathematical symmetry may have no factual counterpart.

Bayes's theorem is mathematically unproblematic, but its interpretation is anything but straightforward. In the factual sciences—particularly in statistical mechanics and quantum physics—A and B often denote states of a thing. In this case, $P(B \mid A)$ is taken to be the probability that the said thing will make a transition from state A to state B. Shorter: $P(B \mid A)$ measures the propensity that the thing concerned will undergo the $A \to B$ transition. Clearly, the transition probability or propensity $P(B \mid A)$ is a property of the thing concerned: radioactive atom, gene, biopopulation, society, or what have you. So far so good. A difficulty would seem to arise when attempting to compute the propensity of the converse transition $B \to A$ using Bayes's theorem. Indeed, the original process $A \to B$ may be irreversible, as in the cases of radiation, radioactive decay, and evolution. In these cases, the inverse probability $P(A \mid B)$ cannot be interpreted as the propensity of the converse transition $B \to A$, because this process just does not occur. In other words, every transition propensity can be exactified as a conditional probability, but the converse is false. This should not be surprising, because pure mathematics knows nothing about reversibility or irreversibility.

A genuine difficulty does arise when a conditional probability $P(B \mid A)$ is interpreted as the propensity of a *cause* A to produce an *effect* B. Thus, Earman and Salmon (1992, p. 80ff) claim that, while $P(E \mid C)$ may be interpreted as the tendency of the cause C to produce the effect E, the inverse probability $P(C \mid E)$ cannot be correctly interpreted as a "posterior propensity", namely as the propensity of E having been caused by C, for this would amount to reversing the causation arrow. This objection is correct, but it does not affect the propensity interpretation. Indeed, a propensity is a property of a *thing*, not of its changes, especially causes or effects. In particular, we may be justified in talking about the propensity $P(B \mid A)$ of a *thing* to undergo the transition $A \to B$ or its converse, but it makes no sense to talk about the propensity of either transition in itself. In short, the Earman and Salmon paradox is dissolved by unearthing the referent, i.e., the thing in question. This caution suffices to save the propensity interpretation, as well as to save us from falling into the old trap of the probability of causes (see Bunge 1973c).

In conclusion, not all the terms of Bayes's formula—or, for that matter, of any other mathematical formula—are interpretable in factual terms. This situation is not new. For example, mathematical economics contains certain equations relating quantities to prices which have negative solutions. The latter are obviously meaningless and must therefore be discarded. Likewise, fully half of the wave-like solutions to Maxwell's equations must be dropped because they would have to be interpreted as representing waves coming from the future.

Let us now examine briefly the main alternative views of probability: the logical, subjectivist (or personalist or Bayesian), and empiricist (or frequency) views.

1.10.2.3 The Logical Interpretation

According to the *logical* interpretation, probability is a certain relation between propositions: it explicates the notion of confirmation of a hypothesis by the empirical evidence relevant to it. For example, if $P(h)$ denotes the (prior or initial) probability of a hypothesis h, then $P(h \mid e)$ is the probability of the hypothesis h given some piece of evidence e. (What is actually meant is the plausibility or verisimilitude of the corresponding propositions.) The main trouble with this view is that there are no objective procedures for assigning probabilities to propositions. In science, probability assignments are made on the strength of measurements or of (hypothetical) random mechanisms, such as blind shuffling, and in science, only facts (not propositions) are assigned probabilities. Given the logical view, what probability should be assigned to a probabilistic hypothesis such as "The probability of a (fair) coin landing heads up is 0.5"? Probability 1? And what is meant by 'probability' in this proposition, if the concept of probability refers to propositions, not to facts? In short, what we need here is a concept of partial truth, not that of probability. (More on this in Sect. 3.8.)

Similarly, probability should not be conflated with *plausibility* —a property of propositions. A hypothesis is plausible or implausible only in the light of the background knowledge as long as it has not yet been tested. But once the tests have been carried out, and provided they are reasonably conclusive, we say that the hypothesis has been confirmed (or disconfirmed), so that—at least for the time being—we are justified in believing that it is (partially) true (or false). That is, after a conclusive test we do not need the concept of plausibility any longer because the hypothesis can be assigned a truth value, and neither a probability nor an *a priori* plausibility. In sum, probability quantitates neither truth nor plausibility.

Incidentally, as with the propensity interpretation, Bayes's theorem also gives rise to paradox with regard to the logical interpretation. Since the prior probability $P(h)$ of a hypothesis h is unknown, it must be assigned arbitrarily. However, the "inverse" probability $P(e \mid h)$ of finding the empirical result e on the assumption that hypothesis h holds would depend on the prior probability of h. Yet since the latter is unknown, this would be a case of extracting knowledge out of ignorance.

1.10.2.4 The Subjectivist Interpretation

The *subjectivist* interpretation construes every probability value $P(x)$ as a measure of the strength of someone's belief in x, or as the accuracy or certainty of his or her information about x. This view was historically the earliest and it is still very popular because it harmonizes with classical determinism. An objection of a mathematical nature against the subjectivist interpretation is that the formula "$P(x) = y$" makes no room for a subject u and the circumstances v under which u estimates his or her degree of belief in x, under v, as y. In other words, the elementary statements of probability theory are of the form "$P(x) = y$", not "$P(x, u, v) = y$". Such additional variables are supernumerary, but they would have to be introduced in order to account for the fact that different subjects assign different credibilities to one and the same item, as well as for the fact that one and the same subject may change beliefs not just in the light of fresh information but also as a result of sheer changes of mood. In sum, the subjectivist interpretation of probability is adventitious, in the sense that it does not match the structure of the mathematical concept.

To bring home the difference between objective indeterminacy and subjective uncertainty, consider the following example discussed by the famous biometrician M.S. Bartlett (1975, pp. 101-104). Of three prisoners, Matthew, Mark, and Luke, two have already been singled out to be executed, but neither of them knows which. Matthew cannot stand the uncertainty and asks the jailer: "Since either Mark or Luke is doomed, you will give me no information about my own chances if you confide in me the name of one man, either Mark or Luke, who is going to be executed." The jailer agrees and answers truthfully that Mark will be executed. Thereupon Matthew felt happier, as reasoning thus: (a) before the jailer replied, his own chances of execution were 2/3; (b) afterwards there are only two candidates for execution, himself and Luke, so his chance of execution has dropped from 2/3 to 1/2. Is Matthew justified in feeling happier? He would, indeed, if the prison director had decided to pick the two victims at random. But this is not the case: so much so, that even the jailer knows who the designated victims are. Since their fates were sealed from the beginning of the story, no probabilities are involved, even though there is uncertainty on the part of the prisoners. Moreover, Matthew's uncertainty about his own fate did not shrink upon hearing the jailer's answer: there could be no rational consolation in confirming that he was one of the victims. Moral 1: probability (unless it is either 0 or 1) implies uncertainty, but the converse is false. Moral 2: Bayesianism may bring comfort (or distress)—though not more so than lies—but not objective truth. It cannot, for it rests on the false assumption that any event, whether objectively random or not, can be realistically assigned a prior subjective probability.

Consider next the case of a really random process that has already run its course, without, however, our fully knowing its outcome. Suppose a woman is known to have two children, one of them a boy. Obviously, the other child is either a boy or a girl. The subjectivist, who does not know the (genetic) sex of the second child, will say that the probability of the child being a boy is 1/2. By contrast, the objectivist will refuse to assign a probability to the *belief* in question. He will argue that the (objective) probabilities (of the underlying events) make sense only during the period between copulation and fertilization, i.e., during that period when there is a *chance* that either an X- or a Y-chromosome carrying sperm may reach the egg first and fertilize it. After fertilization the probability has vanished. (If preferred, one of the probabilities has expanded to 1 while the other has contracted to 0.) The subjectivist's ignorance of this fact does not change anything about it. He confuses the probability of an event with the degree of certainty of his belief in the occurrence of the event—a case of mistaking physics for psychology.

1.10.2.5 The Frequency Interpretation

Empiricists tend to believe that the correct alternative to subjective probability is *frequency*. This view, first proposed by J. Venn in 1866, is still popular among some scientists. Our objections to this interpretation are as follows. Probability and frequency, though related, are different concepts. For one thing, whereas the former is theoretical, the latter is empirical. For another, they have different mathematical structures. In fact, whereas a probability function is defined on an abstract probability space *S*, a frequency function is defined, for every sampling procedure, on a finite subset of *S*, namely the collection of actually observed events (Bunge 1973b). Consequently, a probability statement does not have exactly the same reference class as the corresponding frequency statement: the former usually refers to an *individual* fact, while the latter is about a whole set (or "collective") of facts. Moreover, every frequency is the frequency of actual observations of facts of some kind, whereas a probability may be interpreted as the quantitation of a potentiality yet to be actualized. Hence, equating probabilities with frequencies involves rejecting real (objective) possibility and thus adopting an actualist ontology, according to which nothing is really possible.

The correct procedure with regard to the relations between probability and frequency is not to equate them but to clarify their differences and relations. In our view, frequency is among the *estimators* or *indicators* of probability. More precisely, some probabilistic models may be tested by enriching them with an indicator hypothesis of the form "The numerical value of the probability p is roughly equal to the long-run relative frequency f". (In so doing, probabilities are given by a dimensionless number such as $\sqrt{2}/2$ or 0.5, whereas frequencies are given in percentages such as 70%.) In other words, there are often frequency *estimates* of a probability, but the frequency *interpretation* of probability is wrong. Consequently, the expression 'statistical probability', used by some statisticians and philosophers, is an oxymoron. However, it is useful to indicate that one is using objectively estimated probabilities rather than subjective (or personal) probabilities.

1.10.2.6 Conclusion

To conclude, we are left with the propensity interpretation as the only interpretation of mathematical probability theory utilizable in factual science and technology. Accordingly, there is no point in engaging in such fashionable academic probability games as musing about the probability of phylogenetic trees, about the probability by which one "cause" is a better explainer than the other, or about the probabilities with which a conclusion follows from its premises.

1.11 Upshot

Philosophers will have noticed that the ontology outlined in the preceding is a continuation of certain philosophical traditions. Let us summarize, then, a few philosophical *isms*, which can be explicitly or implicitly found in our ontology. (Besides the danger of indicating one-sidedness and dogmatism, a single *ism* is just insufficient to account for the variety, richness, and changeability of both the world and our knowledge of it.)

First, our ontology is an obvious instance of *materialism*, for it admits only material existents and discards independently existing immaterial objects such as ideas-in-themselves. We regard ideal objects as constructs, i.e., fictions, thought up by material entities equipped with organs capable of "minding", i.e., brains. (More in Chaps. 3 and 6.) We may call this view about constructs *constructionist materialism*.

However, our ontology does not condone *physicalism* (or reductionist materialism), which takes all things to be physical entities. (See also Sect. 3.6.2.) Our materialism embraces *systemism* and *emergentism*. We claim that at the present stage of the evolution of the universe there are at least five genera of systems or levels of systems: physical, chemical, biological, social, and technical. (Of course, all these levels can be divided into numerous sublevels. We will examine the structure of the hierarchy of system levels with respect to biological levels in Sect. 5.3.) Every system at a given level is characterized by qualitative novelties, i.e., properties which do not occur in entities at the lower levels. This is why the higher levels cannot be reduced to (identified with) the physical level. (Note that this is an ontological thesis. If emergent properties can be explained and predicted within bounds from knowledge of lower-level properties, this is an instance of epistemological reduction.)

To accept systemism and emergentism amounts to adopting *pluralism* as regards the diversity of things and processes, i.e., the plurality of kinds of thing and laws. Yet, in other respects, our ontology is *monistic*, since it acknowledges only one substance that possesses properties and undergoes change (namely matter) and it claims that there is only one world, i.e., the (material) universe.

Furthermore, our metaphysics is *dynamicist* and *evolutionist*, for it assumes that every thing is in flux in some respect or other, and that new systems pop up and old ones break down all the time. However, it is not dialectical because we have no use for the beliefs that every thing is a unity of opposites or that every change consists in, or is caused by, some ontic contradiction, such as a contest.

Finally, our metaphysics embraces *determinism* in the broadest sense: it holds that all events and processes are lawful, and that no thing comes out of nothing or disappears into nothingness. It does not, however, assume causalism, for it admits both spontaneity (i.e., uncaused events) and randomness. Despite randomness and chaotic processes (in the sense of modern chaos "theory"), the world is not really chaotic or indeterminate in the traditional sense of lawless. For instance, computers may break down spontaneously, but they do not pop in or out of existence, and they do not turn into pink elephants. (Not even elephants do so.)

We submit that researchers in the factual (natural and social) sciences study exclusively material things—though, of course, with the help of concepts. They thus behave as materialists. To be sure, in contrast to many philosophers of science, only few scientists realize this tacit commitment to materialism or care to acknowledge it (e.g., Levins and Lewontin 1982, 1985; Mayr 1982). The following reasons may account for this. First, only few people are interested in laying bare their own presuppositions: this is a typically foundational and philosophical task. Second, materialism is not for the faint-hearted (neither is idealism). The latter prefer diluted versions of the stronger stuff. These watered-down doctrines are a mixture of materialist and idealist ingredients: they hold that whereas *some* real objects are material, *others* are immaterial. (Though popular, this is not a viable compromise. If the material and immaterial realms are not supposed to be entirely separate, such ontology must explain how material and immaterial objects may conceivably interact. Needless to say, nobody has ever produced any such theory.) Third, to declare oneself a materialist (hence, implicitly, an atheist) amounts to ringing the leper's bell: convicted materialists are either promptly isolated or, worse, joined by undesirable company (e.g., ethical materialists, radical reductionists, unreconstructed Marxists, etc.).

It should come as no surprise, then, that many scientists prefer the label 'naturalism' to 'materialism'. Yet naturalism is, first of all, the opposite of supernaturalism: it does not automatically preclude the existence of immaterial objects inasmuch as they are not supernatural entities. (Incidentally, the term 'natural' here has nothing to do with the distinction between the natural and the artificial: both natural and artificial things are nonsupernatural things, hence natural in the sense of ontological naturalism.) Rejecting supernatural entities is, of course, a necessary step toward a scientific ontology, but it is not sufficient. Unless somebody comes up with a consistent ontological theory of, as well as scientific evidence for, the existence of immaterial objects, such as disembodied souls and spirits, numbers and theories, poems and symphonies, and so on, we cannot grant them any ontological status in a scientific ontology. So Ockham's rule *entia non sunt multiplicanda praeter necessitatem* calls for the most parsimonious metaphysics in tune with contemporary science to begin with, to wit, materialism or (materialist) naturalism.

2 Semantical and Logical Fundamentals

Semantics is popularly seen as just a matter of choice of words. Yet semantics is a rigorous discipline. In fact, it is the family of three research fields: linguistic semantics, which studies the meanings of signs; mathematical semantics, which investigates the models (or examples) of abstract mathematical theories (see also Sects. 3.5 and 9.3.2); and philosophical semantics, which studies the concepts of sense, reference, meaning, truth, and their kin. These three fields have little in common, however.

Here we shall deal only with some of the key concepts of philosophical semantics and, more particularly, with those relevant to the philosophy of factual science. These are the constructs that, unlike those of pure mathematics, refer to real or at least putatively real things, such as the concept of evolution and the theory of selection. We are primarily interested in the concepts of meaning and factual (as different from formal) truth. For example, we want to know how to find out the connotation (sense) and denotation (reference) of the concept of evolution. However, in the present chapter we shall introduce only a few elementary notions insofar as they are of general interest, or as they are relevant to subsequent topics, such as the conception of scientific theories. (More on the semantics of factual science in Bunge 1974a, b.) As the notion of truth involves also epistemological considerations, it will be examined later (Sect. 3.8.1).

2.1 Concept and Proposition

The units of meaning and thus the building blocks of rational discourse are *concepts*, such as "all", "biosystem", "is composed", and "protein". Concepts come in two basic kinds: *logical* and *nonlogical*. Logical concepts are , for instance, "not", "and", "all", and "entails" (or its dual "follows logically from"). They hold (nonlogical) concepts or propositions together, as in "All biosystems are composed of proteins" and "*p* entails *p* or *q*". Examples of nonlogical concepts are "biosystem", "protein" and "is composed"; more on the latter in a moment.

Concepts are used to form propositions (statements). A proposition "says" something about some item or items, that is, it is an assertion or a denial. Consequently, propositions can be true or false, and only they can be the subjects of tests. Since concepts do not assert or deny anything, they can be neither true nor false, and thus the category of testability does not apply to them. Concepts can only be exact or fuzzy, applicable or inapplicable, fruitful or barren.

Note the following points about propositions. For one, propositions should not be confused with sentences. In fact, one and the same proposition, such as "I love you", may be expressed by many sentences, such as 'You are loved by me', 'Ich liebe dich', and 'Je t'aime'. Moreover, a grammatically correct sentence, such as Heidegger's 'The world worlds', need not designate a proposition. For another, propositions should not be confused with thoughts: the former are conceptual objects, hence fictions, the latter are brain processes. Thus 'This book is silly and boring', 'This book is boring and silly', 'Silly and boring is this book', 'Boring and silly is this book' are different sentences expressing so many different thoughts, but they are all lumped into the same proposition. That is, we choose to ignore their linguistic and material differences, feigning that they are one (proposition). (Due to our thing-construct distinction, we distinguish concepts and propositions by double quote marks, whereas signs, symbols, words, and sentences are put in simple quote marks.) Finally, propositions should not be mistaken for proposals, such as "Let's check this proposition". Proposals are invitations to action, so they can be accepted or declined, but not tested to find out whether they are true or false.

Let us now elaborate the nonlogical concepts. The nonlogical concepts can be partitioned into *individuals* such as "Darwin"; *collections* of individuals (sets, classes, or kinds) like "humankind"; and *predicates* such as "lives" or "is alive", and "human" or "is human". In traditional logic "is", as in "Darwin is human", was called the *copula* and treated as a separate logical concept, though never defined. Its function was said to join or glue the predicate (e.g., "human") to the individual or subject (e.g., "Darwin"). However, in modern or mathematical logic 'is' and its cognates ('are, 'was', etc.) designate five different concepts, only one of which involves predication, such as "is human". Thus the proposition "Darwin is human" is analyzed into two concepts—"Darwin" and "is human"—not three. (In fact, it is symbolized as "Hd", where H designates the predicate "is human" and d the individual "Darwin".) On the other hand, the equivalent statement "Darwin is a member of [or belongs in] the human species" is indeed made up of three concepts, one of which, "belongs in" (designated by $\in$), was unknown to traditional logic. (In symbols, "$d \in \mathcal{H}$", where $\mathcal{H}$ designates the collection of all humans, i.e., the species *Homo sapiens*.) The same holds for class inclusion (designated by $\subset$), as in "Humans are mammals", or "$\mathcal{H} \subset \mathcal{M}$" for short. (See also Chap. 7.) Finally, the innocent-looking word 'is' designates two further concepts in mathematical logic, namely that of identity (=), as in "1 is the successor of 0", and that of equality (:=), as in "The cosine of 0° is 1", or "cos 0 := 1".

Returning to predicates, it will be recalled from Section 1.3 that a predicate can be unary such as "lives", binary such as "descends", ternary such as "mediates", and so on. A unary predicate denotes an intrinsic property of a (simple or complex) individual; a binary predicate denotes a relation between two items; a ternary predicate a relation between three individuals, and so on.

One particular kind of relation should be singled out here because it is important and because we shall meet it on occasion: that of mathematical function. A mathematical function matches every member of one class to a single member of another. More precisely, a *function* from a set A to a set B assigns to every element of A a single member of B. One writes $f: A \rightarrow B$, or $y = f(x)$, where x is in A and y is the image of x in B. A is called the *domain*, and B the *codomain* of f. For example, age is representable as a function from the set of things to positive real numbers (in obvious symbols, $a: \Theta \rightarrow \mathbb{R}^+$ and, for a particular thing $\vartheta \in \Theta$, $a(\vartheta) = t$, where $t \in \mathbb{R}^+$). If an arbitrary member y of B is a number, or an n-tuple of numbers, one calls y a *(numerical) variable*. (An n-tuple is an ordered list of n items, where n is a positive integer.)

From an epistemological or methodological viewpoint we can distinguish two kinds of variable and three kinds of function. Variables can represent observable or unobservable properties. Functions may relate (a) observable variables, such as the inputs and outputs of a system, (b) unobservable properties, such as population density and competition, or (c) unobservable to observable variables, such as adaptedness to number of offspring. The functions of the third class relate theory to data, and some of them function as indicator hypotheses (see Sects. 3.2.3, 3.5.5, and 3.5.7.1).

As we saw above, the relation between predicates and individuals is that of *predication* or *attribution*. Predicates are attributed to individuals, couples, triples, or n-tuples. (Note that not all individuals need be irreducible: some of them may be analyzable into collections. In other words, the individual-collection distinction may depend on the level of analysis or on the context. Note further that we are talking about the logico-semantical, not the ontological, concept of individual.) The attribution of predicate F to individual b results in the proposition Fb, which reads "b is an F" (recall Sect. 1.3). The attribution of predicate G to the ordered pair $\langle a, b \rangle$ results in the proposition Gab, and the result of attributing predicate H to the n-tuple $\langle a, b, \ldots, n \rangle$ is the proposition $Hab\ldots n$. Alternatively, we may say that the proposition Fb is the value of the function F at b; and, likewise, Gab is the value of the function G at $\langle a, b \rangle$, and so on. In general, a predicate may be construed as a function from individuals to propositions. (More in Sect. 2.2.)

The relation between individuals and collections is that of belonging, designated by the symbol $\in$. For example, the proposition "Darwin is a human being" can be analyzed as "Darwin belongs in (is a member of) the class of human beings", or "$d \in \mathcal{H}$" for short. Note that in mathematics the terms 'set' and 'collection' are synonymous, and every collection has a fixed membership. This is not so in the factual sciences, where one often studies collections with a variable membership: think of any biological taxon. Such a variable collection is a set proper only at any given time, for it may contain different members at a different time. Moreover, as in the case of extinct taxa, the membership of a variable collection may, at some time, even equal the empty set. (More on variable collections in Sect. 7.2.1.3.)

Finally, we have the relation between predicates and classes. Yet this topic deserves a section of its own because it paves the way to the elucidation of the semantical notion of meaning.

2.2 Extension and Reference

Every predicate determines a class called the *extension* of the predicate. This is the collection of individuals (or couples, triples, etc.) that happen to possess the property designated by the predicate in question. For example, the extension of the unary predicate "metabolizes", as it occurs in the proposition "All living things metabolize", is the class of living things (or biosystems). In obvious symbols, $\mathcal{E}(M) = \{x \in \mathcal{S} \mid Mx\}$, i.e., the extension of M is the collection of individuals in the set $\mathcal{S}$ (of systems) that have the property M. The extension of the binary predicate "descends", as it occurs in the propositional schema "x descends from y", or "Dxy" for short, is the collection of ordered pairs ⟨ancestral organism, descendant organism⟩. Shorter: $\mathcal{E}(D) = \{\langle x, y\rangle \in \mathcal{A} \times \mathcal{D} \mid Dxy\}$, where $\mathcal{A} \times \mathcal{D}$ is the collection of pairs ⟨ancestor, descendant⟩, also called the *Cartesian product* of $\mathcal{A}$ by $\mathcal{D}$. (In the case of sexual reproduction the extension is more complex, namely ⟨maternal ancestor, paternal ancestor, descendant organism⟩). It goes without saying that the extension of a ternary predicate is a set of triples, that of a quaternary predicate a set of quadruples, and so on. However, some predicates have an empty extension, i.e., they apply truthfully to nothing. In other words, some predicates do not correspond to substantial properties of things (recall the property-predicate distinction from Sect. 1.3). Examples: "vital force", "ghost", "immortal". When the extension of a predicate F is empty one writes: $\mathcal{E}(F) = \varnothing$.

Nominalists, such as Woodger (1952), admit only individuals. Hence they mistrust concepts, especially predicates. So they believe that every property must be understood as the collection of individuals possessing it. That is, they conflate predicates with their extensions. This *extensionalist* approach is open to the following objection. Let P_1 and P_2 denote two properties of entities of a certain kind K, e.g., 'feathered' and 'possessing an intertarsal joint'. Since all K's (Recent birds) possess both P_1 and P_2, according to nominalism P_1 is identical to P_2 — which contradicts the hypothesis that P_1 and P_2 are different (see also Bunge 1974a, 1983a; Sober 1981).

In sum, the ordinary language expression 'b is an F' can be construed either as 'Fb', where F is a function, or as 'b is a member of the extension of F', i.e., $b \in \{x \mid Fx\}$. Ordinarily the latter construal of predicates, which is called *extensional*, presupposes the former, which is called *intensional*. This is because we must know what predicate we are talking about and what property it conceptualizes before we can inquire into its extension. However, both construals are mathematically equivalent.

What does a proposition of the form "b is an F", or "Fb" for short, refer to? Obviously, Fb is about b: it attributes F to b. Now, we saw above that Fb may be construed as the value of the function F at b. However, the individual b may belong to more than one collection: it may be an organism, an animal, an insect, a parasite, an ancestor, and so on. (Note that a collection may contain a single member. Such a collection is called a *singleton*; but a singleton is not the same as its solitary member, i.e., $\{b\} \neq b$.) Hence the predicate F may be construed as a function from a collection D of individuals to the set P of all the propositions of the form Fb; in short, $F: D \to P$.

We assume next that the (unary) predicate F refers to any and all of the members of the domain D of F. In other words, the *reference class* of the predicate F equals its domain D, or $\mathcal{R}(F) = D$ for short. For example, the reference class of the energy concept is the collection of all actual and possible material things, and that of ontogenesis (development) is the totality of organisms. Note that, since all (material) things possess energy, the extension and the reference class of "energy" coincide. By contrast, since apparently not all organisms develop (e.g., some unicellular organisms seem to undergo no development proper: see Chap. 8), the extension of "ontogenesis" (O) is properly included in its reference class, i.e., $\mathcal{E}(O) \subset \mathcal{R}(O)$.

To find out the reference classes of higher-order predicates, such as "eats" (binary) or "mediates" (ternary), we must identify their respective domains. Since it is animals that eat, and since what they eat is other organisms or parts of them, the predicate "eats" (E) applies to any ordered pair $\langle$animal, (food) organism$\rangle$. In technical terms, the domain of the function E is the Cartesian product of the collection A of animals by the collection F of food organisms. We stipulate that the reference class of E is the union of the factors A and F, i.e., $\mathcal{R}(E) = A \cup F$. In general, the *reference class of an n-ary predicate* P with domain $A \times B \times \cdots \times N$ will be $\mathcal{R}(P) = A \cup B \cup \cdots \cup N$.

Our next assumption is that the reference class of a proposition, or any other construct in which predicates occur, is the union of the reference classes of those predicates. For example, the reference class of the proposition "Some organisms live in fresh water" is the union of the collection of organisms and the collection of fresh water bodies. Note that the denial of the given proposition has the same reference class. The same holds for the reference classes of "P and Q", "P or Q", "If P then Q", and the other combinations of P and Q by means of logical connectives: they all have the same referents. In short, the reference function R is insensitive to the logical connectives.

Finally, we stipulate that the reference class of a system of propositions, such as a theory, equals the union of the reference classes of all the predicates occurring in the theory. As in building a theory one may introduce and define as many predicates as needed, the task of finding the reference class of the theory looks at first sight open-ended and therefore hopeless. This is, indeed, the case with untidy theories. In the case of axiomatized theories, however, one can easily identify the set of basic or defining predicates, which is a small subset of the collection of all the predicates occurring in the theory.

The reference class of a theory or model is sometimes called its 'ontology'. We warn against this usage of 'ontology'. For one, we take the term 'ontology' either to denote a philosophical discipline, namely metaphysics, or to designate a metaphysical theory. Thus, neither the world as a whole nor the collection of objects in the world (i.e., its composition) is an ontology. Moreover, if we used 'ontology' also for 'reference class', we would have to speak of the ontology of an ontology. For another, concepts with an empty extension, such as "unicorn", do refer, namely to fictional objects, although the latter have no ontological status.

Let us re-emphasize the importance of distinguishing the extension of a predicate from its reference class, although this distinction is hardly made by philosophers. (After all, most philosophers do not care for the property-predicate distinction.) A biologist who criticizes the notion of *élan vital* for having no real counterpart, i.e., for having an empty extension, refers to the *élan vital* while regarding it as pure invention. And a scientist who hypothesizes the existence of an object that has not yet been found assigns a nonempty reference class to the defining predicate(s), even while admitting that, so far, the corresponding extension has proved empty.

The main differences between the extension and the reference class of a predicate are the following. Firstly, the concept of extension presupposes that of (factual) truth, whereas the notion of reference class does not. That is, we include in the extension of a predicate only the items for which it actually holds. Secondly, while the extension of a binary predicate is a set of ordered couples (and in general that of an n-ary predicate a set of n-tuples), the corresponding reference class is a set of individuals. Thirdly, while the extension function is sensitive to negation, disjunction, conjunction and the remaining logical connectives, the reference function is not. For example, the extension of "not photosynthesizing" is the complement of that of "photosynthesizing", while the reference class of both predicates is the same, namely the entire set of organisms (present, past, and future). Again, the extension of "P or Q" is the union, i.e., $\mathcal{E}(P \vee Q) = \mathcal{E}(P) \cup \mathcal{E}(Q)$, and that of "$P$ and Q" is the intersection of the extensions of P and Q, i.e., $\mathcal{E}(P \,\&\, Q) = \mathcal{E}(P) \cap \mathcal{E}(Q)$. By contrast, the reference class of both compound predicates is the same, namely the union of the partial reference classes, i.e., $\mathcal{R}(P \vee Q) = \mathcal{R}(P \,\&\, Q) = \mathcal{R}(P) \cup \mathcal{R}(Q)$. (For details see Bunge 1974a, b.)

Although biologists need not bother about the formalization of the concepts of reference and extension, they should keep in mind the distinction between them, and this for two reasons. One is that sometimes it is not at all obvious what the reference class, let alone the extension, of a given theory or model is. For example, is the theory of selection about genes, genotypes, organisms, groups, or populations, or perhaps all of them together? Thus, the careful semantical analysis of theories and models is an important task of theoretical biology. Another reason for the pertinence of the distinction in question is that it helps us detect the falsity of certain subjectivist and instrumentalist claims, such as the one that "[the theory of selection] is about flora and fauna, *and* about cognitive agents who theorize about them" (Rosenberg 1994, p. 15, italics in the original). The fact that all theories are theories of some cognitive agents, as well as the fact that theories may contain conventions, simplifying assumptions, or statistical artifacts, does not entail that they are therefore *about* cognitive agents; in other words, that cognitive agents are among their referents.

2.3 Meaning

The word 'meaning' has many meanings in both ordinary language and biology. There is talk about the meaning of genetic information, the meaning of evolution, or the meaning of life. We shall avoid such equivocations and adopt only the semantical notion of meaning. That is, we shall admit only constructs, as well as signs symbolizing constructs, as meaning bearers. (Hence, neither molecules nor processes have meaning; and most things and processes also have neither purposes nor goals: see Chap. 10.) In so doing, we shall analyze the concept of meaning into "sense" and "reference", or what is being said about what.

Since the concept of reference has been elucidated previously, we proceed to analyze the notion of sense (or connotation, or intension). Rather than referring directly to an individual, a proposition can "say" something in an indirect way. For example, the statement "Darwin is an English biologist" presupposes the existence of England and of biology. Furthermore, it implies the statement "Some Englishmen are biologists". Thus a proposition "contains" potentially all its logical consequences, so that these must be counted as belonging to the full sense of the given proposition. We call this the *import* of the proposition. Moreover, it makes complete sense only in relation with the propositions that entail it. We call the generators or logical precursors of a proposition its *purport*. (Note that, if a proposition is an initial assumption, i.e., a postulate or an axiom, of a scientific theory, or if a predicate occurs as a primitive, i.e., undefined, concept in a theory, it has no purport other than itself.) In sum, the *full sense* of a proposition is the set of all the propositions it entails or is entailed by, i.e., the union of its purport and its import. However, we must be cautious here because one and the same formula may have somewhat different senses (or none) in different contexts. For example, a proposition about the rate of bacterial growth makes no sense in systematics; and the sense of the predicate "is alive" depends on whether or not it is defined in the given theory. If undefined, its sense equals its import, whereas in a context where it is defined, its sense equals the union of the set of defining predicates (e.g., "metabolizing") with the set of predicates that it entails (e.g., "mortal"). Hence, whenever there is risk of ambiguity, explicit indication of the context should be made.

Having defined the reference and sense of an arbitrary construct, we can now introduce the semantic concept of meaning. We define the *meaning* of a construct *c* as its sense together with its reference: in obvious symbols, $\mathcal{M}(c) =_{df} \langle \mathcal{S}(c), \mathcal{R}(c) \rangle$. In other words, two constructs have the same meaning if, and only if, they are cointensive and coreferential. We further stipulate that every construct has a meaning, i.e., a sense (even if known only in part) and a nonvoid (though possible indefinite) reference class. Its extension, on the other hand, may be empty.

Furthermore, we stipulate that a sign or symbol is *significant* provided it either designates a construct proper or denotes an actual or possible fact. (Note that we call the semantical relation between a sign and a construct 'designation', and the semantical relation between a sign and a factual item 'denotation'. Thus, road signs denote but do not designate, and punctuation marks neither designate nor denote.) Signs or symbols can have meaning only vicariously by designating a meaningful construct. In other words, a sign or symbol acquires meaning indirectly by standing proxy for a construct.

Since names are symbols, they have (vicarious) meaning only if they stand for constructs. If a name just denotes a concrete individual, such as a proper name denoting a particular person, it has no meaning. Thus, 'Jones' is meaningless. This point is relevant to the neonominalist theses in taxonomy according to which (a) biological taxa would be concrete individuals (clades) rather than classes or kinds of organisms, and (b) the names of taxa would thus be proper names rather than names of classes (i.e., of concepts). Yet at the same time, some neonominalist authors (e.g., de Queiroz 1994) speak of the meaning of taxon names, the intension of taxon names, and the definition of taxon names. All this is mistaken: if taxon names do not designate constructs but do denote concrete individuals, they have neither an intension nor a meaning, and they cannot be defined. One can only assign or attach a symbol to an object, but one does not thereby define the former—nor, of course, the latter: definitions are sign-sign or concept-concept *identities*, and there can be no identity between a name and its nominatum. (More on definition in Sect. 3.5.7.1, and more on neonominalism in Sect. 7.3.)

According to the positivist theory of meaning, a proposition is meaningful only in the case of its being testable or, more precisely, verifiable (see, e.g., Carnap 1936-1937.) This view is usually called the 'verification theory of meaning'. Now, although testability is certainly a sufficient condition for the meaningfulness of propositions, it is not necessary. Thus, the proposition "Life is a divine gift" may make perfect sense in a theological context even though it is untestable. Moreover, constructs other than propositions have meaning although they are not testable. For example, the order "Pass me the sugar!" and the moral maxim "We must not be sexist" are perfectly meaningful although they are not propositions and therefore not testable. Consequently we turn upside down the verifiability theory of meaning to read: if a proposition is testable, then it is meaningful. In other words, meaningfulness is necessary for testability. It is not sufficient, though, because tests require test means, such as observation devices and methods of some kind. (In other words, it appears that "is testable" is at least a binary predicate, not a unary one, for it occurs in propositions of the form "*p* is testable by means *m*".)

Finally, we must distinguish between factual and empirical meaning. Obviously, a construct has *factual meaning* if, and only if, it refers to factual items. On the other hand, a construct has *empirical meaning* only in the case when it refers at least partially to human experiences of some kind, such as perceiving, thinking, or doing. Thus, while the statement "The first organisms self-assembled from abiotic precursors" is factually meaningful (and moreover true), it is empirically meaningless, because we have no experience of the past self-assembly of living things. Since experience is a proper part of the real world (see Postulate 1.10), every empirically meaningful construct is also factually meaningful, but not conversely. If necessary, we may call a construct that is factual but not empirical *strictly factual* or *objective*. For example, "temperature" is a strictly factual concept, whereas "hot" is an empirical one. The distinction between factual and empirical meaning or content will come in handy when we examine the problem of whether scientific theories have either factual or empirical content, or both (see Sect. 3.5).

2.4 Logic

Whereas semantics studies the content of constructs, logic studies the *form* of concepts, propositions, systems of propositions (theories), and deductive arguments. (An accessible introduction to elementary logic for biologists is van der Steen 1993; more extensive and technical textbooks are, e.g., Suppes 1957 and Copi 1968.) From a logical point of view, propositions can be simple (atomic) or composite (molecular). A proposition containing one or more logical operators, like "or" and "not", is said to be composite or molecular. An example is "p or not-p", or "$p \vee \neg p$", for short, where p is an arbitrary proposition. Another example is "not-(p and not-p)", or "$\neg(p \& \neg p)$". The first of these propositions is called the *excluded middle principle* (or *tertium non datur*), and the second the *principle of noncontradiction*. Both hold in ordinary or classical logic for all propositions, regardless of their content and truth value. Likewise, the basic rule of deduction, "From p, and If p then q, infer q" (or, in symbols, "$p, p \Rightarrow q \therefore q$"). This principle, called *modus ponens*, holds for any propositions p and q regardless of what they "say" and whether or not they are (formally or factually) true: logic is noncommittal with regard to the truth value of nonlogical formulas.

The laws of excluded middle and noncontradiction are epitomes of *tautologies*, i.e., propositions that are (logically) true by virtue of their form, so that they hold regardless of the state of the world. Tautologies are thus radically different from mathematical, chemical, or biological truths, every one of which depends on the nature of its referents and on the context. While all tautologies are formal truths, not all formal truths are tautologies. For example, "2 + 3 = 5" is true in arithmetic but it does not belong in logic, which does not involve numbers.

Tautologies say nothing in particular about the world, although they are necessary to reason correctly about it. Indeed, logic, the canon of valid reasoning, consists of infinitely many tautologies together with a handful of inference rules. A major use of logic is as a tool for identifying logical platitudes, i.e., tautologies, and their negations, i.e., contradictions or logical falsities. A logical falsity either includes a self-contradictory predicate or it contains a pair of mutually contradictory propositions. An instance of the former is "carnivorous herbivore", and "Aggression is innate and acquired" exemplifies the latter. Being logically false, no empirical operations are required to dismiss them.

Since maximal generality and independence from subject matter are peculiar to logic, no other science has such breadth—or, if preferred, such shallowness. This is why logic can be used to analyze all manner of discourse. And this is why such expressions as 'logic of the process' and 'logic of the situation', sometimes used in scientific discourse, may refer to anything but logic. Logic does not handle processes or situations any more than it tackles photosynthesis or evolution. In sum, logic is the basic theory of rational discourse. It is the study of the form of concepts, propositions, and deductive arguments. It teaches us how to tell correct from incorrect argument, not how to explore the world, let alone how to change it; similarly for mathematics, which is also not committed to any factual subject matter. So mathematics, too, is portable from one field of inquiry to another.

To conclude: however necessary, logic and mathematics are insufficient to study the world. This is because logic is about form and consequence: it is an *a priori* science, which is in no need of empirical operations such as observation, measurement, or experiment. And mathematics deals with constructs and, hence, has no need for empirical procedures either. This is why we call logic and mathematics *formal* sciences as opposed to the *factual* sciences, which study the world. (More on the philosophy of logic and mathematics in Carnap 1939; Bunge 1985a, 1997.)

3 Epistemological Fundamentals

Epistemology is the philosophical discipline concerned with knowledge in general—whether ordinary or scientific, pure or action-oriented. It is partly descriptive, partly normative. Some classical epistemological questions are: What can we know? How do we know? What, if anything, does the knowing subject contribute to her knowledge? Methodology (or normative epistemology) is the discipline that studies the principles of successful inquiry—whether in ordinary life, science, technology, or the humanities. Some methodological problems are: Is there a single best way of producing knowledge and, if so, which is it? What is the scientific method, if any? Is empirical confirmation (or else falsification) necessary and sufficient for evaluating theories?

Many epistemological and, in particular, methodological questions can be dealt with as though knowledge were independent of knowing animals. This allows us to proceed in the traditional way, when we analyze methodological notions such as those of datum, hypothesis, or testability. Yet, as will be obvious from Chapter 1, the materialist claims that there is no knowledge in itself: the learning process happens to occur in and among animals endowed with nervous systems of a certain complexity. In other words, we submit that epistemology must mesh in with biology, psychology, and social science. The immaterial and isolated knowing subject of traditional epistemology must be replaced with the inquiring animal possessing a complex brain, or a team of such animals, embedded in a society. In short, epistemology must be biologized and sociologized. However, since this is not a sociological treatise but a biophilosophical one, we shall take into account only some biological or, rather, psychobiological aspects of epistemology. (More in Chap. 6. For details see Bunge 1967a, b, 1980, 1981a, 1983a, b; Bunge and Ardila 1987.)

3.1 Cognition and Knowledge

3.1.1 Cognition

If we take the biological basis of cognition seriously, our initial assumption must be:

POSTULATE 3.1. Every cognitive act is a process in some nervous system, whether human or not.

Since cognition is an activity of the nervous system, we must turn to neuroscience and psychobiology for its explanation. These disciplines suggest that the acquisition of knowledge, i.e., learning, consists in a change in the connectivity of some neuronal system or other. Now, the mode of connections of neurons and neuronal systems can either be *constant*, i.e., the connections do not change once established, or *variable*. Variable connections, in turn, can be *regular*, i.e., a neuronal system x is connected now with a system y, now with another system z, according to a definite time pattern. Or they can be *random*, i.e., a neuronal system x is connected successively with different systems in a random fashion. The *connectivity* of a neuronal system, then, is the set of couplings among its subsystems. We are now ready to propose some basic definitions and postulates.

DEFINITION 3.1. A neuronal system is *plastic* (or *uncommitted*, or *modifiable*, or *self-organizable)* if, and only if, its connectivity is variable throughout the animal's life. Otherwise, i.e., iff its connectivity is rigid or constant either from the beginning of its formation or from a certain stage in its development on, the system is *committed* (or *wired-in*).

Note that this is a neuronal, not a behavioral, definition of plasticity. Whereas neuronal plasticity entails behavioral plasticity, the converse is not true. A behavior can appear plastic but may be due to the activation of different committed neural systems reacting to different stimuli. Therefore, matters of plasticity cannot be investigated by ethology alone. Our next assumption is:

POSTULATE 3.2. All animals with a nervous system have neuronal systems that are committed, and some animals have also neuronal systems that are plastic.

It is a task of comparative neurophysiologists to test this hypothesis, that is, to find out the members of which animal species possess plastic neuronal systems. Those animals possessing such plastic neuronal systems will be said to be capable of learning.

POSTULATE 3.3. Learning is the specific function of some plastic neuronal systems.

DEFINITION 3.2. Every neural activity (function) involving a plastic neuronal system that has acquired a regular connectivity is said to be *learned.*

Note that, according to the preceding definitions and postulates, the formation of committed connections among neurons during development does not count as learning even if developed through the influence of environmental stimuli. In other words, our peculiar neurobiological construal of learning surmounts the obsolete innate-acquired or genetic-environmental dichotomies. (More on this in Chap. 8.)

It will be helpful, therefore, to adopt T.C. Schneirla's proposal to call "the contribution to development of the effects of stimulation from all available sources (external and internal)" (Lehrman 1970, p. 30) *experience* instead of 'learning'. Thus, for instance, imprinting does not count as learning. Furthermore, our construal of learning also excludes habituation as a form of learning. (Habituation occurs not only in organisms that lack plastic neuronal systems, but also in organisms that lack nervous systems to begin with, such as protists.) We are aware that this narrow neurophysiological construal of learning will not satisfy many an ethologist who prefers an operational notion of learning in terms of observable changes of behavior. However, we have no use for operational definitions, as we shall explain in Sections 3.2.3 and 3.5.6.

3.1.2 Knowledge

Although we cannot detach the outcome (knowledge) from the corresponding process (cognition), we may distinguish them. We thus propose:

> DEFINITION 3.3. The *knowledge* of an animal at a given time *is the set of all items it has learned and retained* up until that time.

In other words, the knowledge of an animal is the collection of changes (processes) in its plastic neuronal supersystem, including the dispositions to replay these processes. According to the preceding definitions and postulates, organisms lacking nervous systems, such as plants, protists, and sponges, cannot know anything. This holds, *a fortiori*, for nonorganisms such as computers. The same holds for animals possessing a nervous system but lacking plastic neuronal systems. For example, most likely neither coelenterates nor nematodes can know anything. We must say, then, for example, that a jellyfish is able to swim but it does not know how to swim. An immediate consequence of Definition 3.3 is:

> COROLLARY 3.1. There is no inherited knowledge.

We do not speak of *innate* (or inborn) knowledge here because not all animals can be said to be born. The term 'innate' can be properly applied only to animals that are viviparous or that hatch from some eggs or pupas, or whatever. (For the different and confusing senses of 'innate' and 'inherited' see Lehrman 1970.) In principle, such animals may learn something in the womb or in the egg provided plastic neuronal systems are involved. The problem of discovering from what developmental stage on such plastic systems are functional is one for comparative embryology. (For a discussion of prenatal stimuli of behavioral development see, e.g., Gottlieb 1970, 1991.) In other words, there may indeed be such a thing as innate or inborn knowledge, though there is no inherited knowledge in the sense of genetically transmitted knowledge (see Chap. 8). However, even if there is innate knowledge, it can be only of the sensorimotor or perceptual kind, but certainly not of the propositional kind, for the latter calls for fairly complex plastic neuronal systems, which become organized along with experience (see Sect. 3.1.4).

3.1.3 "Knowledge in Itself"

An important metaphysical consequence of our view of cognition is that there is no knowledge in itself, that is, separate from the cognitive processes occurring in some nervous system or other. This, of course, is at odds with the idealist tradition. We adopt the materialist view that Platonic ideas, objective knowledge without a knowing subject (Popper 1972), and objective mathematical reality (Putnam 1975) are so many figments of the metaphysician's brain. Just as there is no motion apart from moving things, so there are no ideas in themselves. However, abstracting from the real animals that think up such ideas as well as from the personal and social circumstances under which they ideate is convenient and useful for many a logical or epistemological analysis, and there is no harm in it as long as it is understood as a *fiction*, not as an ontological thesis on the autonomy of ideas.

It might be objected against this materialist thesis that, since we obviously can exchange information with one another, there must be such a thing as a *content* of cognitive processes that can be not only transferred, transmitted, or communicated to other brains but also externalized in the form of artifacts, such as inscriptions and tapes. We claim that there is actually no such content and *a fortiori* no such transfer. Let us explain why.

For most of prehistory, knowledge existed only in individual brains. With the invention of drawing, painting, sculpting, and particularly writing, knowledge could be encoded and "externalized" in cultural artifacts that could circulate in the community. This facilitated the storing, sharing, and enriching of "knowledge". However, it also fostered the myth of the independent "content" of knowledge—independent, that is, from the inquiring subject. It is easy to see how this myth can be generated and maintained. When somebody finishes a book, this piece of structured matter can be detached and seen by somebody else. Even its creator can stand back and contemplate it as if it were self-existing, while, in fact, its "content" depends on its being perceived and understood by some brain. This creates the illusion that we are in the presence of three separate items: the neural (and motor) process resulting in writing, the cultural artifact, and the knowledge or feeling encoded in the latter.

The next step is to collect all such bits of knowledge detached from brains and endow such collection with a life of its own. The final step is to give a name to such a collection of items allegedly hovering above brains and society—e.g., the "realm of ideas" (Plato), "the objective spirit" (Hegel), or "World 3" (Popper). In this way, the illusion is created that such "worlds" of knowledge and feeling persist and subsist once they have been created by concrete individuals in specific social circumstances, and, moreover, that they interact with living beings. However, those collections do not actually constitute worlds (concrete systems), for there is no way in which totally heterogeneous objects, such as statements and disks, can combine to form a system exhibiting emergent properties. Worse, there can be no empirical evidence whatsoever for the hypothesis that such ideal "worlds" lead an existence separate from living brains. Consequently, Popper's (1972) hypothesis of the existence of "World 3" does not comply with his own methodology of conjecture and refutation (for further arguments see Bunge 1980, 1981a).

Having done away with the "information content" of cognitive processes and cultural artifacts, how do we account, then, for communication? To understand communication we must realize that exchanging "information" is not like trading goods, but is interacting with another animal (directly or via artifacts) in such a way that each party elicits certain learning processes in each other's brain. In other words, successful communication consists in the *construction* or *(re)creation* of similar processes in the brains of the animals involved in the interaction. If those (re)constructed brain processes are too dissimilar, we misunderstand each other, and if there is no equivalent construction in the other's brain we do not understand each other at all. If communication were really to consist in the transmission of immaterial information, no such non- or misunderstanding should be possible, except perhaps for disturbances in the communication channel (whatever this may be). In particular, teaching and learning, i.e., education, should not be such difficult and arduous activities as they actually are.

In sum, there is no immaterial content of cognitive processes and of cultural objects. Thus, a sculpture that nobody looks at is just a chunk of matter, and so is a biological paper that nobody reads. Only when such material objects elicit processes of re-construction, re-creation, re-feeling, re-thinking, or re-enacting upon perception by some animal do they exist at, and only at, that time when these processes occur and as long as they can be replayed, i.e., remembered.

In the light of the preceding it will be obvious that the notion of a *meme*, introduced by Dawkins (1976) in analogy to the concept of gene, is nothing but a metaphor. Indeed, whereas pieces of DNA (i.e., genes) are actually passed on in the production of offspring, there are no such things as pieces of knowledge (i.e., memes or ideas) that are literally transmitted to other brains. Therefore, Dennett's (1995) attempt to render the notion of a meme philosophically respectable fails, for it is not true to say that "memes restructure a human brain" (p. 365): if anything, the (re)structuring of the brain (perhaps induced by sensory signals) "produces" a meme—not the other way round.

Finally, if, unlike organisms, ideas (or memes) are neither alive nor self-existing, it should be clear that they neither replicate nor evolve by themselves. We need to emphasize this, because a number of philosophers and scientists, notably Spencer, von Helmholtz, Peirce, Mach, Toulmin, Popper, and, more recently, Hull (1988), have drawn a parallel between the history of ideas and biological evolution. However, this variety of "evolutionary epistemology" consists just of metaphors and analogies, because it rests on the reification of cognitive processes. To be sure, it is legitimate and convenient to study the "history" and "change" of ideas *as if* they were self-existing entities. Yet what is objectionable are the underlying metaphysical theses that there are such things as "knowledge in itself", or "objective knowledge", or that ideas are "historical entities". In particular Hull's selectionist account of scientific knowledge (1988) is remarkably unbiological, for he entirely neglects the constructive aspect of knowledge, which is mandatory if we take neurobiology seriously. If we do so, it becomes obvious that there can neither be such things as the transmission of ideas nor such things as conceptual replicators (memes), conceptual descent, conceptual lineages, and conceptual interaction, not even "via physical vehicles" (p. 436). All these are just metaphors and ellipses, which are attractive, however, because they help us understand in an intuitive way. But this understanding is illusory because it is not based on genuine scientific explanation. (See also Chap. 6; more on the use of inadequate analogies in this kind of evolutionary epistemology in Ruse 1986 and Bradie 1991. For a genuine evolutionary epistemology dealing with the evolution of cognitive abilities see Vollmer 1975, 1983, 1985, 1987a, 1995, and Ruse 1986. See also Riedl 1980, whose exposition is, however, a little vague, as well as Riedl and Wuketits eds. 1987; Bradie 1994a.)

3.1.4 Kinds of Knowledge

Although we have rejected the ontological thesis about knowledge in itself, for many epistemological purposes it is sufficient to study only the "products" of cognitive processes regardless of the learning subject and her social environment. (Note that since we explicitly acknowledge this to be an exercise in simplification for the sake of analysis, we do not violate our own ontology.) To begin with, a classification of knowledge items may be a useful starting point.

First, we can distinguish

1. *Sensorimotor* knowledge—e.g., knowing how to dance or type (but, e.g., being able to breathe or urinate does not count as knowledge).
2. *Perceptual* knowledge—e.g., knowing the song of the nightingale or being able to distinguish a termite from an ant (note that, like all knowledge, these abilities must be learned to qualify as knowledge in our narrow construal).
3. *Conceptual* or *propositional* knowledge—e.g., knowing that the Earth revolves around the Sun (or conversely), or knowing the role of respiration in organisms. (For obvious reasons, in the following we shall be concerned only with propositional knowledge.)

A second useful distinction is that between first-hand knowledge and second-hand knowledge. *First-hand knowledge* is acquired by personal experience, such as research. *Second-hand knowledge* is knowledge about first-hand knowledge: it is "communicated" by word of mouth, books, films, disks, or whatever. Note that this partition does not involve any valuation. First-hand knowledge may be rather worthless (e.g., knowing that one's desk is brown), and second-hand knowledge may be insightful (e.g., having read this book).

A related partition is that between private and public (or intersubjective) knowledge. We may say that an animal b has *private* knowledge of X iff there is nobody except b who has knowledge of X. Otherwise, i.e., iff an item of knowledge is shared among at least some members of a society, it is *public* or *intersubjective* in that society. Private knowledge can be further partitioned into knowledge of one's own states (or, rather, processes), in particular brain processes, or secret knowledge, i.e., knowledge kept classified.

Another distinction is that between tacit (or unconscious, or know-how) and explicit (or conscious, or know-that) knowledge. More precisely, we can say: if subject s knows p, then (a) s has *explicit* knowledge of p iff s also knows that s knows p or knows how to express p in some language; (b) otherwise s has *tacit* knowledge of p.

We must finally caution against mistaking public or intersubjective knowledge for objective knowledge. We define *objectivity* as follows:

DEFINITION 3.4. Let p designate a piece of explicit knowledge. Then p is *objective* if, and only if,
(i) p is public (intersubjective) in some society, and
(ii) p is testable either conceptually or empirically.

Thus, magical rules and religious dogmas may be intersubjective in a given society but they are not objective in the above methodological sense. Note that truth is not involved in objectivity. A statement may be objective and false, or true and nonobjective. For example, "Our planet is hollow" is objective but false, and "I am not sure I'll ever finish reading this fastidious book" may be true but is not objective.

3.1.5 Knowledge and Belief

Many epistemologists have defined knowledge in terms of belief or, more precisely, as a special kind of belief, namely as *justified, warranted, or true belief.* (More explicitly: s knows that p iff s believes p, and s is completely justified in believing p, e.g., because p is true.) However, since knowledge as well as belief must be seen as brain processes, we shall define justified belief in terms of knowledge, instead of defining knowledge as justified belief. Given a "piece" of knowledge (i.e., a thought), we submit that belief is the degree or strength of assent we assign to that thought, i.e., another brain process related to the thought in question. Accordingly, we may grade belief between –1 (maximal disbelief or rejection) and +1 (maximal belief or acceptance). The value 0 is assigned to indifference or suspension of belief. In other words, we must know something, whether true or false, before we can believe it. (Incidentally, we can know as many truths as falsities, and this for the trivial reason that we can always know the negation of any proposition.) We therefore suggest:

DEFINITION 3.5. Let s denote a subject and p a piece of knowledge. Then
(i) *s believes p* $=_{df}$ s knows p and s gives assent to p;
(ii) *s is justified in believing p* $=_{df}$ s knows p and s knows that p is reasonably well confirmed;
(iii) *s is justified in disbelieving p* $=_{df}$ s knows p and s knows that p has been disconfirmed.

To conclude, knowledge involves the notions of neither truth nor objectivity: it can be subjective or objective and it can range from complete truth to utter falsity as well as from practical uselessness to usefulness. Therefore, the vulgar opposition between knowledge and error is not part of our epistemology, because much of our knowledge about facts is at best only partially true. Error is the dual or complement of truth, not of knowledge. (More on this in Bunge 1983a. For the notion of truth see Sect. 3.8.)

Let us now move on to some modes of knowledge acquisition.

3.2 Perception and Observation

To speak of perception and observation as distinct from introspection makes sense only if we assume that the world external to the inquiring subject can also be known by inquirers. In other words, we have to supplement our postulate of ontological realism (Postulate 1.1) with an axiom of *epistemological realism*. This axiom is:

POSTULATE 3.4. We can get to know the world, although only partially, imperfectly (or approximately), and gradually.

Although this postulate may appear rather trivial to the scientist, we state it explicitly here because it clearly belongs to the philosophical presuppositions of scientific research, and because it is at variance with both radical skepticism (no knowledge) and intuitionism (instant knowledge). We can now turn to the question of how some such knowledge of the world may be obtained.

3.2.1 Perception

To begin with, we distinguish perception from sensation. For example, we submit that one may feel cold, or hungry, or pain, but one does not perceive coldness, hunger, or pain. On the other hand, an animal may be able to sense light or sound without perceiving them as light or sound, respectively. That is, feeling or sensing is detecting in an automatic way: it is what sensors do. Perceiving, by contrast, is deciphering or recognizing a sensory message. For example, it is seeing a dark shape in the sky *as* a hawk, or hearing certain sounds *as* the call of a cuckoo. While sensing takes only detectors or sensors, perceiving takes, in addition, neuronal systems capable of "interpreting" what is felt or sensed. Thus, perceiving is always perceiving some thing (or rather events occurring in a thing) in a given way. The schema is always: *animal x in state y perceives object z as w*. (See also Hooker 1978. Note that what we call 'perception' is sometimes called 'cognition': see, e.g., Dretske 1978. Yet, in our view, cognition also comprises conception and evaluation.)

Since percepts are "interpreted" sensations, the states of the perceiving animal's brain, in particular memories and expectations, determine what is perceived as what. Thus, to perceive is to construct, not just to copy. The senses do not give us a *picture* of the world but only *signs* of it, which must be interpreted before they can become cognitive items. What is sensed is some event in a sensory organ. This sensation distorts the ongoing activity of the perceptual system. In other words, the environment does not produce the activity of the CNS but only causes (triggers) changes in its activity: it enhances or dampens, i.e., *modulates*, the incessant activity of the CNS. Accordingly, we have no direct knowledge of the external world. Even the assumption that environmental items trigger events in some sensory system is a brain construction, i.e., either a perception or a hypothesis, which is sometimes correct and at other times incorrect.

Although this is a constructivist thesis, we do not embrace radical constructivism but realism: we maintain that we do not construct the world but only map or represent it (to some extent) with the help of more or less adequate constructs. More precisely, we suggest:

> DEFINITION 3.6. An animal *b* has acquired some (partially true) *perceptual knowledge of some items in its environment E* if, and only if, *b* possesses a plastic neuronal system *n* such that some events in *E* are mapped into events in *n*.

Lest the notion of mapping be misunderstood as pictorial resemblance, we emphasize that any knowledge about the world, whether perceptual or conceptual, is symbolic rather than pictorial (von Helmholtz 1873). In particular, there can be no isomorphism between a brain process (and, *a fortiori*, a construct) and a fact outside the knowing brain. We spell this out in:

> POSTULATE 3.5. Any knowledge of factual items is not direct or pictorial but *symbolic*.

The moral for a realist epistemology is clear: although perception triggered by an external object may give us some knowledge about reality, we do not perceive real things as they really are but only as they appear to us. Hence, naive realism is wrong. The same holds for causal theories of perception, as adopted by most empiricist philosophers. Yet this is not to embrace phenomenalism. Disentangling the object (thing in itself) from the subject (percipient subject) calls for a different kind of brain process, namely hypothesizing and theorizing about things in themselves. The realist, then, assumes that, though appearances or phenomena *are* real, they are only *part* of reality (see Sect. 1.8.2 and Postulate 1.10). Not so the phenomenalist.

3.2.2 Phenomenalism Versus Realism

Phenomenalism is the philosophical school which holds that we can know only phenomena, i.e., appearances to some observer. We distinguish two kinds of phenomenalism: ontological and epistemological. According to *ontological phenomenalism*, only phenomena exist: every thing would be a bundle of appearances, and every change would be a human experience. This view is neatly summarized in Berkeley's famous formula *esse est percipi* (to be is to be perceived). *Epistemological phenomenalism*, by contrast, is the view that only phenomena can be known. While Kant (1787) thought that there must be things in themselves (*noumena*) behind and apart from phenomena, for there must, after all, be something that appears, others regard the existence of things in themselves as undecidable. Needless to say, the first kind of phenomenalism implies the second.

In either version, phenomenalism is incompatible with modern science. (For an early critique of phenomenalism in biophilosophy see Woodger 1929.) Indeed, phenomena or perceptual appearances are only the starting point of empirical inquiry. Even in everyday life, most of us seek realities behind appearances, for we know that looks can be deceptive. *A fortiori* scientists seek reality beneath appearances. This search takes them beyond perception, into conception and, particularly, into theory. Sometimes, we succeed in explaining appearances in terms of hypotheses about imperceptible (or transphenomenal) things or processes. Well-known examples are the explanation of the external appearance (phenotype) of an organism in terms of genes (genotype), and the neurophysiological explanation of overt behavior, after-images, and perceptual illusions.

Although phenomenalism is incompatible with genuine science, it is far from dead. At present, for instance, epistemological phenomenalism survives in a version called 'constructive empiricism' (van Fraassen 1980). It should come as no surprise that this view is avowedly antirealist. After all, phenomenalism is the last-ditch stand of anthropomorphism (see also Giere 1985). As the aim of this school is to come up with "empirically adequate" models of the phenomenal world, not with truth, we submit that the scientist has hardly any use for it. Incidentally, the notion of empirical adequacy as a substitute for truth is an old ideal of phenomenalists. Already Cardinal Bellarmino argued against Galileo, an outspoken realist, that the heliocentric model could not be regarded as true but, at best, as just as empirically adequate as the Ptolemaic one (Duhem 1908). (The same held, by the way, for Tycho's model. For further criticisms of constructive empiricism see Churchland and Hooker eds. 1985.)

We submit that, regardless of their philosophical declarations, scientists behave like *realists*. That is, they presume that there are objective (subject-independent) facts besides phenomena, which are (semi)subjective, and that some such facts can be known—though, of course, conceptually rather than perceptually. Realism, then, is an ontological as well as an epistemological doctrine. However, there are two kinds of ontological realism: idealist and scientific. *Idealist realism* (or *Platonism*) identifies reality with the totality of ideas and their material shadows. Moreover, it assumes that ideas exist autonomously, in a realm of their own, whereas concrete things are their shadows or copies. Thus, a white thing would be only a poor copy of the perfect and eternal idea of whiteness. The Romantic naturalists, in particular Goethe, adopted this view when they postulated that all existing organisms are more or less defective copies of a primordial organism built according to a flawless *Urplan*.

Scientific realism opposes idealist realism: it equates reality with the collection of all concrete things, which in turn may be defined as objects capable of changing in some respect or other. (This is what *we* take to be the kernel of scientific realism. As a matter of fact, 'scientific realism' designates a variety of doctrines: see, e.g., Sellars 1963; Smart 1963; Popper 1972; Putnam 1983, 1994; Bhaskar 1978; Boyd 1984; Leplin ed. 1984; Churchland and Hooker eds. 1985; Harré 1986; Hooker 1987; Rescher 1987; Kanitscheider 1989; Suppe 1989; Vollmer 1990.) According to our version of scientific realism ideas, far from being self-existing, are processes occurring in the brains of some animals. This hypothesis makes it possible to study ideation in a scientific manner. However, scientific realism promotes not only the investigation of objective facts but also the study of the way in which animals "perceive" them and, especially, of the way in which they model their environment. This is particularly necessary in ethology, psychology, and social science because perceptions and ideas, whether true or false, can steer behavior.

3.2.3 Observation

We submit that not every act of perception is an observation. Thus, we distinguish spontaneous perception from *directed* and *selective* perception or *observation*. In the former case we look *at* things, in the latter we look *for* them. When we watch the passing crowd we are onlookers; when we look for a friend in the crowd, we are observers.

Being selective, observation depends on our expectations, on our fund of knowledge, and on our value system (or interests). That is, people with different backgrounds, cognitive attitudes, and value systems are bound to see the (only) world there is in more or less different ways. In short, there is no such thing as the immaculate perception stipulated by the empiricists. Since there is no pure observation, it has become popular to say that observation is *theory-laden*. We take this to be an unlucky expression because most observers do not know any theories proper. Therefore, the expression *hypothesis-driven* seems preferable.

Observation can be *direct*, i.e., unaided by any instruments, or *indirect*, i.e., conducted with the help of some instrument, such as a stethoscope or an electron microscope. However, most facts are not directly observable: think of electrons or quasars, of genes or populations, of atomic collisions or historical events, of extinct organisms or neuronal processes. We call such unobservable things or facts *transphenomenal*. The term 'transcendental', though traditionally with the same meaning, is nowadays laden with theological connotations and is thus better avoided. We also do not speak of "theoretical entities", as is often done, because facts are neither theory-dependent nor in any other sense "theoretical" (recall Sect. 1.8.1). Only our knowledge of such facts can be theoretical. In other words, our *hypothesizing* the existence of a certain unobserved or unobservable entity may be motivated or driven by theoretical considerations, i.e., it may occur in a theoretical context, but it does not make the fact to which the hypothesis (truly or falsely) refers theoretical itself.

Scientists have devised literally thousands of means of extending the reach of our senses, because they suspect that reality encompasses more than appearance. It is worth mentioning that only a few such means, such as the scale or the binocular, extend the scope of the human senses. Most others, such as the Geiger counter and the electron microscope, are not extrasensitive eyes, ears, or skins, but tools without biological precursors. Yet all of them supply directly observable facts, such as clicks, pointers on a dial, or pictures on a computer monitor. We can therefore say that all knowledge of factual matters either consists of or involves some direct observation. (This is the empiricist ingredient of scientific realism.)

The historical sciences, such as cosmogony, geology, paleontology, (parts of) evolutionary biology, and archeology, are not exempt from this condition. Since historical scientists cannot observe events in the past, most of their observations are limited to remains such as fossils or prehistoric tools. So they must conjecture the possible use of such tools or the possible morphology of extinct organisms. However, in doing so, they will check their hypotheses with other findings, e.g., compare the fossils to Recent organisms. That is, regardless of the amount of speculation historical scientists may engage in, at some point there is some observation involved, if only to compare the thing or process to be explained with some item that is already known. A field of inquiry that involves no observation at all is either strictly formal, i.e., logical, mathematical, or semantical, or else purely speculative, hence unscientific.

Whereas direct observation is hypothesis-laden, indirect observation (as a scientific activity) is *hypothesis-* or even *theory-dependent*. (Again: our observations of imperceptible facts, not the facts themselves, are theory-dependent.) This is because indirect observation involves not only some technical means but also a set of hypotheses or even some theory employed in its very design and operation. For example, the detection of radio waves employs electromagnetic theory. True, I do not need to know any theory when listening to a radio or using a microscope (Hacking 1985). However, without theoretical background, the raw data supplied by indirect observation means would be *scientifically unjustified*, i.e., they would be personal (subjective) yet not scientific data. In many cases, the raw data would be even unintelligible and thus useless, for they bear no obvious relation to the observed facts. For example, a layperson would not know what to make of an X-ray diffraction pattern or the spots on an electrophoretogram.

Every indirect observation requires an "interpretation" of the data. This "interpretation" consists of hypotheses about the nature of the events of interest and their interaction with the detector. Such unobservable-observable links used to be called *operational definitions* (Bridgman 1927). Actually they are not definitions, i.e., conventions, but hypotheses. If scientific, these hypotheses must be testable, so that they are better called *indicator hypotheses*. An *indicator* is an observable property (or the corresponding variable) that points to an unobservable property (or the corresponding variable). For example, the height of a thermometric column is a temperature indicator; the vital signs of a person, such as heart beat and blood pressure, are indicators of her physiological state; and speech is an indicator of the activity (normal or pathological) of the Wernicke and Broca areas.

In the ideal case, the unobservable-indicator relation is a known function $U = f(I)$. In other words, an unambiguous indicator hypothesis maps an observable fact onto a transphenomenal one, in such a way that, by observing the former, we can directly infer the latter. For example, to a first approximation the length L and temperature t of a thermoelastic body are related by $L = L_0 (1 + \alpha t)$, where L_0 is the length at 0 °C and α is a characteristic of the material. Measuring L, L_0, and α, the formula gives $t = (L - L_0)/\alpha L_0$. It will come as no surprise that, compared to physics, unambiguous indicators are rare in biology. For example, rapid eye movement during sleep seems to be a reliable dream indicator.

For better or for worse, most indicators are ambiguous and therefore fallible, that is, they are one-to-many relations rather than functions. For example, a particular phenotype does not always allow one to infer a certain genotype, or conversely; and low biodiversity may indicate either cold climate at present or in the past (e.g., during an ice age)—or the ravages of industry. Fortunately, the ambiguity in an isolated indicator hypothesis may be removed by using two or more indicators simultaneously, so that unobservables are best ferreted out with the help of whole batteries of mutually compatible indicator hypotheses. (More on indicators in Sects. 3.5.5 and 3.5.7.1.)

3.2.4 Datum

The result of a perception and, in particular, of an observation is a *datum*. A datum is a particular (as opposed to a general) item of knowledge. More precisely, a datum is a singular statement of the form "Thing x is in state y (or undergoes process z)" or "There are things of kind K". In ordinary life, many data, as the etymology of the word suggests, are *givens*, such as the data of memory, perception, and hearsay, which are not necessarily sought for and which are usually, though often wrongly, taken to be true. Moreover, they are regarded as "facts", so that even in the scientific and philosophical literature the words 'datum' and 'fact' are often used interchangeably. However, this usage is incorrect, for data are propositions, not facts (see also Hooker 1978). Data *refer to facts* and thus can be more or less true and, if less than true, corrigible. If a datum does not refer to a fact it is false. Example: reports on sightings of UFOs and Bigfoot. In other words, we should not confuse factual propositions with the facts they refer to. Finally, we must distinguish data from evidence. An *evidence* is a datum relative to some hypothesis or theory: there is no such thing as evidence in itself.

In factual science all data are *empirical*: they are outcomes of observations, measurements, or experiments. Consequently, they are *produced* rather than "given" or "collected". Yet the fact that both observation and data are hypothesis-driven should not be puffed up to the point of epistemological relativism because, at least in science, both incompatible observations and rival underlying hypotheses can be checked against other data or theories or both. (For a critique of epistemological relativism see Siegel 1987 and Suppe 1989.) Hence, they may be shown to be of unequal worth (e.g., truth value, generality, or depth). The same holds for data because many of them can be replicated, and all of them are, at least in principle, checkable by different means or methods. Consequently, even if all of them are hypothesis-driven, many of them are invariant with respect to changes in the driving hypothesis.

Empirical data can be objective or subjective. The former inform about facts in the world external to the observer, whereas subjective data inform about the subject's feelings, perceptions, desires, and intentions. For example, "*X* is a mammal" is an objective datum, whereas "I feel happy" is a subjective datum. Subjective data are inadmissible in the natural sciences, though admissible, nay indispensable, in psychology and social science.

We can also distinguish primary or direct data from secondary ones. The latter are derived from the former by statistical or other techniques. Such derivation is necessary when data come massively. In this case, they must be subjected to statistical elaboration, in order to find percentages, averages, modes, deviations from the averages, correlation coefficients, and other statistical parameters. At least in the advanced sciences, theory determines which the relevant statistical parameters are: they are those the theory can help calculate.

In the less-developed fields of inquiry, scientists spend most of their time and effort in what is often (though wrongly) called 'fact-finding', that is, in producing data. Though undoubtedly indispensable, this task can only be a means rather than an end, for data do not "speak" by and for themselves. Indeed, data are worthless in themselves: they are useful only as inputs to some brain thinking of a hypothesis or a theory, and which is thus capable of supplying understanding. Moreover, interesting data can be collected only in the light of interesting hypotheses, and their collection involves methodological sophistication and careful planning. Let us therefore say a few words about scientific investigation.

3.3 Inquiry

Walking around with the botanist's vasculum and the insect net, and collecting and describing indiscriminately what crosses one's path, is the caricature of the field biologist. Indeed, traditional natural history was purely descriptive, i.e., its explorations consisted merely in collecting data. (This is not to belittle the importance of description in biology or in any other science.) The modern field biologist, however, no longer collects data indiscriminately. He or she collects data in order to solve a problem. A *problem* (a *Fragestellung*; or, in Kuhnian terms, a *puzzle*) may be generated by former observations (data) or by gaps in one's knowledge. For example, the natural historian's description of the geographic distribution of the members of certain species may lead to the problem of why they only occur in certain areas; and the phylogeneticist who wants to show the monophyly of a certain group is led to search for particular features of phylogenetic significance if they are as yet unknown in all species of the putative monophyletic group. In sum, a problem may be construed as the difference between what is known and what one wishes or needs to know. (More on problems in Bunge 1983a.)

(In this construal, problem-solving behavior presupposes knowledge, i.e., cognitive abilities. Furthermore, problem-solving behavior is *purposive* behavior. Thus any automatic exploration conducted by animals incapable of knowing anything is not of the problem-solving kind: not knowing anything, they have no knowledge gaps to fill. *A fortiori*, we cannot side with those biologists who say that all surviving organisms have "solved" successfully the "problems" posed by their environment by means of "adaptations". Neither can biological evolution be regarded as a problem-solving process. We take this to be metaphorical talk. Finally, it is doubtful that we are justified in assigning to inquiring automata the ability to solve problems, for they have no purpose or design of their own, because their activity—even if quasi-intentional—is dependent on their manufacturers and users. They are not really inquiring machines but machine aids to human inquiry.)

Yet how are problems solved in science? We submit that they are solved by intuition and methodical inquiry.

3.3.1 Intuition

When faced with a problem of a familiar kind we can resort to our own fund of knowledge, or to the knowledge stored in other people's brains or in libraries. (Recall that to speak of knowledge stored in a library is a metaphor.) This will not suffice if the problem is of a new kind, because in this case we need additional knowledge and some intuition or flair to guide us in the search for such knowledge. Intuition is that ill-defined ability to spot problems or errors, to "perceive" relations or similarities, to form concepts or hypotheses, to conceive of strategies, to design experiments or artifacts—in short, to imagine, conceive, reason, or act quickly in novel ways. (See also Bunge 1962.) However, the need for intuition in every field of human endeavor does not entail that it suffices, or that we cannot go beyond it, or that it is superior to experience and reason. Indeed, intuition is insufficient: it is not a substitute for work but a guide to it. It must lead to *method* if we want to gain some reliable knowledge about the world.

3.3.2 Method

A *method* is a prescription for doing something that can be formulated in an explicit manner. It is a rule, or set of rules, for proceeding in an orderly fashion toward a goal. Thus, a method can be formalized as an ordered n-tuple every member of which describes one step of the procedure: First do this, then that, and so on. By contrast, contemplation, intuition, and guessing (i.e., trial and error) are not methodical procedures because they are not rule-directed.

Some methods are *general*, that is, utilizable in various research fields. Think of what is called the 'scientific method', the 'experimental method' and the 'statistical method'. Other methods, such as Giemsa or Golgi staining in histology or DNA hybridization in molecular biology, are *specific*, that is, they are restricted to particular fields of research. Specific methods are also called 'techniques'. A technique may be called *scientific* iff (a) it is intersubjective in the sense that it gives roughly the same results for all competent users; (b) it can be checked or controlled by alternative methods; and (c) there are well-confirmed hypotheses or theories that help explain, at least in outline, how it works. We call a method that complies with only one or two of these conditions *semiscientific*, and one that complies with neither *nonscientific*. Note that condition (b) is that of testability, and (c) is that of justification, as opposed to faith or authority.

We mentioned the scientific method as one with a large scope, in contrast with any of the special methods. Now, it is often argued that there is no such thing as *the* scientific method. This skepticism seems to be a reaction against the naive view that the scientific method be a set of simple, invariable and infallible recipes for finding definitive truths. Of course, there are no such simple rules. Nevertheless, we contend that there is a general *scientific method*, which we take to consist in the following ordered sequence of cognitive operations (Bunge 1983a):

1. *Identify a problem* (whether gap or dent in some body of knowledge)—if possible an important bit of ignorance. If the problem is not clearly stated, go to the next step, otherwise to step 3.
2. *State the problem clearly*, if possible in mathematical terms.
3. *Search for information, methods, or instruments* likely to be relevant to the problem. That is, scan what is known to see whether it can help solve the problem.
4. *Try to solve the problem with the help of the means collected in the previous step*. Should this attempt fail, go to the next step; if not, to step 6.

5. *Invent* new ideas (hypotheses, theories, or techniques), *produce* new empirical data, or *design* new experiments or new artifacts that promise to solve the problem.
6. *Obtain a solution* (exact or approximate) of the problem with the help of the available conceptual or material means.
7. *Derive the consequences* of the tentative solution thus obtained. If the solution candidate is a hypothesis or a theory, compute predictions or retrodictions; if new data, examine the effect they may have on existing ideas; if new experiments or artifacts, assess their possible uses and misuses.
8. *Check the proposed solution.* If the solution candidate is a hypothesis or a theory, see how its predictions fare; if new data, try to replicate them using alternative means; if new techniques or new artifacts, see how they work in practice. If the outcome is unsatisfactory, go to the next step, otherwise to step 10.
9. *Correct* the defective solution by going over the entire procedure or using alternative assumptions or methods.
10. *Examine the impact* of the solution upon the body of background knowledge, and state some of the new problems to which it gives rise.

We claim that the scientific method as outlined above can (and should) be applied to all inquiries, whether mathematical or empirical, scientific, technological, or humanistic. (This is a thesis of *scientism*. No surprise, then, that it is denied by many philosophers.)

The scientific method is the most general method in the sciences, hence also in biology. As mentioned previously, there are further general methods such as the experimental and the statistical method, though they are less general than the scientific method. Many biologists claim that there is another very important general method in biology, namely the so-called *comparative method*. However, there is no such thing as the comparative *method*. Indeed, although the comparison between two or more things may be methodical (orderly) rather than erratic, it is not ruled by a method of its own. That is, there is no set of general rules for comparing things in some respects.

Finally, we have to warn against a common confusion of the terms 'method', 'methodics', and 'methodology'. We call *methodics* a set of methods used in a certain field of inquiry. Unfortunately, 'methodics' often is wrongly equated with 'methodology'. However, *methodology* is a metadiscipline studying the processes and the methods of inquiry. In other words, methodology is *normative epistemology*.

3.4 Hypothesis

3.4.1 Conjecture and Hypothesis

Solving problems, whether practical, cognitive, or moral, involves conjecturing: that this fruit is edible and that animal dangerous; that this stone may be suitable for fashioning an ax; and that this action may benefit (or hurt) my kin. Every problem elicits some conjecture or other, and every conjecture poses the further problem of finding out whether it is adequate, i.e., true, efficient, or good.

We submit that not all of our conjectures deserve being called 'hypotheses', but that only the *educated guesses* which are *formulated explicitly* and are *testable* qualify as scientific hypotheses. (See also Bunge 1983d.) We take the expression 'educated guess' to designate a conjecture that, far from being wild, is compatible with some background knowledge. The expression 'explicitly formulated' refers not only to propositions but also to artifacts such as a reconstruction of a stone ax or the historical reconstruction of an extinct organism. Finally, a conjecture can be said to be testable if, and only if, (a) it refers exclusively (whether truly or falsely) to material (real) entities, and (b) it can be checked for truth by contrasting it with data or with the bulk of accepted knowledge. (More on testing in Sect. 3.7.) A conjecture that does not meet these standards will be called *wild* or a *pseudohypothesis*.

3.4.2 The Generation of Hypotheses

According to empiricism, the task of researchers is finished once they have their data. At most, they may summarize and cautiously generalize the information thus obtained. In other words, they may formulate inductive generalizations, i.e., low-level hypotheses. A hypothesis may be said to be *low-level* if it contains no substantive concepts other than those occurring in the data it covers. For example, a data-fitting curve, i.e., the interpolation of a continuous curve from a finite number of data, is of this kind. It should be noted that, although we may form a hypothesis by inductive generalization from data, there is no such operation as the *deduction* of a hypothesis from data. Any such deduction is impossible because a datum d entails infinitely many statements of the form $h \Rightarrow d$ but no particular proposition h. Recall that $h \Rightarrow d$ is true even if h is false; and $d \,\&\, (h \Rightarrow d)$ does not entail h. In short, the first mode of hypothesis generation is jumping to a general conclusion from observed cases (induction).

Besides forming hypotheses by induction, we may form them by noting associations. For example, if we know that A and B often occur together, or one after the other, we are likely to conjecture that A and B are somehow correlated or even functionally related. Such associations may suggest hypotheses of the following types: qualitative ("All A's are B's"), statistical ("f percent of A's are B's"), probabilistic ("The probability that a be at place b at time t equals p"), or causal ("Event a produces change b in thing c"). However, unless trained scientifically, we are likely to see correlations even where there are none. In other words, we tend to care only for occurrences of A and B and to disregard the remaining cases, namely A and not-B, not-A and B, and not-A and not-B. This may lead to the fallacy of nonexistent association or correlation, which is at the root of many superstitions, such as the belief in magic, astrology, and miraculous cures. Moral: when investigating the possible association between two factual items *A* and *B*, examine the full contingency table

$$\begin{matrix} A\,B & A\,\overline{B} \\ \overline{A}\,B & \overline{A}\,\overline{B}\,. \end{matrix}$$

Another source for hypothesis formation is provided by similarities and analogies, whether real or imaginary. Thus, Faraday invented the concept of electromagnetic field by analogy with the concept of an elastic body, and Darwin's hypothesis of natural selection was inspired by his readings on economic competition and overpopulation.

Finally, the deepest and therefore most important hypotheses are not suggested by any of the aforementioned options, but are newly invented. The reason is that the most interesting hypotheses concern facts inaccessible to direct observation. To understand some of the wealth of our complex and messy perceptions, we must imagine transphenomenal things or connections. For example, Harvey, who had just as many anatomical data available as his predecessors, was the first to hypothesize that the heart, the arteries and the veins form a system of closed circuits. Part of this conjecture was to assume the existence of small blood vessels (capillaries), invisible to the naked eye, that connect the ends of the arteries to the beginnings of the veins.

This is, incidentally, one of many examples showing that hypotheses are often generated by thinking about problems, not data. This is because hypotheses of this kind contain concepts that go beyond the data at hand, i.e., concepts referring to transphenomenal facts. Such concepts are often called 'theoretical' (as opposed to observational or empirical concepts). However, since they need not be part of a theory, the term *transempirical* seems preferable. We call conjectures containing transempirical concepts *high-level hypotheses.*

In sum, whereas a low-level hypothesis may be generated from data, a high-level hypothesis is an invention generated by a problem, and its test requires the search for new data.

3.4.3 Scope and Depth of Hypotheses

We distinguish several kinds of hypotheses. Our first distinction will be with regard to scope or extension. The real scope of a hypothesis is not always apparent but must be revealed by logical analysis. Concerning the scope of hypotheses, we have

1. *Singular* hypotheses, such as "This behavior belongs to the animal's courtship display".
2. *Particular* hypotheses:
   a) *Indefinite* particular hypotheses, such as "There is extraterrestrial life" or "There are organisms that can metabolize arsenic", which specify neither precise

place nor time and are therefore hard to check. Such hypotheses can be verified by coming up with at least one of the hypothesized items. Their strict refutation requires that the whole world be searched and no referent of the hypothesis be found —a practical impossibility. From a pragmatic point of view we shall say, however, that the failure to find the hypothesized items despite a long and careful search *confirms* the negation of the hypothesis—until new notice.

b) *Definite* particular hypotheses, such as "There are symbiotic algae in the epidermis of corals".

3. *General* hypotheses:

a) *Bounded universal* hypotheses, such as "All fish in this lake have been poisoned", whose referents are restricted to certain places or times.

b) *Unbounded universal* hypotheses, such as "All protons are composed of quarks". Such hypotheses are falsified by coming up with a single counterexample. However, they are verifiable only if we manage to investigate the whole world —again a practical impossibility. Therefore, we must settle for the weaker notion of confirmation.

Note that there are also mixed hypotheses. For example, the hypothesis "The Earth spins" is singular in one respect but universal in another. More precisely, it is referentially singular but universal with respect to time, for it can be reformulated as "For all times since its formation the Earth spins". This also holds for other astronomical or geophysical law statements concerning our planet.

Hypotheses come not only in different scopes but also in different depths. We define *depth* as follows:

> DEFINITION 3.7. A hypothesis h is *deep* $=_{df}$ h contains at least one transempirical concept (aside from any logical or mathematical concepts). Otherwise, h is *superficial*.

For example, the hypothesis, held for centuries, that syphilis is only a skin disease, is literally superficial. In the wake of the microbiological revolution initiated by Koch and Pasteur, the deep (transempirical) hypothesis was formulated that syphilis was caused by microbes—then transphenomenal entities. The hypothesis proved to be true when the members of *Treponema pallidum* were later identified as the pathogens.

Now, depth comes in degrees. For example, a *black box* model of a system, i.e., one that ignores its composition and internal structure, is superficial, for it represents only the observable inputs and outputs. Epitomes of this kind of hypothesis are the stimulus-response models of behaviorism and the box-and-arrow models found in descriptive ecology. If we allow for internal states of the box, our representation of the system becomes somewhat deeper. For instance, if the classical ethologist assumes that the response to a certain stimulus depends on some internal state, such as a motivation or a drive, he can be said to have built a *gray box* model. Still, a gray box model can be further deepened by specifying a precise internal mechanism, such as stating the neurophysiological processes mediating between sensory input and behavioral output. If this is accomplished we can speak of a *translucent box* model of the system in question. In sum, the deeper hypotheses are those suggesting some mechanism or other, i.e., the *mechanismic* or *dynamical* hypotheses. (Note that 'mechanis*mic*' refers to processes of all kinds, not just mechanical ones: mechanical or electrical, chemical or cellular, organismic or ecological, economic or cultural.)

Nonmechanismic hypotheses, whether or not they contain transempirical concepts, are called *phenomenological*, *kinematical*, or *black box* conjectures. (See also Bunge 1964.) For example, the photosynthesis formula "$6CO_2 + 6H_2O \rightarrow C_6H_{12}O_6 + 6O_2$" is not superficial because it contains only transempirical concepts, but it is phenomenological since it does not suggest any mechanism relating output to input.

The preceding distinctions deserve to be spelled out in:

> DEFINITION 3.8 A hypothesis is called *mechanismic* if, and only if, it conjectures some mechanism. Otherwise, it is called *phenomenological.*

It will be obvious that scientific investigation often starts with phenomenological hypotheses, which are later replaced by mechanismic ones. For example, the hypothesis that biopopulations evolve is phenomenological. If supplemented by concepts such as mutation, natural selection, and reproductive isolation, which suggest the mechanisms by which biopopulations evolve, it may be turned into a mechanismic hypothesis. We submit that only mechanismic hypotheses have explanatory power (see Sect. 3.6.3).

3.4.4 The Methodological Status of Hypotheses

Although we make conjectures all the time, we do not assign the same status to them all. Upon examination, some hypotheses prove to be weaker, or wilder, or less deep, or less testable, or less true, or less useful than others. Therefore, it will be useful to distinguish hypotheses with regard to their methodological status.

A first distinction is that between substantive and nonsubstantive hypotheses. A hypothesis is called *substantive* iff it describes, explains, or forecasts facts of some kind. By contrast, the *nonsubstantive* hypotheses are valuational or prescriptive, i.e., they guide research and action. Two kinds of them may be distinguished: methodological and axiological. A *methodological* (or *instrumental)* hypothesis is a tentative statement concerning the manner of studying a substantive hypothesis, i.e., it is a conjecture about the adequacy of a method, procedure, or instrument to investigate a certain fact. An *axiological* hypothesis is a value judgment, such as the statement that theory *A* explains a given set of facts better than theory *B*.

A hypothesis may or may not belong to some organized body of knowledge, such as a theory. If they do, we speak of *systemic* hypotheses, if not we call them *stray* hypotheses. While we prefer systemic hypotheses in science, most of the hypotheses we formulate in daily life are stray. The advantages of systemic over stray hypotheses are the following. For one, the hypotheses in a system can join to generate (logically) further hypotheses. For another, systemic hypotheses are supported not only by whatever empirical data may be compatible with them but also by the other components of the system.

The stray hypotheses that account only for a small subset of the total set of facts to which they refer are called *ad hoc*. For example, the hypothesis "Dinosaurs became extinct because their brains were too small to cope with a complex environment" is *ad hoc*, because it only "saves the appearances", i.e., the fact that dinosaurs became extinct. However, it does not take into account why they had been successful during the whole Mesozoic era, and why many other species also became extinct at the end of the Cretaceous period.

Other *ad hoc* hypotheses are introduced to save further hypotheses endangered by unfavorable evidence. These are called *protective* hypotheses. The use of protective hypotheses is legitimate if they are testable independently of the hypothesis they are supposed to protect. For example, Darwin hypothesized that there are gaps in the fossil record, in order to save his theory of evolution and, in particular, the hypothesis of evolutionary gradualism, which requires the existence of intermediate forms between the distinct Recent forms of organisms. This *ad hoc* hypothesis, however, is testable independently of evolutionary theory. Taphonomy—the discipline studying the processes of fossilization—shows, for instance, that organisms possessing solid body parts, such as an inner or outer skeleton, will be overrepresented in the fossil record, whereas soft-bodied organisms are only rarely preserved. We call protective hypotheses that are independently testable *bona fide ad hoc* hypotheses.

Yet not all protective hypotheses are *bona fide*. There are also *mala fide* hypotheses, which are not independently testable or perhaps not scrutable at all. *Mala fide ad hoc* hypotheses can often be found in everyday reasoning and in pseudoscience. For example, the psychic's defense that his failures are due to somebody's hostility inhibiting his paranormal abilities, and the psychoanalytic fantasy that people who do not exhibit their Oedipus complex have repressed it, are paragons of *mala fide* protective hypotheses.

Finally, a word on the notion of a *null hypothesis*. This is the hypothesis that the variables under investigation in an experiment are mutually unrelated, so that the data are "due to chance". The formulation of a null hypothesis is indispensable in the preliminary stage of an experimental research, when all one has to go by is the programmatic hypothesis that two given variables may be correlated. The thing to do is to try and refute the null hypothesis. The refutation of the null hypothesis clears the way to the conjecturing of an alternative and substantive hypothesis. Once this is at hand, it should help one design a more sophisticated experiment aiming at answering a more precise question, e.g., of the form "Are the variables in question related in such and such a manner?". For example, if H_0 = "$y = a$", then H_1 = "$y = ax + b$", H_2 = "$y = ax^2 + bx + c$", H_3 = "$a \exp(bcx)$", etc. Only experiment will help identify the truest hypothesis. (See Sect. 3.7.4.2. For the notion of null hypothesis in ecology see Sect. 5.2.)

3.5 Theory and Model

Although the conjecturing of hypotheses is an important scientific activity, it does not automatically provide us with a comprehensive and coherent view of reality. To achieve the latter, we must aim at producing hypotheses which refer roughly to the same domain of objects and which allow us to deduce further hypotheses, or to deduce them from even more general propositions. That is, if we strive for systematic knowledge, our hypotheses must be related both logically and referentially. In other words, we must organize our hypotheses into *theories*, which ought to be further organized into systems of theories.

A theory, then, is a *hypothetico-deductive system*. This conception obviously contradicts some more or less popular views, such as the view that theory is the opposite of hard fact, that theories are the same as hypotheses, that theories are general orientations or approaches or "scientific practices" (e.g., Culp and Kitcher 1989), or that theories are generalizations of observed facts (inductivism). We reject these beliefs for the following reasons. First, theories are not opposed to facts. What is true is that some theories do not fit the facts they purport to represent, i.e., they are false, and still others are irrelevant to them. Second, theories are not single hypotheses but systems of such. That is, even in the special case where a theory contains only a single axiom, the theory consists of this hypothesis *plus* all its consequences. Third, since approaches are no more and no less than ways of viewing and handling things or data, they may at most suggest a type of theory. Fourth, like hypotheses proper, and unlike empirical generalizations, theories contain (higher-level) concepts that may not occur in the data relevant to them.

The conception of a theory as a hypothetico-deductive system embraces theories of all scopes in all fields, from pure mathematics to physics to biology to engineering. This conception was born over two millennia ago in mathematics: recall Euclid's *Elements*. However, since it was endorsed by logical empiricism, which is still influential, it is often believed to be inextricably linked to the so-called *neopositivist* or *standard* or *received* view of theories. This received view of theories has been under attack for several years by philosophers (see, e.g., Suppe 1972, ed. 1974). Some philosophers have proposed alternative views of theories, such as the so-called *structuralist* and *semantic* views of theories. The neopositivist view is indeed inadequate, and we thus have to look for an alternative conception of theories. However, this alternative is neither the "semantic" nor the structuralist conception, as will be shown in Section 9.3.2 when examining evolutionary theory. We will therefore expound in this chapter what we call the *realist* view of theories (as developed by the senior author: see Bunge 1967a, c, 1974a, 1977c, 1983a.)

3.5.1 The Structure or Syntax of Theories

We begin by sketching some examples of theory, emphasizing the structure or formal skeleton. The matter of theory content or interpretation will be tackled later.

Consider a set of data of the form "Fx" or "x is an F". (Read: "individual x [e.g., an organism] possesses property F" or, more rigorously, "individual x is attributed a property designated by predicate F". Recalling the property-predicate distinction from Section 1.3.2, we must keep in mind that one and the same property may be represented by alternative predicates in different theories.) In this schema (or open formula) the variable or blank x may take on such values as the letters $a, b, \ldots, n$, naming particular individuals. We are then faced with a *data base* composed of n propositions, neither of which is deducible from any others in the base. (Recall also that data are not facts but propositions about facts.) These propositions may be conjoined: $Fa \;\&\; Fb \;\&\; \ldots \;\&\; Fn$, which can be abbreviated as "$(\forall x)_S Fx$", where $S = \{a, b, \ldots, n\}$. The formula, read "For all x in S, Fx", is an *empirical generalization*: it compresses all the n data into a single proposition. Note that the universal quantifier $\forall$ is bounded or limited to the set S: our formula does not state that all individuals are F's but only those in the set S, which is finite. This is one of the limitations of inductive generalizations. However, this restriction can be lifted: that is, we may *extrapolate* the given generalization to a superset T of S, as when we conjecture that what holds for laboratory rats holds also for humans. A far more serious limitation of inductive generalizations is that they only contain predicates representing observable properties. If they made assertions about unobservable properties, such as "is as remote descendant", or "is an oncogene", they would not be inductions.

We started from n data and compressed them, without running any risks, into a bounded generalization. So far, no theory. But on other occasions we start from one or two observations, jump tentatively to a general "conclusion", deduce some of its consequences, and perhaps end up by checking them for truth. In this case we proceed in a hypothetico-deductive way: we build a theory—albeit one of the simplest possible kind. We may proceed then in either of two different ways:

	$(\forall x)_S Fx$	$(\forall x)_S Fx$	
Finite induction	$\uparrow$	$\downarrow$	*Finite deduction*
	$Fa, Fb, \ldots, Fn$	$Fa, Fb, \ldots, Fn$	

In sum, the data base $\{Fa, Fb, \ldots, Fn\}$ is not a hypothetico-deductive system, for it contains no hypotheses. On the other hand, the hypothesis "All x are F, where x is in S", together with its n immediate consequences, is a (tiny) hypothetico-deductive system; but it is not a *scientific* theory because it does not contain any law statements (more on law statements in Sect. 3.5.8). Since the hypothesis "$(\forall x)_S Fx$" is the basic statement of this hypothetico-deductive system from which all other statements follow, it is called an *axiom* or *postulate*.

If the initial hypothesis includes numerical variables ranging over infinite sets, it encompasses infinitely many propositions: it generates an infinitely richer system than the above. Consider, for example, the allometric equation

$$y = ax^b, \text{ with } a, b \in \mathbb{R}, \qquad [1]$$

which occurs almost everywhere in biology. (For example, the basal metabolic rate of mammals is such a power function of the body mass.) With the help of elementary mathematics, equation [1] can be rewritten in two equivalent forms:

$$x = (y/a)^{(1/b)}, \qquad \log y = \log a + b \log x. \qquad [2]$$

Moreover, with the help of the calculus we may compute any of the infinitely many derivatives of y:

$$y^{(n)} = ab(b-1)(b-2)\ldots x^{b-n}, \text{ with } n \in \mathbb{N}. \qquad [3]$$

In short, the single equation [1], conjoined with a few mathematical formulas, has generated the infinite set $\{y^n(x) \mid n \in \mathbb{N}\}$, every member of which follows deductively from its predecessor. This, then, is a full theory. However, the higher-order derivatives have no known biological interpretation, so that the theory has no biological interest except for its postulate [1] and its first derivative. In other cases, a single formula, such as a partial differential equation, is so rich that, with the help of mathematics, it generates a bulky theory of scientific interest.

Let us now go from the case of a single axiom to that of two or more postulates. Consider the following two indeterminate universal generalizations:

All A's are B's, or $(\forall x)(Ax \Rightarrow Bx)$
All C's are D's, or $(\forall y)(Cy \Rightarrow Dy)$,

where the predicates C and D are independent of A and B. Because of this independence, nothing follows from conjoining the two hypotheses. Moreover, the difference in notation (x and y) for the arbitrary individuals concerned suggests that they are not of the same kind. The given set, then, is a *set* of hypotheses but not a *system*: it is not a hypothetico-deductive system or theory.

Let us now introduce the following bridge between the two above hypotheses: $C = B$. Because now all the individuals share one property, namely the one designated by B, they belong to the same kind, so we set $x = y$ and write

$$(\forall x)(Ax \Rightarrow Bx), \quad (\forall x)(Bx \Rightarrow Dx). \qquad [4]$$

By virtue of the logical law of the hypothetical syllogism these two hypotheses entail the consequence:

All A's are D's, or $(\forall x)(Ax \Rightarrow Dx)$. [5]

(A consequence of a postulate, such as [5], that is not evident but in need of proof by means of logical rules is called a *theorem*. The immediate or obvious consequences of postulates, definitions or theorems are called *corollaries*.)

Moreover, if the particular individual b happens to be an A, we infer that it is also a D. (Both the general and the particular results are potentially contained in the system [4], but we may not have known them at the time of assuming [4].) Nor are these the only consequences of [4]. Indeed, the logical principle of addition "From p, infer p or q", where q need not be related to p in any way, allows us to deduce infinitely many propositions from either of the above formulas. However, this logical principle must be handled with care: in science we are seldom interested in smuggling in propositions that have nothing in common with the ones we are considering. To prevent such contraband we must add the following proviso: no predicates should occur in the theorems of a theory which do not occur in its axioms or its definitions.

3.5.2 The Semantics of Theories: Population Growth Theory

Let us now study a very modest quantitative theory (or mathematical model) occurring in a number of sciences, among them biology and demography. However, we shall specialize it from the start to biology, i.e., "read" or interpret (most of) its variables in biological terms.

Consider a biopopulation of unicellular asexual organisms endowed with an unlimited food supply, and protected from environmental hazards; assume further that overcrowding does not affect the rate of reproduction. It is well known that these simplifying assumptions are "heroic" (i.e., brutal) idealizations: real populations usually do not find themselves in such ideal conditions. However, the resulting theory "works" (is true) to a first approximation. (As a matter of fact, it is employed in making demographic projections.) If the preceding assumptions are spelled out in quantitative terms, a minitheory or *theoretical model* results. We will formulate it in an orderly, i.e., axiomatic, fashion, stating both the mathematical and the semantical assumptions, as well as a couple of logical consequences.

The minitheory or model to be presented presupposes ordinary predicate logic with identity, elementary number theory and naive set theory, and the elementary theory of finite difference equations. The primitive (i.e., undefined) concepts of our axiom system are designated by P, T, x_t, and k. They are subject to the following axioms:

A1 P is a nonempty finite collection.

A2 T is a subset of the set of natural numbers. (I.e., $T \subseteq \mathbb{N}$.)

A3 x_t is a function from the Cartesian product of P by T to the natural numbers. (I.e., $x_t : P \times T \rightarrow \mathbb{N}$.)

A4 The value of x_t at $t = 0$ is greater than 1, i.e., $x_0 > 1$.

A5 k is a positive real number. (I.e., $k \in \mathbb{R}^+$.)

A6 The value of x_t for any given t in T is proportional to its value at $t - 1$:

$$x_t = kx_{t-1} \quad , \text{ or } \quad x_{t+1} = kx_t. \qquad [6]$$

A7 P represents a population of unicellular asexual organisms.

A8 An arbitrary member t of T represents the time elapsed from the start ($t = 0$).

A9 x_t equals the (numerosity of the) population P at time t, and x_0 represents (the numerosity of) the initial population.

A10 k represents the rate of growth of P.

So much for the foundation of our theory. (The *foundation* of a theory consists in a list of its logical and mathematical presuppositions, a list of its primitive, i.e., undefined, concepts, and its mathematical and semantic postulates.) The rest is either logical consequence or application to particular cases. To indulge in literary metaphor, the first five postulates introduce the main characters, whereas the remaining axioms constitute the plot. The subsystem constituted by postulates A1 to A6 is a self-contained *mathematical formalism*: it specifies the (formal) structure of the theory. However, like any other formalism, ours may be interpreted in alternative ways, e.g., as describing collections of things of a number of different kinds, biological or nonbiological. The remaining axioms, A7 to A10, interpret the four basic concepts in *factual* terms: they are the *semantic assumptions*, which endow the formalism with a biological content (or meaning). (Note that they do not interpret the basic concepts in *empirical* or phenomenal terms, i.e., in terms of observations or experimental operations. Rather, they refer to facts, whether observable or not.) In short, the subsystem constituted by postulates A1 to A6 constitutes the *syntax*, and the remaining axioms, A7 to A10, constitute the *semantics* of the theory. A different set of semantic postulates would produce a different theory with the same structure. Thus syntax and semantics complement each other. However, they are not on the same footing: every interpretation must fit the formalism like a glove to a hand. To pursue this metaphor, an unlimited number of gloves (interpretations) may fit the same hand (formalism). That is, the same mathematical formalism may be interpreted in an unlimited number of alternative ways.

The foundation of our minitheory is now ready for either processing, application, or empirical test. A couple of general consequences of the above postulates follow. The first is a corollary or immediate consequence, whereas the second requires deduction with the help of the principle of complete induction.

COROLLARY. For all t in T,
(i) if $k = 1$, then $x_{t+1} = x_t$, i.e., the population remains constant;
(ii) if $k > 1$, then $x_{t+1} > x_t$, i.e., the population grows;
(iii) if $k < 1$, then $x_{t+1} < x_t$, i.e., the population decreases.

THEOREM. For all t in T, $x_t = x_0k^t$. (*Proof* By mathematical induction.)

Because t is a variable and k an indeterminate parameter, this theorem entails a double infinity of propositions, one for each pair <k, t >:

$$x_1 = x_0k, \quad x_2 = x_0k^2, \quad x_3 = x_0k^3, \dots .$$

Note that, in order to get down to numbers, both the initial population x_0 and the parameter k have to be assigned definite numerical values. These can only be obtained from observation: they are data. Moreover, observation may refute postulate A5 by showing that k is not a constant. In fact, for many populations the rate of growth decreases with increasing population, i.e., overcrowding dampens growth. A common assumption is that k is not a constant but depends linearly upon x_t, i.e., $k_t = a - bx_t$, where b is essentially the reciprocal of the carrying capacity of the habitat. This assumption entails replacing A6 with the famous logistic equation $x_t = kx_t\,(1 - ax_t)$, which, however, is still a poor describer of the growth of most populations (Solbrig and Solbrig 1979; Lewontin 1984.) Yet, whichever population growth equation may turn out to be the truest, it will be the one *basic law statement* contained in the theory. All of its biologically meaningful consequences will be *derived laws*$_2$.

The minitheory we have just sketched is conceptually so unproblematic that nothing of scientific substance was gained by axiomatizing it. Ours has been merely a didactic exercise aiming at exhibiting both the hypothetico-deductive structure and the factual content of scientific theories. However, axiomatics has several advantages. First, it is convenient in all cases for didactic purposes, as it is easier to remember a handful of axioms and a few outstanding theorems than to recall a large and disordered set of formulas. Second, axiomatization is the only way to make sure that a given proposition really is (or is not) part of a given theory. In particular, one may need to check whether a given proposition is a working assumption rather than an idle one. Or one may wish to check whether a given proposition does, in fact, follow from the given premises rather than having been smuggled into the theory. Third, axiomatization is required to distinguish assumptions from their logical consequences as well as to distinguish assumptions from definitions. (More on definitions in Sect. 3.5.7.1.) The latter is sometimes necessary to avoid trying either to prove definitions or pass empirically groundless hypotheses as definitions. In other words, axiomatics clarifies the logical form and methodological status of any theory components. Fourth, it exhibits the referents of a theory. For instance, an adequate axiomatization of a theory of evolution should state explicitly what evolves. Last, but not least, axiomatics turns conceptual disorder into order. (For further virtues of axiomatics see, e.g., Hilbert 1918; Bunge 1973b.)

As axiomatization is occasionally confused with formalization, and the latter with symbolization, it might be useful to point out the differences between these operations. An axiomatization may, but need not, involve a formalization and symbolization of its concepts and statements. For example, occasionally the axioms, definitions and theorems of a theory may just be stated in plain language. If we abbreviate the concepts and statements by symbols, we engage in symbolization, yet not formalization. Only if the constructs involved are endowed with a precise logical and mathematical form, do we speak of formalization. Of course, formalization usually involves symbolization, but it does not amount to axiomatization: an untidy list of formulas is not an axiomatic system.

The above axiom system illustrates what we call the *realist* conception of factual (in particular scientific) theories as mathematical formalisms enriched with semantic assumptions or concept-fact relations. It is realist because it involves interpreting the mathematical concepts in objective or impersonal terms rather than in terms of appearances, observations, or experiments. That is, it views scientific theories as having a factual content but not an empirical one: it refers to things out there, not to our actions to handle them or to check the theory itself. No scientific theory is concerned with its own test. However, a theory can have empirical content if it *is* a theory about experience, e.g., a psychological theory.

3.5.3 Degrees of Abstraction, Generality, and Depth of Theories

3.5.3.1 Degrees of Abstraction

Theories can be partitioned in two different ways: with regard to abstraction, and with regard to generality. Let us begin with abstraction, where the term 'abstraction' is used in its semantic sense, not in the epistemological sense of remoteness from sense experience. As regards its degree of semantic abstraction (or interpretation), a theory can be of either of the following kinds:

1. *Abstract* or *uninterpreted*, such as Boolean algebra and set theory;
2. *Semiabstract* (or *semi-interpreted*), such as the propositional calculus (a model of Boolean algebra);
3. *Interpreted in mathematical terms*, such as the infinitesimal calculus (in which all the functions are interpreted as taking values in the set of the real numbers);
4. *Interpreted in factual terms*: all the theories in science and technology.

To exemplify the interpretation of abstract theories, we make use of one of the simplest abstract theories: semigroup theory. This algebraic theory can be introduced by the following axiomatic definition: a semigroup $G_{1/2}$ is an arbitrary set S together with an associative operation (concatenation) $\circ$ among any two members of S. In symbols:

$$G_{1/2} = \langle S, \circ \rangle \text{ such that, for any } x, y \text{ and } z \text{ in } S, x \circ (y \circ z) = (x \circ y) \circ z.$$

Note that the nature of the elements of both the set and the operation are left unspecified. This why the theory is called 'abstract'. Semigroup theory can be complicated (enriched) in many ways. That is, one can construct any number of abstract semigroups by adding assumptions consistent with the associativity postulate. One of them could be that S contains an identity element e : for all x in S, $e \circ x = x \circ e = x$, in which case the semigroup is called a *monoid*.

An indeterminate number of objects, many known and others yet to be invented, satisfy the above definition of a semigroup. That is, there is an indefinite number of *mathematical interpretations* or *models*$_1$ of a semigroup. One of the simplest is constituted by the natural numbers $\mathbb{N}$ together with the addition operation, i.e., M_1

$= \langle \mathbb{N}, + \rangle$. Indeed, for any three nonnegative whole numbers x, y, and z, $x + (y + z) = (x + y) + z$. (In fact M_1 is a monoid, since zero behaves like the unit element under addition: for all x in S, $x + 0 = 0 + x = x$.) In other words, from the viewpoint of abstract algebra, M_1 results from $G_{1/2}$ by adding the semantic assumptions (i.e., interpretations)

$$\text{Int}\,(S) = \mathbb{N}, \qquad \text{Int}\,(\circ) = +.$$

An alternative model is produced by interpreting S as the set $\mathbb{Z}$ of integers and $\circ$ as multiplication, i.e., $M_2 = \langle \mathbb{Z}, \times \rangle$. But, of course, the addition and multiplication of whole numbers came historically much earlier and they only provided two "concrete" examples or models$_1$ of the abstract semigroup formalism, a twentieth century invention. (For more on logical models and formal truth see Bunge 1974b.)

The relations between an abstract theory A and any of its models$_1$ M_i are those of interpretation and its dual, abstraction:

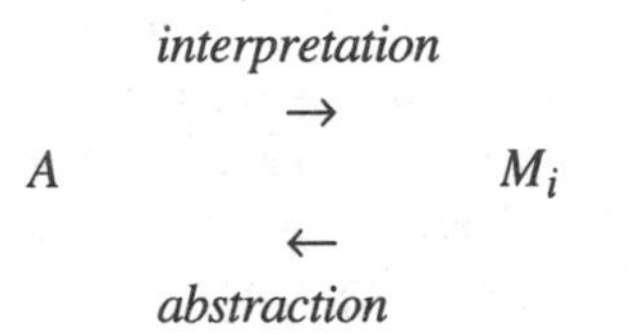

However, the interpretation in question is neither factual nor empirical but *mathematical*. Thus consider the formula "$(\forall x)(\exists y)\,(x \circ y = x)$", which occurs in group theory. It "says" that the group contains an element which, when concatenated with an arbitrary element of the same group, leaves it unaltered: that designated element is, of course, the unit element we called e earlier. The given formula holds (is mathematically true) if the group elements are taken to be integers, in which case the unit element is the zero.

In order to handle such abstract theories, *model theory* was built systematically, particularly from the early 1950s on. This metamathematical endeavor had three major goals. One was to help ferret out the abstract formalisms (uninterpreted hypothetico-deductive systems) common to the members of a large family of mathematical theories. A second aim was to fashion a ready consistency criterion. (An abstract theory is consistent if, and only if, it has at least one model$_1$—i.e., iff its formulas are true under at least one mathematical interpretation, e.g., in numerical terms.) A third aim was to introduce the concept of (formal) truth for formulas of an abstract theory. (A statement is said to be true in A if, and only if, it is true in every model$_1$ of A, i.e., if it holds under every interpretation of the basic concepts of A. Leibniz's familiar metaphorical rendering of this idea is: a proposition is logically true or tautological if, and only if, it holds "in all possible worlds", i.e., under every interpretation. This poetic version is useful in suggesting that logical truth has nothing to do with adequacy to the real world.) The third goal was to provide a tool for investigating global properties of theories, such as consistency and inter-theory isomorphism—or lack of it. (See, e.g., Tarski 1954-1955.)

With regard to our critique of the "semantic" view of theories in Section 9.3.2, it is important to realize that models in the above model-theoretic or logical sense of the word, i.e., models$_1$, are interpretations of abstract formulas or theories *within mathematics*. Thus, Tarski (1953, p. 11), the father of modern model theory, states: "A possible realization in which all valid sentences of a theory *T* are satisfied is called a model of *T* "—where *T* is an abstract (or formalized) theory like the theories of groups, lattices, or rings, not a "concrete" one like trigonometry or the calculus, much less a factual one like mechanics or selection theory. Another builder of the theory describes it thus: "Model theory deals with the relations between the properties of sentences or sets of sentences specified in a formal [i.e., abstract] language on [the] one hand, and of the mathematical structures or sets of structures which satisfy these sentences, on the other hand" (Robinson 1965, p. 1). And in a standard textbook we read: "Model theory is the branch of mathematical logic which deals with the connection between a formal language [abstract theory] and its [mathematical] interpretations, or models" (Chang and Keisler 1973, p. 1). That is, mathematical interpretation is a construct-construct relation and, more particularly, an intertheoretical affair. Therefore, model theory is incompetent to say anything about extramathematical—e.g., biological—contents.

By contrast, when building scientific theories, we are interested in establishing construct-fact relations, i.e., relations between theories on the one hand and some things or events on the other. In other words, we are interested in *factual* interpretations of mathematical formalisms, i.e., in factual models (models$_2$), not mathematical ones. In principle, such a factual model may be obtained by interpreting in factual terms either (1) an abstract theory, or (2) a semi-abstract theory, or (3) an already mathematically interpreted abstract theory, or (4) a "concrete" mathematical theory that does not go back to any abstract theory (see above). To indulge in model talk, in the cases of (2) and (3) a factual model would be a model$_2$ of a model$_1$. However, there are only a handful of examples for the cases (1)-(3) in science and technology. The best-known examples of these kinds are these: (a) Boolean algebra can be interpreted in terms of electric circuits; (b) combinatorics can be used to handle combinations and permutations of objects of any kind; (c) packing theory (or geometric combinatorics) can be used to minimize waste of packaging material. In fact, most scientific theories belong in (4), i.e., they make use of mathematical formalisms not derived from any abstract theories. Typically, they include differential (or finite difference) equations which, when suitably interpreted in factual terms, represent rate equations or equations of motion (e.g., the spread or "diffusion" of a population over a territory).

To give an example, we make use of the abstract theory of semigroups again. The interpretation of *S* as the set of all sentences of a language, and $\circ$ as sentence concatenation, results in the simplest possible factual model of an arbitrary (natural or artificial) language. Indeed, the sentences in any given language concatenate associatively: If *x*, *y*, and *z* are sentences of a language, so is their concatenation $x \circ y \circ z$, which satisfies the associative law: $x \circ (y \circ z) = (x \circ y) \circ z$.

An alternative $model_2$ is obtained by interpreting S as the collection of all concrete things (or substantial individuals), and $\circ$ as physical addition (or juxtaposition). This model exactifies the intuitive notion of physical addition, occurring in propositions such as "The biomass of two biopopulations taken together (or "added") equals the sum of their individual biomasses". Clearly, this model is of interest to all the factual sciences and their underlying ontology, if only because it allows one to define the part-whole relation in an exact fashion. Indeed, if x and y are substantial individuals, then x is a *part* of y, or $x \sqsubset y$ for short, if and only if $x + y = y$, where + stands here for physical addition, an ontological interpretation of the concatenation operation $\circ$ (Bunge 1977a).

The above (linguistic and ontological) $models_2$ are merely factual interpretations of a ready made abstract theory. That is, each of them resulted from enriching an abstract theory A with a set I of factual interpretations of the basic (undefined) concepts of A. In obvious symbols, $T = Cn(A \cup I)$, where '$Cn\ X$' is to be read "the set of all logical consequences of X". We call the $models_2$ that result from assigning the basic concepts of an abstract theory a factual interpretation *bound models*.

However, this situation is rather exceptional in biology: in this field most models are not bound, i.e., they must be crafted from scratch. Moreover, their mathematical formalisms are built with components borrowed from several extant mathematical theories, such as high school algebra, Euclidean geometry, and the calculus, none of which is abstract. We call the $models_2$ that do not result from assigning the basic concepts of an abstract theory a factual interpretation *free models.*

3.5.3.2 Degrees of Generality

Theories come not only in different degrees of abstraction, but also in different degrees of generality. (The more abstract a theory, the more general it is, but not necessarily conversely.) For example, a theory of population growth is more general than a theory of the growth of a population of *Drosophila*, and a theory of protein synthesis is more general than a theory of myoglobin synthesis. The former are *general theories* whereas the latter are *special theories* or *theoretical models* ($model_3$). The relation between the two is this: a theoretical model follows from a general theory G when the latter is enriched with a set of subsidiary (or auxiliary) assumptions S —in our cases those that individuate the members of the population, or the protein respectively, in question. In obvious symbols, $M_i = Cn(G \cup S_i)$, read "M_i equals the set of logical consequences of the union of G and S_i".

The subsidiary assumptions S may be said to sketch the specific features of the referents of the theory in question. For example, to a first very crude approximation a cell may be sketched as a sphere; to a second approximation, as a sphere containing a smaller sphere—the nucleus—and so on. An alternative sketch S' will, when conjoined with G, produce a different $model_3$ M' of either the same things or of different things of the same general kind. Since the ingredient S of M is an idealization of the things of interest, M may be said to describe directly the sketched model object, and only mediately its real referent(s). Moreover, since any sketch S is only an idealized "picture" of the real referents, while M describes S accurately, it describes only its real referent(s) to some approximation. If the specific theory M turns out not to match satisfactorily its real referent(s), then G, S, or both must be repaired. If G has a good track record, S will be the suspect, and a different sketch S' referring to the same real referent(s), will be tried out. Otherwise, G as well as M may have to be changed. Thus, $models_3$ or idealizations of real referents, such as ideal gases, free electrons, and ideal populations, are not idle fantasies but conceptual sketches of real things, introduced with the sole aim of getting theorizing started. Complications are added only as needed, i.e., as the discrepancies between $model_3$ and reality become serious. As science advances, its models are supposed to represent their real referents with increasing accuracy.

Similarly as with $models_2$ of abstract theories, a special theory constructed by enriching a general theory will be called a *bound* $model_3$. If no general theory is available, a (mini)theory or $model_3$ M must be crafted from scratch. Such a theory will be called a *free* $model_3$. Since in the case of free models $G = \emptyset$, a free model consists only of special assumptions and their logical consequences, i.e., $M_i = Cn(S_i)$.

As regards (actual or intended) generality (or coverage, or range), a factual theory can fall into one of the following classes:

1. *Specific* (= theoretical model = $model_3$), such as the theory of the linear oscillator, and a model of the action of a given drug on a certain organ. Semantic properties: (a) all the concepts have a factual content; (b) the reference class is a rather narrow kind—typically, a species (in the ontological, not biotaxonomic, sense). Methodological properties: (a) conceptually testable (i.e., one may check in a fairly direct manner whether the model is consistent with the relevant parts of the bulk of antecedent knowledge); (b) empirically fully testable (i.e., both confirmable and disconfirmable provided it is enriched with some empirical data D. That is, what is subjected to tests is not the theory T itself but $T \cup D$.)

(Specific theories or $models_3$ should not be mistaken for subtheories. A theory T_2 is a *subtheory* of a theory T_1 if, and only if, T_1 entails T_2 or, equivalently, every formula of T_2 is included in T_1. In other words, a subtheory is part of another theory. A specific theory, by contrast, contains assumptions that do not occur in the general theory, so that it cannot be part of the latter. Rather, it is a specialization or application of a general theory.)

2. *General*, such as classical particle mechanics, population genetic theory, and the theory of organismal selection (see Sect. 9.2.2). Semantic properties: (a) all the concepts have a factual content, and (b) the reference class is a genus (in the logical sense), every member of which is representable by some theory of type (1). Methodological properties: (a) conceptually testable; (b) empirically testable provided it is enriched with subsidiary assumptions S, indicator hypotheses I, and data D. That is, what is subjected to tests is $T \cup S \cup I \cup D$.

3. *Hypergeneral fully interpreted*, such as continuum mechanics, the Turing-Rashevsky theory of morphogenesis, and general selection theory (see Sect. 9.2.2.2).

Semantic properties: (a) all of the concepts have a factual content; and (b) the reference class is a large family, every member of which is representable by a type (2) theory. Methodological properties: (a) conceptually testable; and (b) empirically untestable by itself but can become empirically testable if suitably enriched with subsidiary assumptions so as to become a type (2) theory.
4. *Hypergeneral semi-interpreted* (or *scaffolding* theory), such as information theory, general systems theory, game theory, automata theory, general control theory, and the ontological theory of selection (see Sect. 9.2.2.1). Semantic properties: (a) only some concepts are assigned factual interpretation; and (b) the reference class is a broad family, every member of which is representable by a type (3) theory. Methodological properties: (a) conceptually testable; and (b) empirically untestable by itself but may become vicariously testable upon being fully interpreted and enriched with subsidiary assumptions.

3.5.3.3 Degrees of Depth

Like hypotheses, theories can be more or less deep. For example, quantum electrodynamics, which construes light beams as streams of photons, is deeper than classical electrodynamics, and this, in turn, is deeper than wave optics, which is deeper than ray optics. Molecular genetics is deeper than classical genetics, for it explains genetic facts in molecular terms, whereas classical genetics treated genes as black boxes. And a biopsychological learning theory is deeper than a behaviorist learning theory, for it explains learning as the reinforcement of neural connections.

Since all that is said about the depth of hypotheses in Section 3.4.3 also holds for theories, we need not repeat it here. Suffice it to recall that a shallow theory treats its referents as black boxes with invisible innards. It is often called *phenomenological*, as believing that it represents only phenomena or appearances. Actually, this is a misnomer, because even "phenomenological" theories contain concepts denoting imperceptible properties, such as energy, entropy, and temperature in the case of thermodynamics, the paragon of "phenomenological" theories. For this reason, *black box* or *empty box theory* are better names. By contrast, a *translucent box theory* is one that accounts for the composition, structure, and dynamics of its referents. Finally, a *gray box theory* is one that represents the innards of its referents in a schematic way.

The deeper theory is one that postulates some mechanism at a lower level of organization: it is a *mechanismic multilevel* theory, in contrast to a phenomenological single-level theory. The mechanism need not be mechanical: it may be electromagnetic, chemical, biological, economic, political, or what have you. In most cases, the mechanism is imperceptible, so it must be conjectured before it can be found. It goes without saying that if the theory is scientific the mechanism must be experimentally accessible, however indirectly. A deep theory tells us not just (part of) what happens but also by virtue of what mechanism it happens. Hence, it has explanatory power. Moreover, it may prove to have practical interest, for if we know a thing works we may tamper with its mechanism to our convenience.

3.5.4 Formal and Factual Models and Theories

In the preceding sections we saw that theories may deal with mathematical or factual items. It is therefore useful to distinguish between formal and factual theories. More precisely, we say that a theory is *formal* iff it concerns exclusively conceptual objects, such as numbers or philosophical ideas. By contrast, if the domain or reference class of a theory contains factual items, such as molecules or organisms, it is said to be *factual*. Note that, whereas a formal theory contains no reference to factual items, a factual theory may contain not only factual statements but also statements about attributes of its own concepts. For example, a theory about the change of factual items will contain mathematical statements concerning the (formal) attributes of the functions representing the change of the things in question, such as continuity and monotonous increase.

Furthermore, we must distinguish (at least) two different acceptations of the word 'model': the mathematical—model$_1$—and the scientific—model$_2$ and model$_3$ (Bunge 1969):

1. *Model-theoretic* (or *logical*) *model*: a theory *M* about mathematical objects of a certain kind, which results from interpreting an abstract theory *A* in mathematical terms. One says that the propositions of *A* are true (or are satisfied) under the given interpretation. (For example, group theory is satisfied by the rotations on a plane.) This is the concept of mathematical truth, which has nothing to do with that of correspondence with reality.
2. *Factual (epistemological* or *theoretical) model*: a scientific, technological or humanistic theory *T* about concrete objects of some restricted kind. *T* may be construed as the result of interpreting a mathematical formalism *F* in factual terms. (To worsen the terminological model muddle, factual models couched in mathematical terms are often called *mathematical models*.) Unlike a model in the model-theoretic sense, a theoretical model refers to real (or putatively real) things, and it need not be true to fact: it may be only approximately true or even utterly false. In other words, unlike model-theoretic models, which are subject to a coherence theory of truth, factual models are subject to a correspondence theory of (partial or approximate) truth (see Sect. 3.8.)

Since all the referents of a formal theory are mathematical constructs, formal theories (or abstract models$_1$) are by definition unrelated to reality. (Recall the thing-construct distinction from Sects. 1.2 and 1.6.) Factual theories, by contrast, are doubly related to reality, namely semantically and methodologically. That is, they *refer* to (certifiedly or putatively) real things, and they are *tested* by confrontation with experience, which is part of reality. What has to be done in order to prepare a factual theory for testing will be explained in the following.

3.5.5 Theory Operationalization

Only low-level empirical generalizations, such as "All birds have feathers", can be directly confronted with empirical data. Any hypothesis containing concepts that fail to represent directly observable features of things must be joined with indicator assumptions in order to face reality. Such *indicator hypotheses* link unobservables to observables (see also Sect. 3.2.3). For example, the vital signs, such as heart beat and blood pressure, are health indicators whose values allow one to check hypotheses about internal (in particular pathological) processes in the human body. What holds for hypotheses holds, *a fortiori*, for hypothetico-deductive systems or theories. A few examples should help clarify these matters, as well as distinguish the methodological problems of operationalization and test from the semantic matter of interpretation or meaning. They should also lay to rest the neopositivist tenet that theories are operationalized by enriching them with "operational definitions". Indeed, to begin with, a definition is a purely conceptual operation, one that involves no empirical operations (see Sect. 3.5.7). Second, an indicator hypothesis is a conjecture that may fail empirical tests.

Consider the hypothesis that stress (or rather strain) gives headache or, more precisely, that stress is a (sufficient) cause of headache: $S \Rightarrow H$. Headache is directly felt, but stress is not. However, there are standard methods for measuring stress levels, for instance to measure corticosterone levels in blood. That is, one uses the indicator hypothesis that the higher the corticosterone level, the higher the stress level, or $C \Rightarrow S$. We have then a system of two hypotheses:

Substantive hypothesis	$S \Rightarrow H$
Indicator hypothesis	$C \Rightarrow S.$

The conjunction of the two entails $C \Rightarrow H.$

The intervening or unobservable variable S has disappeared from the final result, which contains only predicates (variables) denoting observable properties—the ideal positivist formula.

It must be stressed that not all indicator hypotheses are of equal worth. Only those that are justified theoretically and enjoy empirical support can be used with any confidence. The hypothesis that relates the clicking of a Geiger counter to radioactive disintegration is of the first kind: the theory of the instrument shows what the clicking mechanism is, and measurement with independent techniques show that the Geiger method is reliable. This is not the case of the classic Mendelian hypothesis that the genotype-phenotype relation is a one-to-one mapping, so that phenotypic features are reliable indicators of genotypes. In fact, pleiotropy, polygeny, position effects, and phenocopies render the phenotype an unreliable genotype indicator. Yet today, molecular biological techniques allow one to determine the genotype without recourse to (macro)phenotype. In fact, the genotype is now conceived of as a subset of the phenotype (see Sect. 8.2.3.3).

Let us now show in some detail how to operationalize a theory, i.e., how to prepare it for empirical tests. Call T the general theory to be tested, and S the set of subsidiary assumptions specifying the salient particular features of the referent—e.g., the composition, environment, and structure of the system in question. From T and S we build the (bound) model M of the referent, which is to be subjected to tests. We now introduce the set I of indicators crafted with the help of T

and scraps of antecedent knowledge in any of the relevant branches of knowledge (e.g., optics and chemistry). Enter also the set D of available empirical data concerning the referents of the theory and relevant to it. Such data may be somewhat remote from the theory: for example, they may consist of DNA sequences revealed by electrophoretograms, while the theory is about the inheritability of genes leading to certain diseases. This theory-experience gap is bridged by the indicator hypotheses I: they allow one to "read" such macroproperties in terms of observable facts. We may call this a *translation* of the available data into the language of the theory. These translated data D', which follow logically from D and I, are next fed into the model M to yield the translated model M'. Finally, this result is translated back into the language of experience by means of I. That is, M' is joined to I to entail M^*, the operationalization of the theory or, rather, of the (bound) model M. Hence, not T itself but some consequences of T together with the subsidiary assumptions S, the data D and the indicator hypotheses I, face whatever fresh empirical evidence may be relevant to T: see the following diagram.

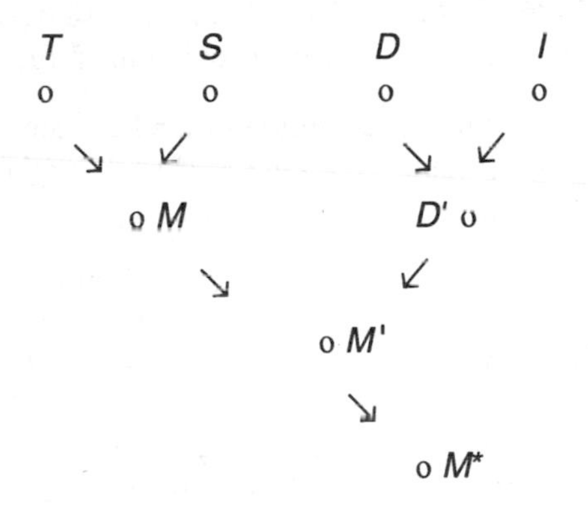

In sum, indicator hypotheses (also misnamed 'operational definitions' and 'correspondence rules') are indispensable to prepare hypotheses and theories for empirical tests. We emphasize, however, that they usually do not belong to theories, i.e., they are extratheoretical assumptions. Hence, whereas all scientific theories have a factual content (interpretation), most of them do not have empirical content. Only a psychological theory dealing with experience of some kind, such as a color perception theory, may have an empirical content.

The search for reliable indicators is a task both theoretical and empirical. It is a theoretical task, because only theories can tell us whether a given observable trait does, in fact, point to one particular unobservable rather than another, by disclosing the objective relation between the phenomenal and the transphenomenal. And only empirical checks can validate such a hypothesis. So much for theory operationalization. (See Bunge 1973b for applications to physics.) Our view of theory testing is at variance with the so-called received view, to which we turn now.

3.5.6 The Neopositivist or Received View of Scientific Theories

The so-called *received view* of scientific theories originated in the Vienna Circle (1926-1936), the cradle of neopositivism. This was a version of positivism updated with the help of mathematical logic: it has accordingly been called 'logical empiricism'. According to it, scientific theories (a) are hypothetico-deductive systems built out of data via induction, and (b) consist of mathematical formalisms interpreted in empirical terms, i.e., in terms of observations or experiments rather than facts independent of the knowing subject. (That is, the received view is empiricist and thus conflates interpretation with test.) Such interpretation is allegedly performed via *correspondence rules*, or pairings of concepts to observations. Bridgman (1927) called them 'operational definitions'.

According to this view, the concept-experience pairing would be direct in the case of observable properties, such as dissolving or breaking, and indirect in the case of unobservable ones, in particular dispositional properties, such as solubility and fragility. Carnap (1936-1937) and Hempel (1965, p. 188ff) analyzed the rules of the second kind as "bilateral reduction sentences" of the form

> For all x, if x is subjected to test condition of kind T, then x has property P if and only if x exhibits a response of kind E. (In self-explanatory symbols: $(\forall x)(Tx \Rightarrow [Px \Leftrightarrow Ex])$.)

Here P is supposed to designate a theoretical (or transempirical) predicate and E an empirical one. For example, a substance will be said to be soluble in water (P) only in the case that, when poured in a glass of water (T), it is observed to dissolve (E).

Unfortunately for positivism, this is not the way theoretical predicates are actually interpreted in science: the bilateral reduction formulas are just philosophical artifacts rooted in a crude empiricist epistemology. Consider, for example, the allometric equation that we met in Section 3.5.1, and which we now rewrite in the form

$$\mu = aM^b.$$

This hypothesis is (approximately) true to facts under several interpretations, among them the following:

Int (μ) = basal metabolic rate (minimal energy turnover) of mammals;
Int (M) = body mass.

These two variables are measured with instruments found in any well-equipped hospital. However, even though one can read their values in dials, the instruments containing these dials are designed with the help of certain indicator hypotheses, which are actually theoretical formulas. Some of these are "Mass = Weight/gravitational acceleration", "$\mu = c \cdot$ total heat production", "$\mu = d \cdot$ oxygen consumed", and "$\mu = e \cdot$ carbon dioxide production", where c, d, and e designate empirical constants. Every one of the last three formulas is an indicator hypothesis: it links the unobservable μ (basal metabolism) to variables that are measurable in fairly direct ways. Each of these indicator hypotheses is involved in the design and operation of a measuring instrument of a given type. If only because of this multiplicity, empirical tests cannot confer meanings. In sum, *interpretation precedes test*, not the other way round. The reason is obvious: before measuring something we must know what we intend to measure. (For criticisms of both operationalism and the received view see Bunge 1967a, b, 1974a; and for an application of the realist view to physical theories see Bunge 1967c.)

In sum, according to logical empiricism, a scientific theory is a hypothetico-deductive formalism plus a set of correspondence rules, some or all of which may be bilateral reduction sentences. By contrast, according to our version of scientific realism, a scientific theory is a hypothetico-deductive formalism plus a set of objective interpretation (or semantic) hypotheses. (These are hypotheses, not rules or definitions: indeed, they may have to be changed in the light of new evidence.) So, the neopositivists were right concerning the syntax (structure) of scientific theories, but wrong concerning their semantics (content or interpretation).

Some philosophers have criticized the positivist conception of scientific theories for either or both of the following reasons: because it construes them as hypothetico-deductive systems, or because it includes in them descriptions of test procedures. These critics have proposed correcting either or both of these alleged flaws by advancing a formalist view, which involves no construct-thing relations. Formalism comes in two versions: the structuralist and the "semantic" views. First applied to physical theories, both versions have now also entered the philosophy of biology under the common label 'semantic view of theories' (Beatty 1981; Lloyd 1988; Thompson 1989). As in the philosophy of biology the "semantic" view is usually discussed with regard to the structure of evolutionary theory, we shall examine it in more detail in Section 9.3.

3.5.7 Theories and Conventions

3.5.7.1 Definitions

Our simple theory of population growth (Sect. 3.5.2), exemplifying the syntax and semantics of a theory, contained only postulates (or axioms), from which we derived a theorem and a corollary. It contained no definition. Thus, all the concepts in our theory of population growth are undefined or *primitive*. As a matter of fact, the most important concepts in any context are the primitive ones, for they help define all the other concepts in the context. (However, the defining concepts in a given context may turn out to be defined in another.) For example, in number theory "one" is usually defined as the successor of zero. Here the primitive or defining concepts are, of course, those of zero and successor. Another example: the concept of life is usually undefined in evolutionary theory, but it is definable in ontology (see Chap. 4).

The problem with talking about definitions is that the word 'definition' has many popular connotations (see any textbook of logic, e.g., Copi 1968). To avoid ambiguity, we adopt a rather narrow construal of definitions as *identities* of the form "the defined object = the defining object", where the object in question is either a sign or a construct (Peano 1921). Such definitions are sometimes called *stipulative* or *nominal* definitions. Obviously, according to this view, concrete objects cannot be defined: they can only be described. Thus the expression 'real definition' is a misnomer for 'description', and ought to be avoided.

As far as logical form is concerned, there are two main kinds of (nominal) definition: explicit and implicit. An *explicit* definition is of the form "$A =_{df} B$". For example, "spermatozoon $=_{df}$ male gamete". The symbol '$=_{df}$' is read "identical by definition". The symbol '*df*' plays only a methodological role: it indicates that A is the defined concept or *definiendum*, and B the defining one or *definiens*. From a strictly logical point of view, then, explicit definitions are identities. So much so that, from this point of view, "$A = B$" is the same as "$B = A$". If A and B are propositions, a definition of the form "$A =_{df} B$" entails the equivalence "A if, and only if, B", but not conversely.

The definition of a class is of the same kind as the definition of any other concept: it is an identity. Although there are *epistemic* differences between, say, "$C =_{df} \{x \mid Px\}$" and "$1 =_{df}$ the successor of 0", there are no *logical* differences between them. In the former case we construct a new set, in the latter we identify a known concept with a combination of two old concepts ("0" and "successor").

In an *implicit* definition the defined concept (definiendum) cannot be expressed in terms of other concepts: it occurs in combination with other constructs. In other words, an implicit definition is an identity where the defined concept does not occur alone in the left-hand side. Example: "X is a bird $=_{df}$ X possesses feathers". Another example is provided by the concept "if..., then" of logical implication, symbolized $\Rightarrow$, which may be defined thus: "For any propositions p and q, $p \Rightarrow q =_{df} \neg p \vee q$".

A subset of the class of implicit definitions is constituted by the so-called *axiomatic* definitions, which occur in axiomatized theories. An axiomatic definition is of the form "A is [or is called] an $F =_{df} A$ satisfies the following axioms: ... ". Axiomatic definitions work well in mathematics, because they are ideal for characterizing mathematical objects, which are constructs. However, if used in the factual sciences, they can easily be misunderstood as "definitions" of concrete objects (witness the defenders of the so-called structuralist view of theories). Still, our theory of population growth, for instance, could be stated in the form of an axiomatic definition and would thus read: "Definition 1: A relational system (structure) $\mathcal{P} = \langle P, T, x_t, k \rangle$ *represents* a population of organisms, if and only if..."—and here our ten axioms would be listed. However, it would be mistaken to conclude that our theory would then *define* a population rather than *describe* it. In fact, every axiomatic definition can be broken down into one or more postulates and a definition proper. For instance, we could as well state our theory of population growth in the form: "Postulate 1: There are concrete systems of kind $\mathcal{P}$ *representable* by the mathematical structure $\langle P, T, x_t, k\rangle$ such that..."—and here the ten axioms would be listed, followed by "Definition 1: The systems of kind $\mathcal{P}$ satisfying Postulate 1 are called *organismal populations* ". (A further example is provided by Postulate 4.1 and Definition 4.1.) Thus, what is defined is the *concept* of population (growth), but not any concrete population.

Since definitions are identities, any proposition in a theory containing the symbol or construct in the definiendum must be exchangeable *salva veritate*, i.e., without any change in truth value, with a proposition containing the symbol or construct in the definiens. For example, exchanging the term 'spermatozoon' with the expression 'male gamete' in the proposition "The spermatozoa of nematodes have no flagella" leaves its truth value unaltered. Another requirement for a correct definition is that of noncreativity. That is, the introduction of a definition into a theory must not allow one to deduce any new theorems.

We can now do away with a popular mistake, namely the belief that definitions specify the meaning of words or concepts. Since definitions are identities, the definiens must be equated neither with the meaning nor the intension of the definiendum. In fact, the meaning of the definiens, if any, is prior to the meaning of the definiendum, which, by virtue of definition, acquires the meaning of the definiens. For example, "male gamete" is not the meaning of the word 'spermatozoon', but the term 'spermatozoon' is identified with the expression 'male gamete', and thus acquires the same meaning as the latter.

Since definitions are stipulations or conventions, not assumptions, they are neither true nor false, but only useful or useless, or practical or impractical. Thus, they are in need of neither proof nor empirical evidence or testing. To be sure, the definitions in factual science have real referents: they used to be called "real" definitions and thus smack of truths of fact. Example: 'Multicellular organism $=_{df}$ Biosystem composed of cells'. Still, even such definitions are conventions. What happens is that we may become so used to employing the defined concepts, that we find them "natural". In principle, nothing but practical convenience stands in the way of changing the conventional name for a thing, property, or process.

Definitions are conventions bearing only on concepts or their symbols, not on facts. We can do many things with the concrete things accessible to us, except define them. For example, a system of equations, such as in Newton's mechanics, does not define its referents, i.e., bodies and their behavior, but *describes* them. By the same token, if someone says that life is "defined" by metabolism, self-regulation, and self-reproduction, or what have you, it must be taken to mean that these properties are regarded as peculiar to, or characteristic of, living beings. Moreover, a conjunction of such properties may be used as a *criterion* or *test* for distinguishing living beings from nonliving ones, much as *aqua regia* was used as a test for gold.

If the formulation of criteria, tests, or indicators for unobservable facts is systematically conflated with definition, as is the case with operationalism, we arrive at what is called 'operational definitions'. The main problem with operational definitions is that *there just are no* operational definitions (Bunge 1967a, 1974a, b,

1983a). In fact, operational definitions are indicator *hypotheses*, not identities. For example, the conditional "*X* is an acid solution iff litmus paper turns red in *X*", tells us not what an acid *is* but how to *recognize* it. (The modern concept of acidity is condensed in the identity "Acid = proton donor". Neither of the two definiens concepts stands for a directly observable item.) Likewise, '*X* is homologous with *Y* iff *X* occupies the same position in system *A* as does *Y* in system *B*' does not tell us what a homology is but, at best, how to recognize it (see also Mahner 1994a.)

Another example of such a mistake is the popular definition of "natural selection" in terms of differential reproduction. Of course, differential reproduction may be the result of natural selection which allows us to recognize that selection may have occurred in the first place. The process of selection can consist only in a biosystem-environment interaction, that is, it can only be regarded as an ecological process (see, e.g., Bock and von Wahlert 1965; Brady 1979; Bradie and Gromko 1981; Damuth 1985; Brandon 1990). That this process may lead to or result in differential reproduction must be formulated separately as a criterion, not a definition. (Further examples in Mahner 1994a; for an earlier analysis of operationalism in biology see Hull 1968.)

Another common mistake is the belief that one has set up an operational definition just because the terms in the definiens are observational or nearly so, or because we can (directly or indirectly) observe their referents. For example, the identity "Fertilization $=_{df}$ Fusion of the nuclei of a male and a female gamete" is not an operational definition just because we can (sometimes) observe this fusion under the microscope. In sum, beware of operational definitions, for they are pseudodefinitions.

Finally, we must point out that whether a formula is a definition, i.e., a convention, or not depends on the context in which it occurs. For example, the formulas "Water is a substance composed of H_2O molecules", or "A gene is any segment, whether continuous or not, of a DNA molecule that occurs as a template in the synthesis of a polypeptide" are hypotheses about the identity of factual items. They are thus true or false and in need of tests. However, in a different context (or theory) they may occur as definitions. For example, we may wish to replace the expression 'substance composed of H_2O molecules' by the term 'water' in a certain theory. This is now a convention, not a hypothesis, because we could as well suggest the term 'aqua' (or 'eau', or whatever) rather than 'water' in the definition.

3.5.7.2 Notational Conventions, Units, and Simplifying Assumptions

Every theory couched in mathematical terms contains a set of (explicit or tacit) notational conventions, which assign nonlinguistic objects to signs. We shall distinguish two kinds of notational convention: designation rules and denotation rules. A *designation rule* assigns a construct to a symbol. Example: "$\mathbb{N}$ designates the set of natural numbers". A *denotation rule* assigns a factual item to a symbol. Example: "*w* denotes the fitness value of an organism with a certain genotype".

Theories may involve *units* such as those of length, mass, or time. Units are conventional and they are involved in the very characterization of magnitudes such as physical quantities. Although units are conventional, they are usually not wholly arbitrary. For example, the second or, more precisely, the mean solar second used to be defined as 1/86400 mean solar day, which, in turn, results from astronomical observations and calculations. (The latest definition of "second" involves concepts occurring in relativistic quantum theory.)

All except fundamental theories contain some *simplifying assumptions*. For example, in certain calculations or measurements we may pretend that the value of π is 3.14; that the system of interest is a black box (i.e., has no internal structure); that it is perfectly isolated or has only the properties we happen to have discovered so far; that certain discontinuous variables are continuous, or conversely, and so forth. All such simplifying assumptions are introduced as conventions to expedite modeling or inference, or even to make them possible.

In sum, all theories contain conventions and, paradoxically, factual theories contain the most. The epistemological moral is obvious: although factual theories may represent their referents in a fairly true manner, such representations are not pictures or copies. In fact, they are *symbolic constructions* bearing no resemblance to the objects they represent (see Postulate 3.5). In particular, theories are in no way isomorphic to their referents. (More on this in Sect. 9.3.2.)

3.5.8 Theories and Laws (Law Statements)

In Section 1.3.4 we noted that the word 'law' is highly ambiguous, because it designates four different concepts: objective pattern (or natural regularity), formula purporting to represent an objective pattern, law-based rule (or uniform procedure), and principle concerning any of the preceding. To avoid confusion, we had called these concepts *law*$_1$, *law*$_2$, *law*$_3$, and *law*$_4$—or *pattern*, *law statement*, *rule* (or *nomopragmatic statement*), and *metanomological statement*—respectively.

For example, the metabolic rate equation "$\mu = aM^b$" is a law$_2$. It represents, to a good approximation, the actual metabolism (law$_1$) of mammals in ordinary conditions. Alternative equations for the basal metabolic rate—involving age or other variables in addition to mass—are different laws$_2$ representing the same objective pattern or law$_1$ to better or worse approximations. One of the rules (laws$_3$) based on the above law statement prescribes the average caloric (joule) intake requirement of an adult mammal. Another such law-based rule allows one to prescribe the dosage of a drug required to produce a desired effect. An example of a law$_4$ or metanomological statement is "All biological laws$_2$ must be invariant with respect to changes in reference frames". Another is "Every biological process fits some law(s)".

A biological law$_1$, or objective pattern, is a constant relation among two or more (essential) properties of a biological entity. In principle, any such pattern can be conceptualized in different ways, i.e., as alternative laws$_2$. As a matter of fact, the history of any advanced science is to a large extent a sequence of laws$_2$. It is hoped that every one of these constitutes a more accurate representation of the corresponding objective pattern or law_1, which is assumed to be constant and, in particular, untouched by our efforts to grasp it. Likewise, the history of technology is to some extent a sequence of $laws_3$, or law-based rules of action. Note that, in principle, there are two $laws_3$ for every law_2: one for getting a thing done and another for avoiding a given effect. Thus, a law_2 of the form "If *A*, then *B*" is the basis for two law-based rules: "In order to obtain *B*, do *A*", and "In order to avoid *B*, refrain from doing *A*". Note further that, since rules are precepts for doing something, they—unlike law statements—do not describe, explain, or forecast: they prescribe. Therefore, statistical law statements, such as in the schema "70% of all *x* possess property *y*", should not be called 'rules'.

Since law statements are sometimes construed as prohibitions (e.g., by Popper 1959), we emphasize that law statements are supposed to represent objective patterns of being and becoming, so that they first of all tell us what is and what is possible. If we know what is possible, we can of course *infer* what is impossible. In other words, by delimiting the set of lawfully possible states and events of things we *eo ipso* characterize the set of conceivable but factually impossible states and events. Still, we insist that, since there are no negative facts (Sect. 1.8.1), scientific statements (including $laws_2$) are supposed to refer to facts, not to nonexistents. For this reason, $laws_2$ should be formulated as assertions, not prohibitions or counterfactuals. For example, the competitive exclusion principle of ecology "No two species can have exactly the same ecological niche" should be stated as "Every species has its own peculiar niche". (This formulation is still in need of improvement: see Sect. 5.4.)

As for the $laws_4$, or laws of laws, they are of two kinds: scientific and philosophical. The first example offered above is of the first kind: it states that the biological $laws_2$ should be stated in such a way that they do not depend on the reference frame (in particular the observer) or on the type of coordinate chosen. It is an objectivity requirement. The second example is the principle of lawfulness (or no-miracle) restricted to biology: it is a philosophical thesis (see Postulate 1.4). Unlike the former, whose truth can be checked with pencil and paper, the principle of lawfulness is irrefutable, though extremely fertile, for it encourages the search for pattern. By the same token, it succors the biologist who despairs of ever "making sense" of his figures, i.e., of finding the patterns they fit.

Not all formulas deserve being called biological $laws_2$. For example, the regularities found by curve fitting are called *empirical formulas*. They may also be called *quasilaws*, for they are candidates for promotion to the rank of full law_2. In advanced theoretical science a formula is called a *law statement* iff it is general, systemic, and well confirmed. More precisely, we adopt:

DEFINITION 3.9. A factual statement is a *law statement* if, and only if,
(i) it is general in some respect (e.g., it holds for a certain taxon);
(ii) it is part of a (factual) theory (hypothetico-deductive system); and
(iii) it has been satisfactorily confirmed (for the time being).

For example, the equations of motion of physics, and the rate equations of chemical kinetics, are $laws_2$, and so are their logical consequences. If a law_2 occurs among the postulates of a theory it is called a *basic* law_2 (sometimes also called a *principle*). On the other hand, any logical consequence of basic law statements and definitions are called *derived* $laws_2$. Note that the basic-derived distinction is contextual: what is a principle in one theory may be a theorem in another. For example, Newton's second law of motion is a theorem in analytical dynamics, and the second principle of thermodynamics is a theorem of statistical mechanics. The former is the paragon of a dynamical law, and the latter that of a statistical law, i.e., one referring to a large aggregate of entities, every one of which behaves roughly independently of the others. Most physiological laws are of the first kind, and most of the laws of population genetics are of the second type.

3.6 Understanding

3.6.1 Explanation

Like every other scientific discipline, biology aims at *understanding* its subject matter. Now, understanding is not an all-or-none operation: it comes in several kinds and degrees. For example, we may understand a fact or construct with the help of empathy, metaphors, analogies, or hypotheses. In any of its modes understanding involves systematizing, that is, we either fit the given item into our preexisting cognitive or epistemic framework, or we transform (e.g., expand) the latter to accommodate the new item. In doing so, our epistemic framework need be neither factual nor scientific. We can, for example, understand the adventures of fictional entities such as Don Quixote or Superman. After all, understanding is a *psychological* category, not a methodological one.

Obviously, here we are only interested in a special and historically rather recent mode of understanding, namely understanding by scientific means. Since understanding comes in degrees, so does scientific understanding. We thus can analyze three epistemic operations: description, subsumption, and explanation proper (Bunge 1967b, 1983b).

3.6.1.1 Description

From a logical viewpoint a *description* or *narrative* is an ordered set of factual statements. For example, "This petri-dish initially containing one bacterium of species *A* contains 2 bacteria after 20 minutes, 4 after 40 minutes, and 8 after 1 hour". Given this description alone, i.e., without using further tacit knowledge, we may understand that some cell multiplication seems to take place in the petri-dish but we do not know why there are just 8 bacteria after 1 hour. Only further data and hypotheses can answer the why-question.

3.6.1.2 Subsumption

Investigating the matter further, we count and sample the population every so often and come up with a table or a graph population vs. time. Suppose that we find that the growth pattern is 1, 2, 4, 8, 16, 32, 64, etc., at 20-minute intervals. This, now, accounts for the fact to be understood, i.e., the eightfold increase of population size over 1 hour, because we recognize it as the fourth member of the sequence. We can say that the fact has been subsumed under a general pattern, namely: "The given population of bacteria of species *A* doubles every 20 minutes". In other words, we have turned the fact to be explained into a particular case of a law statement.

The inference has been this:

Premise 1. For every x and every t, if x denotes a population of bacteria of species *A* in condition *B*, then the population of x at $(t + 20)$ min is twice the population of x at t. (In general, $N_t = N_0 2^t$, where N_0 is the initial population and t is the number of 20 min periods.)
Premise 2. b at $t = 0$ is a population of N_0 bacteria of species *A* in condition *B*.
Conclusion. b at $t = 60$ min is a population of $N_3 = 8N_0$ bacteria of species *A* in condition *B*.

A *subsumption* then is also an ordered set of statements, but one such that the last statement follows from the preceding ones. Generalizing, the basic subsumption schema is this:

Pattern	For all x, if Px then Qx [i.e., $\forall x\ (Px \Rightarrow Qx)$]
Circumstance	Pb
Given fact	Qb

The factual statement that refers to the given fact that is to be explained is usually called the *explanandum*, and the explaining premises the *explanans*.

Sometimes, especially in biology, the pattern occurring in a subsumption is merely a systematic or classificatory statement. For example, the fact that organism b possesses a certain feature can be "explained" in terms of its belonging to a certain systematic taxon. For example, the deductive inference may be:

Premise 1. All mammals possess a squamoso-dentary joint.
Premise 2. b is a mammal.
Conclusion. b possesses a squamoso-dentary joint.

Although this is a logically valid inference, it has no explanatory power, since the feature "possessing a squamoso-dentary joint" has been used to define "mammal" in the first place. More on this in Sect. 7.2.2.5.

Subsumption is sometimes regarded as a "top-down" approach (Kitcher 1989b; Salmon 1989). We believe this to be a misnomer because subsumption does not involve several levels of organization. In a subsumption, we do not "explain" some aspect of a higher-level entity by reference to some lower-level entities, i.e., parts of the higher-level system. In other words, subsumption is a same-level account. All that is involved is the demonstration that some fact is a special case of a general pattern. We see no reason to call a deduction a 'top-down approach', just as we do not regard an induction as a "bottom-up approach". (For genuine upward and downward explanations see Sect. 3.6.1.4.)

3.6.1.3 Explanation Proper

Subsuming allows us to understand a given fact as a particular case of a general pattern. However, we still do not understand the pattern itself, and therefore the fact to be accounted for remains half understood. We are told *how* things are, not *why* they should be the way they are. Returning to our microbiological example, we should recall that, from a logical point of view, bacteria might grow in many different ways or in none. However, cytology and microbiology reveal that bacterial populations happen to grow by division. In our example, every bacterium of species A divides at the end of 20 minutes; in turn, every daughter bacterium divides after 20 minutes, so that the original bacterium has been replaced by four, and so on. This, then, is the growth mechanism in the case of populations of bacteria: cellular division. (To be sure, it is a neither a causal nor a mechanical mechanism.) We have now attained an *explanation* proper, or *mechanismic* account.

The logical reconstruction of the explanation process is this. From the observation and timing of cell division we hypothesize (and check) the law of growth, i.e., $N_t = N_0 2^t$. The rest follows as in the case of subsumption.

The difference between subsumption and explanation is not logical, because both are deductions from statements referring to regularities and circumstances, in particular law statements and data. The difference is another. Subsumption only answers how-questions, explanation how- or why-questions. Both answers are given in terms of patterns, such as trends, empirical generalizations, or law statements: the given fact to be accounted for is shown to be a particular case of such pattern. But whereas in the case of subsumption the pattern itself remains unaccounted for, in the case of explanation it is a mechanismic hypothesis or theory.

The basic pattern of subsumption was:

(i) For all x: if Px then Qx; Pb $\therefore$ Qb.

The corresponding explanation pattern is:

(ii) For all x: if Px then Mx; For all x : if Mx, then Qx, Pb $\therefore$ Qb,

where 'M' symbolizes some mechanism such as cell division. (For example, the first premise could be "For every x, if x is a bacterium, then x reproduces by cell division", and the second "If x reproduces by cell division, then the offspring of x grows geometrically".) Now, the two general premises of (ii) jointly entail the general premise of (i).

Phenomenalists and conventionalists might say that this shows that M is dispensable. A realist, however, concludes instead that *explanation subsumes subsumption*, logically, epistemologically, and ontologically: logically, because every explanation is a subsumption, but not conversely; epistemologically, because explanation presupposes more knowledge than subsumption; and ontologically, because explanation goes deeper into the matter than subsumption, in pointing to some (conjectured or established, perceptible or imperceptible) mechanism. In other words, the quest for explanation is a quest for knowing deeper and deeper levels of reality. Consequently, as with hypotheses and theories, explanations come in various *depths*: an explanation E_1 is *deeper* than an explanation E_2 iff E_1 involves more system levels than E_2. (For the concept of level see Sect. 5.3.)

As *mechanismic* is a neologism, some remarks may be in order. Descartes and his followers required that all mechanisms be strictly mechanical. Field physics, evolutionary biology and other scientific developments have relaxed this condition. We now understand that mechanisms—processes in things—need not be mechanical or mechanistic: they may be physical, chemical, biological, psychological, social, or mixed. They may be natural or artificial, causal, or stochastic, or a combination of the two. The only condition for a mechanism to be taken seriously in modern science is that it be material, lawful, and scrutable (rather than immaterial, miraculous, and occult). In short, we ought not to confuse 'mechanismic' with 'mechanistic'.

We have called 'subsumption' what most philosophers call 'explanation'. The classical account of that operation is that of Hempel and Oppenheim (1948), usually called the *deductive-nomological model of explanation*, or *D-N-model* for short, which has been criticized in various ways since its inception. (See also Hempel 1965; for a [somewhat biased] history of the notion of scientific explanation see Salmon 1989.) Let us note only two criticisms. The first is that the D-N--model is concerned only with the logical aspects of explanation. However, as we stated previously, explanation involves extralogical aspects as well, in particular epistemological and ontological ones. Hence, we should avoid explanations in open contexts; that is, the explanatory premises must not be stray conjectures but should belong to theories (hypothetico-deductive systems). The second shortcoming is that it is concerned with theoretical subsumption, not genuine explanation. We submit that a genuine explanation is one that invokes some mechanism or other, whether causal or stochastic (probabilistic).

To qualify as scientific, an explanation must satisfy three conditions: (a) *logical:* it must be a formally valid (nonfallacious) argument; (b) *semantical*: at least one of its premises must refer to some mechanism or other; (c) *methodological*: its premises and conclusion(s) must be testable and preferably reasonably true. A further desideratum is that the explanans generalization(s) should not be omniexplanatory, i.e., they should not purport to explain just about any fact. An example of an omniexplanatory hypothesis is one that resorts to God's will, as is the case with creationist "explanations" (see, e.g., Mahner 1989; Sober 1993). *Dictum de omni, dictum de nullo.*

To conclude, we can say that explanation is an epistemic process involving three components: (a) an explainer or animal doing the explaining; (b) the object(s) of the explanation, i.e., that which is being explained; and (c) the explanatory premises or explanans. All the explainers we know are human persons. If it is said that a hypothesis or a theory explains such and such facts, this is an ellipsis. Only a knowing subject is able to explain facts with the help of a theory and data. The objects of explanation are facts; and the explanatory premises are the hypotheses and data involved in the explanation.

From a historical point of view, it might be interesting to note that, although the preceding distinction between subsumption and mechanismic explanation was introduced by the senior author in 1967 (Bunge 1967b), it has resurfaced under the name 'ontic conception of explanation' (see, e.g., Salmon 1989, p. 182ff).

3.6.1.4 Kinds of Explanation

Several kinds of explanation have been, rightly or wrongly, distinguished in the scientifico-philosophical literature. Let us examine the most common ones.

Statistical Explanation. Occasionally (e.g., Hempel 1965), the following inference patterns are believed to constitute statistical explanations:

Almost every *A* is a *B*.	*f* % of *A*'s are *B*'s.
<u>*c* is an *A*.</u>	<u>*c* is an *A*.</u>
c is most probably a *B*.	The probability that *c* is a *B* is *f*.

None of these is a statistical or probabilistic explanation. The reason is that no probability statement can be deduced from an empirical generalization and a singular statement, because the concept of probability does not occur in the premises. Thus, both arguments are invalid. At best, this inference is an inductive statistical syllogism according to the rule: try to conjecture a probability (an individual property) from an observed frequency (a collective property). Another misuse of the concept of probability occurs when it is claimed that the conclusion in such an argument would only follow from the premises with a certain probability. This only makes sense under a subjectivist interpretation of probability. The plausibility of an inductive argument has nothing to do with probability (recall Sect. 1.10.2.3).

A genuine statistical or probabilistic explanation can only be attained if one of the premises is a statistical or probabilistic law statement. For example, the fact that (roughly) half of all newborn babies are males, or females respectively, may be (partially) explained with the help of the probabilistic law statement: "The probabilities of the formation of an XX-zygote and an XY-zygote are roughly equal, about 0.5." The corresponding statistical law statement could be: "About half of the zygotes in a large population are of the XX-type and the other half of the XY-type." (For the sake of simplicity we only refer to genetic sex and disregard complications such as XO- or XYY-types, or developmental events that may result in a shifted sex ratio, or in XX-males, or XY-females, or what have you.)

These and other premises entail the conclusion, i.e., the statement about the sex ratio of babies, deductively. That is, the explanandum statement follows with certainty and not with a certain degree or probability. Neither does it make sense to say that if *c* is a child the probability its being a girl would be 0.5. For a child is either a boy or a girl—hermaphrodites aside. Randomness, hence the legitimate use of the probability concept, was (presumably) involved when X- and Y-sperms were on their way to the egg. After they united to form a (viable) zygote, no randomness, hence probability, is involved any longer that could have any bearing on the genetic sex of the child. (The phenotypic sex or gender is, of course, another matter.) The use, or rather misuse, of probability talk in cases like this refers to our expectation to meet a boy or a girl when, for instance, waiting in front of the delivery room. Again, this presupposes a subjectivist, hence illegitimate, probability interpretation (recall Sect. 1.10.4).

Upward and Downward Explanation (Subsumption). In order to elucidate the terms 'upward' (or 'bottom-up') and 'downward' (or 'top-down'), we need to recall the notions of microfact and macrofact from Section 1.8: For any system *x*, a *macrofact* is a fact occurring in *x* as a whole, whereas a *microfact* is a fact occurring in, or to, some or all of the parts of *x* at a given level.

We can now say that a *top-down* (or *microreductive*) *explanation* of a macrofact is the deduction of the proposition(s) describing the latter from propositions describing (micro)facts in components of the system in which the macrofact occurs. By contrast, a *bottom-up* (or *macroreductive*) *explanation* of a microfact is the deduction of the proposition(s) describing the latter from propositions describing macrofacts occurring in the system as a whole. As biology studies both microfacts and macrofacts, we need both microexplanation and macroexplanation, depending on whether we want to explain macrofacts or microfacts.

However, both top-down (or microreductive) explanations and bottom-up (or macroreductive) explanations are incomplete. The reason is that in both cases we start by assuming that we have to do with a *system* and its parts. The only difference lies in the *problem* in question: in the case of microexplanation the problem is to explain the whole by its parts, while in the case of macroexplanation the problem is to explain the parts by reference to the whole. A few examples will help bring this point home.

Let us start with some examples of alleged triumphs of microreduction. Although ferromagnetism is said to be explained in terms of the alignment of the atomic spins and their associated magnetic moments, one starts by assuming that one is dealing with a macro-object such as a piece of steel. Although genetics explains heredity in terms of DNA molecules, the latter are assumed to be cell components and moreover components of one of the regulatory subsystems of the cell. Although physiological psychology explains learning in terms of the reinforcement of interneuronal connections, it starts by considering a large system of neurons. Consequently, none of these examples is a case of pure microreduction.

The case of macroreduction or bottom-up explanation is analogous. For example, the behavior of a molecule in a liquid depends on whether or not it is on the surface. The function of a gene depends on its position within the chromosome. The roles of social animals within their group depend on their place (e.g., their rank) in it. None of these is a case of pure macroreduction.

In sum, since both macroreduction and microreduction are necessary, but neither is completely satisfactory, we must try and combine them. From a systemic viewpoint, a satisfactory explanation of any fact involves two or more levels: at least that of the system or whole and that of the components or parts. And, in addition to the components of the system, it takes the environment and the structure of the system into account. In other words, it analyzes any system in terms of a CES model, as introduced in Section 1.7.2. We call an explanation that involves all this a *systemic explanation*. It combines and subsumes both bottom-up and top-down explanation. In top-down explanation the explanandum concerns a system, and the explanans premises refer to the composition and structure of the system. In bottom-up explanation the explanandum refers to system components while the explanans premises refer to the structure or the environment of the system.

Narrative Explanation. Since evolutionary biology deals (partly) with historically unique events, it has been claimed that evolutionary explanations do not fit the D--N model of scientific explanation (subsumption). Rather, explanations in evolutionary biology would be historical narratives, i.e., so-called narrative explanations (Goudge 1961; Mayr 1982). After initial criticism (Ruse 1971), this view has flared up again in the wake of the "historical entity" metaphysics in the philosophy of biology (Hull 1989).

Insofar as a so-called narrative explanation only *chronicles* historical events and processes, it is a description, not an explanation proper. Indeed, "historical narratives are viewed as descriptions of historical entities as they persist through time" (Hull 1989, p. 181). As descriptions occur in all scientific disciplines, the existence of such historical descriptions is not a case for a methodological difference between biological and nonbiological disciplines. They would only be such if accompanied by the claim that no subsumptions and explanations whatsoever do occur in evolutionary biology; but this thesis is clearly wrong.

Viewed at a closer range, many narrative explanations are not purely descriptive, for they either tacitly imply or explicitly invoke laws, causes, and mechanisms, even though they are not stated in the form of proper deductive-nomological arguments. (See also Ruse 1973; M.B. Williams 1986.) For example, narrative explanations often conjecture some adaptive scenario. The use of the notion of adaptation, however, implies the application of the theory of natural selection, which, in turn, refers to a (general) mechanism of evolution. Moreover, reference to mechanisms and causes, in our strict as well as in the broad sense, presupposes the existence of laws, although they may not be explicitly referred to in the narrative. (See also M.B. Williams 1986). For example, a narrative explanation of—pardon the metaphysically ill-formed expression—the vestigiality of the eyes of some cave-

-dwelling species may involve the law statement "Organs that are no longer used will be reduced", which holds under the condition that the organ in question is neither genetically nor developmentally correlated to another feature with a positive selective value. The degree of reduction is then either an indicator of the evolutionary time elapsed or of the degree of correlation to other, still adaptive features.

In sum, whereas some historical narratives may be nothing but descriptions, others, or at least certain parts of them, should be statable as subsumptions and explanations proper. Therefore, the occurrence of historical narratives does not render biology methodologically unique. (See also Ruse 1971, 1973; Hull 1974; Rosenberg 1985; van der Steen and Kamminga 1991.)

3.6.2 Prediction

Predictions are inferences from the known to the unknown. For example, what can we conclude from the knowledge that (a) strychnine in dose d is lethal, and (b) individual b took a dose d of strychnine? Of course, we can conclude that b will die. Since this argument can be spelled out in the form "For all x: If Px then Qx, $Pb \therefore Qb$", it has the same logical structure as an explanation of the subsumption kind (see Sect. 3.6.1.2). Indeed, there is no logical difference between a scientific prediction and a subsumption: both are deductions from generalizations (particularly law statements) and singular statements (data or "circumstances"). By analogy to the terms 'explanans' and 'explanandum', we shall call the premises in a predictive argument the *projectans*, and the conclusion, i.e., the forecast, the *projectandum*.

The differences between a prediction and a subsumptive explanation are not logical but extralogical. First, in a subsumption we start with the explanandum and search for the explanans. In a prediction, on the other hand, the projectans is available whereas the projectandum is sought for. Second, instead of referring to a circumstance or a condition, the singular premise in a prediction may refer to an indicator (or symptom) only. Think of the famous barometer example: a drop in the barometer reading allows us to predict a weather change without its being a condition, let alone a cause, of it. Moreover, unlike explanations proper, forecasts need not invoke any mechanism. For example, the regularity occurring among (the referents of) the premises of a scientific forecast can be of the black-box kind (such as an input-output relation as in the strychnine example), or even a trend or a statistical correlation. Third, the projectandum is always a singular statement, never a generalization or a law statement.

Being singular statements, projectanda are special cases of descriptions. They describe, first of all, *unknown* facts. Whether these facts are past, present, or future, or whether they are actual or just possible, is irrelevant from a methodological point of view (though, of course, not from a pragmatic point of view.) The importance of predictions for the methodologist consists in their providing the ultimate check of factual hypotheses and theories. The terms 'prediction' and 'forecast' therefore are misleading because they etymologically suggest a reference to future events. However, a projectandum may refer to past events, in which case we speak of a *retrodiction* (or *postdiction*) or *hindcast*. (The term 'prediction' in the broad sense subsumes both prediction proper and retrodiction.)

Scientific predictions may be either empirical or theoretical. *Empirical* (e.g., statistical) forecasts are based on unexplained correlations. For example, if we have found that *X* correlates highly with *Y*, then every time we find *X* we may expect to find *Y* as well—provided the correlation is not spurious. *Theoretical* predictions, on the other hand, involve the knowledge of laws. That is, the general premise in the projectans is a law statement, not an empirical generalization. From a methodological point of view, theoretical forecast is in principle superior to empirical prediction because $laws_2$ belong to theories and are deeper than correlations. Practically, however, a theoretical prediction may score lower than an empirical forecast if it relies on incomplete or inaccurate data.

Until recently, this was the case with weather forecasting: empirical generalizations ("rules of thumb") performed better than theoretical models. As the latter became more sophisticated, and satellites provided more accurate data, scientific weather forecasting became more precise. Even so, global and long-range weather forecasting is far more precise than local and short-term prediction. This may be so because local atmospheric turbulence is "chaotic", and, as a consequence, even a slight perturbation may result in a large unpredictable change. On the other hand, if a process is stochastic and we know its probabilistic law, we should be able to assign a definite probability to each possible future. (In addition, we should be able to predict averages and variances.) By contrast, if the process is chaotic in the sense of nonlinear dynamics, a branching out may occur but, since the underlying laws are not stochastic, there is no way of assigning a probability to every branch. (We should emphasize, by the way, that predictability is an epistemological category, so that one must not define a "chaotic" system, i.e., an ontological category, in terms of unpredictability. See Sect. 5.5.3.)

What holds for prediction proper need not hold for retrodiction. Indeed, in the case of an irreversible process approaching a state of equilibrium, one and the same final state may be reached from different initial states. Hence, in this case, retrodiction will be impossible while prediction will be possible, because all or nearly all trajectories will converge to a single state.

To conclude, while we can uphold the optimistic thesis that all facts can be explained, if not now, then later, we cannot sustain the faith in the predictability of everything. Thus, though prediction is *a* mark of science, it is neither its only peculiarity nor a characteristic of every bit of it.

3.6.3 Unification

Scientific understanding is provided not only by description, subsumption, explanation, and prediction, but also by unification. For example, since the genetic material was identified with DNA molecules, genetic variation has been explained as change in the composition and structure of such molecules. That is, part of genetics has been reduced to biochemistry (molecular biology). Yet reduction is not the only key to understanding: sometimes integration is. For example, biotic evolution is not understandable at the molecular level alone, but calls for a merger of genetics (both molecular and populational), morphology, systematics, paleontology, biogeography, ecology, and developmental biology, i.e., actually almost all the branches of biology. In sum, unification is brought about either by reduction or by integration (Bunge 1983a, 1991a, b; Bechtel ed. 1986).

3.6.3.1 Reduction

Reduction is a kind of analysis (i.e., an epistemic operation) bearing on concepts, propositions, explanations, or theories, or on their referents. The reduced object is conjectured or proved to depend on some other, logically or ontologically prior to the former. If *A* and *B* are both either constructs or concrete entities, to reduce *A* to *B* is to identify *A* with *B*, or to include *A* in *B*, or to assert that every *A* is either an aggregate, a combination, or an average of *B*'s, or else a manifestation or an image of *B*. It is to assert that, although *A* and *B* may appear to be very different from one another, they are actually the same, or that *A* is a species of the genus *B*, or that every *A* results somehow from *B*'s—or, put more vaguely, that *A* "boils down" to *B*, or that "in the last analysis" all *A*'s are *B*'s.

The following are instances of reduction, whether justified or not, in the field of basic science. The heavenly bodies are ordinary bodies satisfying the laws of mechanics; heat is random molecular motion; chemical reactions are inelastic scatterings of atoms or molecules; life processes are complex combinations of chemical processes; humans are animals; mental processes are brain processes; and social facts result from individual actions—or conversely.

Let us begin with *concept* reduction. To reduce a concept *A* to a concept *B* is to define *A* in terms of *B*, where *B* refers to a thing, property, or process on either the same or on a lower (or higher) level than that of the referent(s) of *A*. Such a definition may be called a ***reductive definition***. For example, a reductive definition reduces the concept of heat (in thermodynamics) to the concept of random atomic or molecular motion (in statistical mechanics). In other words, a reductive definition identifies concepts that had been treated separately before. This is, incidentally, the reason that reductive definitions are usually called 'bridge formulas' or 'bridge principles' in the philosophical literature. They are also called 'bridge hypotheses' (Nagel 1961), presumably because they are often originally proposed as hypotheses.

We may distinguish three kinds of reductive definition of a concept: (a) *same level*, or $L_n \rightarrow L_n$; (b) *top-down*, or $L_n \rightarrow L_{n-1}$, or *microreductive*, and (c) *bottom-up*, or $L_n \rightarrow L_{n+1}$, or *macroreductive*. Example of the first ($L_n \rightarrow L_n$): "Light $=_{df}$ Electromagnetic radiation". Examples of the second ($L_n \rightarrow L_{n-1}$): "Heat $=_{df}$ Random atomic or molecular motion", "Gene mutation $=_{df}$ Change in the structure of a DNA molecule". (This type of reduction has actually enabled geneticists to distinguish gene mutations of several kinds, such as substitution mutations, frameshift mutations, etc.) Example of the third ($L_n \rightarrow L_{n+1}$): "Alpha animal $=_{df}$ The animal occupying the top rank in a group of conspecifics".

The reduction of a *proposition* results from replacing at least one of the predicates occurring in it with the definiens of a reductive definition. For example, the proposition "Organisms inherit genetic material" is reducible to the proposition "Organisms inherit DNA molecules" by virtue of the reductive definition "Genetic material $=_{df}$ Aggregate of DNA molecules". The given proposition is said to be reduced. Whereas truth values are preserved under such transformations, meanings are not, because they are contextual.

The analysis of *theory* reduction is somewhat more complex. To begin with, we distinguish reduction from restriction. We thus propose:

> DEFINITION 3.10. Let T_1 designate a theory with reference class $\mathcal{R}$. Then a theory T_2 is the *restriction* of T_1 to $\mathcal{S}$, where $\mathcal{S} \subset \mathcal{R}$, if, and only if, T_2 is a subtheory of T_1 and refers only to the members of $\mathcal{S}$.

For example, particle mechanics is a subtheory of classical mechanics (Bunge 1967c; for the concept of subtheory recall Sect. 3.5.3.2).

Unlike theory restriction, theory reduction involves reductive definitions acting as bridges between the new (or reducing) and the old (or reduced) theory. For example, ray optics is reducible to wave optics by way of the reductive definition "Light ray $=_{df}$ Normal to light wave front". We shall say that this is a case of *strong* reduction. Yet reduction may also involve additional assumptions not contained in the reducing theory. For example, the kinetic theory of gases is reducible to particle mechanics by enriching it, in addition to the reductive definitions of the concepts of pressure and temperature, with the subsidiary hypothesis of molecular chaos (or random initial distributions of positions and velocities). We shall call this an instance of *weak* or *partial* reduction. More precisely, we formulate:

> DEFINITION 3.11. Let T_1 and T_2 designate two theories with partially overlapping reference classes, D a set of reductive definitions, and A a set of subsidiary hypotheses not contained in either T_1 or T_2. Then we say that
> (i) T_2 is *fully* (or *strongly*) *reducible* to $T_1 =_{df} T_2$ follows logically from the union of T_1 and D;
> (ii) T_2 is *partially* (or *weakly*) *reducible* to $T_1 =_{df} T_2$ follows logically from the union of T_1, D and A.

(For the classic treatment of reduction see Nagel 1961. For later studies see, e.g., Bunge 1977b, 1983b, 1991a; Causey 1977.)

In the philosophy of biology it is still an issue whether biology is reducible to chemistry or even physics, or whether it is a science of its own (e.g., Ayala 1968; Simon 1971; Mayr 1982; Rosenberg 1985, 1994). For example, the discovery that the genetic material consists of DNA molecules is sometimes regarded as proof that genetics has been reduced to chemistry (Schaffner 1969). However, chemistry accounts only for DNA chemistry: it tells us nothing at all about the biological functions and roles of DNA, e.g., its role in morphogenesis. That is, the functions and roles of DNA in a living cell cannot be accounted for by chemistry, because the concept of a living cell is alien to chemistry. Similar things could be said for other cases. (See also Beckner 1959.) Thus, when Schaffner (l.c., p.346) admits that, in order to account for living beings, one needs not only chemistry but also to regard the organisms' structure (organization) and environment, then no full reduction of biology to chemistry has been achieved. In sum, biology is at best weakly reducible to chemistry, which in turn is only weakly reducible to physics (Bunge 1982, 1985a). Moreover, not even theories within physics are reducible to one "basic" theory, such as quantum theory (Bunge 1973b, 1983b, 1991a). Even the quantum theory contains some classical concepts (e.g., those of mass and time), as well as hypotheses about macrophysical boundaries, so that it does not effect a complete microreduction. If this holds for electrons, it should hold, *a fortiori*, for organisms. (For more on reduction in biology see Kitcher 1984b; Rosenberg 1985, 1994; Hoyningen-Huene and Wuketits eds. 1989; Gasper 1992.) Likewise, psychology and social science are only weakly (partially) reducible to the corresponding lower-level disciplines.

3.6.3.2 Reductionism

Whereas reduction is an epistemic operation, *reductionism* or, rather, *microreductionism* is a research strategy, namely the adoption of the methodological principle according to which (micro)reduction is in all cases necessary and sufficient to account for wholes and their properties. The ontological partners of microreductionism are physicalism and atomism (or individualism). According to physicalism, things differ only in their complexity, so that wholes can be understood entirely in terms of their parts. Therefore, all the sciences are thought to be reducible to physics, and such reduction to physics would result in the unity of science—one of the illusory programs of logical positivism. (See also Causey 1977; Rosenberg 1994.) As we shall submit in a moment, the unity of science can be only achieved by a combination of moderate reduction with integration.

The dual of microreductionism is macroreductionism, which is often called 'antireductionism'. (The ontological counterpart of macroreductionism is holism.) The microreductionist thesis is that we know a thing if we find out what it is "made" of, while the macroreductionist thesis is that we know it if we figure out its place in "the scheme of things" (i.e., the larger system). Yet to explain how systems of any kind work we need to combine *microreduction* with *macroreduction*. The reason is that a system is characterized not only by its composition but also by its environment and structure (recall Sects. 1.7.2 and 1.8.1).

We thus recommend a strategy of *moderate reductionism*, that is, the strategy of reducing whatever can be reduced (fully or partially) without ignoring variety and emergence. Moreover, moderate reductionism must aim at accounting for variety and emergence. After all, the ontological counterpart of moderate reductionism is emergentist materialism, according to which wholes have properties not shared by their components but which, far from being self-existent, result from the latter (recall Sect. 1.7.3). Therefore, reductionism is not founded on a metaphysical materialism *tout court*—*pace* Rosenberg (1985, p. 72). Both physicalism and our version of emergentism belong in metaphysical materialism. But whereas physicalist materialism is associated with radical reductionism, emergentist materialism can only admit a moderate form of reductionism. Moreover, the epistemology and methodology associated with emergentist materialism aims at attaining the unity of science not only through (moderate) reduction but also through inter- and multidisciplinarity, that is, through the integration of theories and disciplines, to which we turn now.

3.6.3.3 Integration

Things cannot always be explained by reduction: quite often they can only be explained by placing them in a wider context. For example, the life history of an individual organism is explained not only in terms of genetics, physiology, and developmental biology, but also in ecological and evolutionary terms.

We need the integration of approaches, data, hypotheses, and even entire fields of research not only to account for those things interacting strongly with items in their environment. We need epistemic integration everywhere because there are no perfectly isolated things, because every property is related to other properties, and because every thing is a system or a component of some system (Recall Sects. 1.3.4 and 1.7.1.) Thus, just as the variety of things requires a multitude of disciplines, so the integration of the latter is necessitated by the unity of the world (Bunge 1983b, 1991b).

Suffice it to analyze briefly the simplest case of the merger of two theories. Examples of such amalgamations are analytic geometry (the synthesis of synthetic geometry and algebra), celestial mechanics (the union of mechanics and the theory of gravitation), electromagnetic theory (the merger of the theories of electricity and magnetism), and the Synthetic Theory of evolution (the union of Darwin's theory of selection with genetics). Now, not every union of two theories (or rather of their sets of formulas) is meaningful. For instance, no meaningful set of statements, i.e., theory, results from joining genetics with the theory of plate tectonics. Therefore, the precursor theories must share referents and thus also some specific concepts (variables, functions). Furthermore, in the vast majority of cases, the precursor theories have to be supplemented with formulas connecting concepts of the two theories and thus constituting a glue between them. For example, the theories of electricity and magnetism could not have been synthesized into the electromagnetic theory without the addition of Faraday's law of induction and Maxwell's hypothesis of displacement currents. And the Synthetic Theory of evolution required not only the theory of selection and genetics, but also glue formulas such as "Phenotypic variations are the result of genic changes".

We summarize the preceding in the following:

DEFINITION 3.12. A theory T is a *merger* of theory T_1 and T_2 iff

(i) T_1 and T_2 share some referents and some concepts;
(ii) there is a (possibly empty) set G of glue formulas relating some concepts of T_1 to concepts of T_2; and
(iii) the formulas in G are sufficiently confirmed.

Condition (i) excludes theories that have nothing to do with another. Condition (iii) is added because, in principle, there are infinitely many possible glue formulas. Note finally the methodological difference between glue formulas and the reductive or bridge formulas mentioned in Section 3.6.3.1: the latter are definitions, the former postulates.

Integration is brought about not only at the theoretical level, but also at the level of entire scientific disciplines. A discipline that connects two or more scientific disciplines is called an *interdiscipline*. More precisely, an interdiscipline (a) overlaps partially with two or more disciplines, as containing problems, referents, concepts, hypotheses, and methods involving specific concepts of the given disciplines, and (b) contains hypotheses (or, possibly, theories) bridging the original disciplines. Examples of such interdisciplines are biophysics, biochemistry, biogeography, physiological psychology, and neurolinguistics. (More on integration and interdisciplinarity with regard to the biological sciences in Darden and Maull 1977, as well as Bechtel ed. 1986.) Indeed, the success of the existing interdisciplines justifies the stipulation that any two scientific research fields can be bridged by one or more research fields. Every such successful interdiscipline contributes to cementing the unity of science.

3.7 Test and Evidence

3.7.1 Some Methodological Principles

In the preceding we occasionally mentioned the notions of indicator, evidence, and test. In so doing, we tacitly assumed that it is a characteristic of science and technology that everything in them is checkable: every datum, every hypothesis and theory, every method and artifact is supposed to be able to pass some test or other. Indeed, we may formulate the following *testability principle*, which is supposed to hold for all the sciences, formal or factual, basic or applied, as well as for technology and the modern humanities:

RULE 3.1. Every datum, hypothesis, technique, plan, and artifact must be checked for adequacy (i.e., either truth or efficiency).

Obviously, this principle is at odds with any uncritical attitude such as resorting to authority, intuition, self-evidence, revelation, or blind faith. To be sure, it would be impossible for us to test every proposition, method, or artifact that comes our way, because our life and our resources are too short for that. Therefore checking is a social endeavor. That is, we submit our findings to the examination of our peers, and every time we borrow or quote a result obtained by fellow researchers we hope that our source is authoritative, i.e., competent and responsible, and put our trust in it. We thus cannot, even in science, dispense with a modicum of authority. However, this trust is neither blind nor unshakable: we do so only provisionally, that is, we are ready to give it up the moment it is shown to be incorrect. In short, we must also adopt a *fallibilist principle*:

> RULE 3.2. Regard every cognitive item—be it datum, hypothesis, theory, technique, or plan—as subject to revision, every check as recheckable, and every artifact as imperfect.

However, there are degrees of adequacy, hence of inadequacy: some propositions are truer than others, some methods more accurate or powerful than others, and some checks more rigorous than others. Consequently, we may be confident that, in many cases, if not always, there is room for the improvement of our knowledge and our methods. We shall spell out this (unprovable) optimistic belief in the *meliorist principle*:

> POSTULATE 3.6. Every cognitive item, every proposal, and every artifact worth being perfected can be improved on.

The proviso 'worth being perfected' is added because there may be a point where any further investments in even serviceable ideas, procedures, and artifacts would by far outweigh the returns on them.

3.7.2 Evidence and Testability

As we have no direct access to the world, we can only grasp it through experience and reason. To indulge in metaphor, experience is at the interface between ourselves and our external world. (Your external world includes me, and mine includes you.) Experience, i.e., perception and action, mediates between the world and our ideas about it, providing us with raw material for reasoning. The resulting elaboration is a set of ideas, such as images, concepts, propositions, diagrams, models, and theories. We check these ideas about reality by contrasting them with empirical data, not with the world itself. In particular, we do not confront a proposition p about fact(s) f with f itself but with some datum (or data) d relevant to f. We can do this because both p and d are propositions, which f is not. (Recall from Sect. 3.2.4 that data are not facts, but that a datum is a particular proposition of the form "Thing x is in state (or undergoes process) y", and "There are things of kind K", whereas an empirical datum is a datum acquired with the help of empirical operations, such as observation, measurement, experiment, action, or a combination thereof.)

Now, not all data constitute evidence for or against an epistemic item: only relevant data may do so. An empirical datum may be said to be *relevant* to an epistemic item only if it refers to the latter. If preferred, a datum is relevant to a proposition if both share at least one predicate. Moreover, the properties represented by the predicates in question should possibly be lawfully related. For example, if we want to use a barometer reading as evidence for or against a hypothesis about the occurrence of certain weather processes involving atmospheric pressure, a lawful relationship of the readings of a barometer to atmospheric pressure must be assumed.

In addition, every datum must be "interpreted" as possible evidence in the light of some body of knowledge. For example, whereas the traditional Chinese pharmacists "read" certain fossil teeth as dragons' teeth, the paleontologist, armed with the theory of evolution, sees them as remains of the extinct ape *Gigantopithecus*. Sometimes, however, such interpretation is already built into the very construction of measuring instruments, or it is explicitly laid out in their operating manuals. Thus, we have learned to read temperature (not time or weight) from a thermometer. We compress the preceding considerations into the following definition:

DEFINITION 3.13. An empirical datum *e* constitutes *empirical evidence* for or against a proposition *p* if, and only if,
(i) *e* has been acquired with the help of empirical operations accessible to public scrutiny;
(ii) *e* and *p* share some referents (or predicates);
(iii) *e* has been interpreted in the light of some body of knowledge;
(iv) it is assumed (rightly or wrongly) that there is some regular association (e.g., a law) between the properties represented by the predicates in *e* and *p*.

In short, refined and exact empirical data are anything but theory-free perceptual reports. Furthermore, data may not be error-free. Such errors can be either random or systematic (i.e., deriving from bias or defective experimental design). Therefore, data must be checked instead of being taken at face value. The new (checking) run of empirical observations may be carried out with the same technique or, preferably, with an alternative (equivalent or better) method. Ideally, it is done by independent workers in order to minimize personal bias. In sum, scientists not only check their hypotheses by means of evidence, but they also check the latter. Thus, checking and rechecking is of the essence of science and technology, just as it is alien to pseudoscience and pseudotechnology.

Having elucidated the notion of empirical evidence, we are ready to propose some definitions concerning testability. The first is:

DEFINITION 3.14. A proposition *p* is said to be
(i) *empirically confirmable* $=_{df}$ there is direct or indirect, actual or potential empirical evidence for *p*;
(ii) *empirically disconfirmable* (or "*refutable*") $=_{df}$ there is direct or indirect, actual or potential empirical evidence against *p*.

DEFINITION 3.15. A proposition *p* is said to be
(i) *testable* $=_{df}$ *p* is either only confirmable or only disconfirmable;

(ii) *strongly testable* $=_{df}$ p is both confirmable and disconfirmable;
(iii) *untestable* $=_{df}$ p is neither confirmable nor disconfirmable.

The following points should be noted. First, only propositions are testable for truth. Concepts are perhaps testable for relevance and power, but not for truth. Second, testability is not an intrinsic attribute of propositions but is relative to the available or conceivable empirical means, because empirical data are obtainable by some means but not others. Moreover, a proposition may be better testable by data of one kind rather than another. Thus testability, too, comes in degrees. So we must be prepared to deal with sentences of the form 'The testability of *p* relative to means *m* equals *t*', '*p* is better testable with means *m* than with *n*', and '*p* is more testable than *q* with the help of *m*'.

Third, we have made room for potential data or, what amounts to the same, for testability *in principle* alongside actual testability. More often than not we do not possess, at a given time, the proper technique to test a hypothesis. Fourth, we have also included *indirect* empirical evidence, i.e., evidence through some intermediary body of knowledge. For example, the hypothesis "There were dinosaurs" is not directly testable, because there are no living dinosaurs nowadays but only certain fossil bones, eggs, and footprints. However, the hypothesis, together with the vast body of comparative morphology, systematics, evolutionary biology, paleontology, and geology, allows one to interpret such data as evidence for it. In other words, whenever the fact(s) to which a hypothesis refers is (are) not directly observable we resort to some intermediaries, namely indicators or diagnostic signs. As a matter of fact, most scientific hypotheses and theories are only testable via indicator hypotheses, because they refer to unobservable facts.

Consider now what can be inferred from indicator hypotheses of the forms *If U then O*, and *If O then U*, where *O* stands for an indicator of the unobservable property or fact *U*. The first says that *U* is sufficient for *O*, and *O* necessary for *U*. If *U* is assumed, *O* follows by *modus ponens* ($U \rightarrow O, U \therefore O$). Though valid, this inference is unhelpful, because we want to have access to *U* through *O*, not the other way round. Moreover, since *U* is sufficient but not necessary for *O*, the latter might be imputed to a different unobservable. So indicator hypotheses of the form $U \Rightarrow O$ are ambiguous.

Let us now probe indicator hypotheses of the second type, namely *If O then U*. Now we are allowed to conclude that *U* is the case if, in fact, *O* is observed. However, for this logically valid inference to be methodologically correct, the hypothesis $O \Rightarrow U$ must have been invented to begin with and subsequently empirically confirmed by observation or experiment, neither of which is an easy task. The difficulty lies not so much in the unobservability of *U* as in the circumstance that observables are manifestations of unobservables. Consequently, the natural thing to do is to imagine and postulate *U*, attempting to guess its observable concomitant or consequence *O*. In other words, the world is such that the natural thing to do is to make conjectures of the form *If U then O*, rather than of the form *If O then U*. This is exactly what scientists and technologists have been doing since the Greek and Indian atomists.

However, before rushing to designing indicators or making observations, we should ascertain whether the hypothesis to be checked is testable to begin with. A first necessary condition for a proposition to be empirically testable is that it refer to facts of some kind, that is, it must have factual content or meaning. This is, for instance, not the case with formulas in pure mathematics, for they refer to conceptual objects only, and hence are only conceptually testable. A second necessary condition is that the proposition be not a logical truth (or tautology), because such a statement holds no matter what may be the case. For example, the statement "Organism *x* put into habitat *y* will either adapt to *y* or die" is true under all circumstances. (Note that the statement has a factual reference. Therefore nontautological or synthetic statements should not be called 'empirical [read: factual] statements' just because their truth depends on facts, not form.) The same holds for analytic statements, which are true by definition, i.e., by virtue of the meaning of the concepts involved. For example, "Every prey species has at least one predator" is true by definition, hence in no need of testing. A third condition of testability is that the proposition does not contain a proviso rendering it untestable in real time. One such proviso is "ultimately" or "in the last analysis", occurring, for instance, in the propositions "All phenotypic features are ultimately caused by genes", or "In the last analysis all behavior is selfish". Such propositions are untestable because it is not specified what "ultimately" or "in the last analysis" is, let alone who is to perform the last analysis or when. At first sight, the *ceteris paribus* (other things being equal) condition would seem to perform a similar function. Indeed, it allows one to explain away negative evidence as an effect of changes that the model does not contemplate. However, the *ceteris paribus* condition is quite legitimate in reference to open systems. It only points to a limitation of the model, and it does not make it invulnerable when, in fact, the other factors do not vary. A fourth testability condition is that the given proposition be not shielded from disconfirmation by a *mala fide ad hoc* hypothesis, that is, by a second proposition which is not independently testable, and whose only function is to protect the first.

Assuming that the preceding conditions are met, we still face the problem that hypotheses (and theories) come in different degrees of generality. In principle, singular propositions are better testable than universal ones. For example, "This fish has been poisoned" is easier to test than "All fish in this river have been poisoned". However, there may also be extraordinary difficulties in testing singular hypotheses, such as "This astronomical object is a black hole". Concerning general propositions, it is true that, in principle, a single unfavorable case refutes the claim to universality—provided the evidence is reliable. Nevertheless, such a result does not dispose of generality. In fact, though false in a given range, the generalization may hold in another. For example, a hypothesis about vertebrates in general may, if refuted, still hold for some subtaxon. Yet this can be established only by continuing the tests after the first disconfirmations are in.

Paradoxical as it may sound, stray generalizations are, in general, harder to test than generalizations included in theories. The reason is that, while an isolated generalization can count only on the evidence bearing directly on it, a hypothesis belonging to a hypothetico-deductive system can also count on whatever evidence favors other components of the system. In other words, a systemic hypothesis can enjoy both direct and indirect empirical support. Such indirect support is multiplied if the theory in question agrees with other theories in the same or different fields.

Besides generality, there is another problem with theories: a theory can never be exhaustively tested because it is composed of infinitely many propositions. Therefore, one must always confine oneself to testing a finite subset of its infinite set of propositions. These propositions must somehow be selected by the scientist, e.g., because they are of interest or of relevance, because they are new (or, rather, newly derived), or just easy to test. This tends to produce clusters of frequently tested hypotheses, along with propositions that are seldom, if ever, checked.

Worse, if theories in general are hard to test, the most general among them are the hardest: if testable at all, they are confirmable, but not necessarily refutable as well. In order to test a general theory, we must enrich it with specific assumptions concerning the precise composition, environment, and structure of the system in question, since we have to contrast the theory with data about particular objects of certain kinds. That is, we must conceive of a theoretical model or specific theory representing the object in question. This model, not the general theory to which it is bound, is put to the test. However, if the test result is negative, we do not know whether the general theory or the special assumptions are wrong. Fortunately, uncertainty shrinks (but does not completely vanish) if a variety of models of the same general theory either succeed or fail.

Finally, what about metaphysics or ontology? Ontology is so general that it is certainly not directly empirically testable. Worse, much of traditional metaphysics is even unintelligible. However, we believe that a *scientific* ontology (such as the one sketched in Chap. 1) is *indirectly* testable by virtue of its compatibility and coherence with science.

3.7.3 Confirmation Versus Falsification

The two famous philosophical schools of thought concerning the status of evidence may be labeled *confirmationism* (or *inductivism*) and *refutationism* (or *falsificationism* or *deductivism*). The former is associated with Carnap and Reichenbach, the latter with Popper. So, confirmationism belongs to empiricism, falsificationism to rationalism. And neither fits actual scientific practice.

Confirmationists claim that a few exceptions hardly matter, whereas refutationists hold that they are decisive. The former think of the degree of confirmation of a hypothesis as the ratio of the number of favorable cases to the total number of data. And, because of the formal analogy between this ratio and Laplace's definition of probability, they identify the two, and these, in turn, with degree of truth. Refutationists care only for negative evidence. They argue that, while no number of confirmations of the consequent *B* of hypothesis "If *A* then *B*" suffices to confirm it, a single negative case suffices to refute it according to the modus tollens inference rule "If *A* then *B*, and not-*B*, then not-*A* ".

Real scientific practice fits neither confirmationism nor refutationism. For one thing, the two philosophies share the empiricist belief that empirical data are firm, whereas scientists know that these are nearly as fallible as hypotheses. (Recall again that data are not facts but propositions.) This is why they test data of a new kind against well-tried hypotheses. This is also why they protect well-tried hypotheses with (*bona fide*) *ad hoc* hypotheses when negative data seem to refute them. Shorter, there is no rock bottom empirical basis, and not all hypotheses are equally flimsy. In fact, some of them are supported by other hypotheses which, in turn, have been satisfactorily confirmed. Thus, support for a hypothesis comes partly from empirical data and partly from the extant body of relevant knowledge; so much so, that hypotheses are checked against the latter, that is, they are conceptually tested, before being subjected to empirical tests.

To be sure, refutationism is a useful alert call against naive confirmationism, but it is not a viable alternative to it. Firstly, it exaggerates the importance of criticism: after all, theories must be created before they can be criticized. Secondly, it denies the importance of confirmation, which is a powerful motivational driving force of scientists. (See also Hull 1988.) After all, scientists are credited for proposing good data, hypotheses, and theories, not for refuting them. (There are no Nobel Prizes for the falsification of theories.) Thirdly, it restricts the function of observation and experiment to the refutation of theories, while actually they are also necessary to produce data (which are necessary to operationalize theories), to "discover" problems, and even to suggest (modest) hypotheses. Fourthly, if refutationism were correct, we would be justified in upholding all the testable hypotheses that have not yet been refuted. For example, we should be justified in believing in heaven and hell as well as in the divine creation of the universe. The methodological skeptic, such as the scientist, rejects these myths not because they have been refuted by observation or experiment, but because they are not supported by any positive evidence. Finally, if falsification were the sole arbiter of the scientific status of theories, we would have to regard overwhelmingly refuted pseudoscientific theories, such as astrology, as scientific. Yet we do not, and not because they are false, but because they are externally unsound, that is, incompatible with the bulk of our well-confirmed knowledge. (Besides, many of them also have internal defects.) However, there is one context in which refutationism works: the testing of null hypotheses (Sect. 3.4.4).

In sum, real-life scientists, unlike those imagined by some philosophers, are equally interested in positive and negative empirical evidence. Besides, scientists also value the compatibility of the hypothesis or theory under trial with the bulk of the background knowledge. Their verdict is favorable only in the case of compatibility and reasonable confirmation—and even so only until new notice.

3.7.4 Empirical Operations

To gain empirical evidence for or against a hypothesis, we must engage in empirical operations, i.e., actions involving sense experience or perception of some kind. The outcome of such perception is an empirical datum. Since we studied observation, the basic mode of data generation, in Section 3.2.3, we shall now examine two further empirical operations, namely measurement and experiment.

3.7.4.1 Measurement

Measurement may be characterized as quantitative observation or, more precisely, as observation of quantitative properties. In a broad sense, most organisms are able to "measure", i.e., detect, quantitative properties, such as gradients and deviations from the optimal values of certain parameters. Think of temperature, light, salinity, or acidity. This does not even require the possession of a nervous system. The latter, however, is necessary in order to speak of perception and observation proper. Moreover, measurement in the strict, in particular scientific, sense involves measuring instruments. More precisely, we can say about measurement proper that (a) every measurement presupposes the conceptual operation of quantitation, or assignment of numbers to the degrees of a property (together with some appropriate unit); (b) some measurements also presuppose the construction of indicator hypotheses if the quantitative property, such as a field intensity, is not directly observable; and (c) measuring instruments must be equipped with pointers, digital dials, or some other indicators, as well as with scales, which allow one to "read" them.

Now, what exactly do measuring instruments measure? To be sure, they measure properties of things, not things or facts. As not all properties are measurable, we suggest elucidating the notion of a measurable property thus:

DEFINITION 3.16. A property is *measurable in principle* iff it is
(i) quantitative, and
(ii) either manifest (observable) or lawfully related to a manifest property.

It goes without saying that whether a property is actually measurable at a given time depends on the state of the art, opportunity, and resources. In particular, the concept representing the quantitative property in question must have been quantitated. And if the concept denotes a transphenomenal property, an indicator hypothesis concerning its lawful relation to a manifest property must have been conjectured. This is the well-known thesis that scientific measurement is theory-dependent.

When we observe and, in particular, measure some property of an object, we may or may not disturb the object. For example, the behavior of an animal might be more or less different from its natural behavior when the normal habitat is disturbed by the presence of the ethologist. The same holds, for instance, for physiological measurements. For example, the measurement of body temperature in humans is practically unintrusive, whereas the measurement of glucose turnover in the brain with the help of radioisotopes is somewhat intrusive, and the measurement of action potentials in the limbic system by means of electrodes is extremely intrusive. We must therefore distinguish measurements of two kinds, which are elucidated by:

DEFINITION 3.17. A measurement technique is said to be *intrusive* if, and only if, it changes in any way the state (and, *a fortiori*, the kind) of the object of measurement. Otherwise, it is *unintrusive*.

The intrusiveness of some measurements poses the methodological problem of whether such interference, if it occurs, can be corrected. The answer is: sometimes yes, sometimes no. For example, sometimes we can use different techniques, or we can explain with the help of some theory how and to what degree a measurement operation interferes with the object in question. The problem of whether a measurement interferes with or even creates its own objects is notorious with regard to quantum physics. However, we must leave this debate to the philosophy of physics (see Bunge 1967c, 1973b, 1983b, 1985a). More on measurement in Bunge (1967b, 1971, 1983b).

3.7.4.2 Experiment

Unlike measurement of the unintrusive kind, every experiment involves controlled changes in the object of study. Indeed, an experiment may be defined as a controlled action performed by a person on some other object, or on part of itself, with the aim of solving some cognitive or practical problem concerning the object of experimentation, and such that the person is able to receive, record, and analyze the reaction of the object to that action. Thus, what is typical of all experiment, in contrast to mere observation, is that the experimenter controls the object and its environment (Bunge 1983b).

Therefore, *pace* Mayr (1982) and others, there are no "natural experiments" (*Naturexperimente*). Nature just changes—sometimes gradually and sometimes catastrophically—and scientists may observe and analyze some of these changes. Thus *Naturexperimente* belong to the so-called '*ex post facto* experiments', which are not really experiments but unplanned events analyzed with hindsight. Similarly, thought experiments (*Gedankenexperimente*) are not genuine experiments. A thought experiment consists in imagining what would happen if certain facts were to occur or had, or had not, occurred. For instance, computer simulations are thought experiments. Gedankenexperimente have no validating force but they may spark off interesting hypotheses. In fact, the very design of a real experiment is a thought experiment. And every student of the past is constantly making thought "experiments".

Let us return to real experiments. An *experimental device* (or *setup*) is a concrete system with at least three components: the object x of study, an object supplying a stimulus (input) of some kind to x, and an object monitoring the response (output) of x to that stimulus. (In the case of micro-objects, such as molecules, the measuring instrument is linked to the measured object via an amplifier of some sort.) An *experiment* on x consists in (a) subjecting x to a stimulus and observing its response; (b) observing the output of the same object x or one of the same kind when not subjected to the stimulus; and (c) comparing the responses and determining whether the difference between them is significant, i.e., due to the stimulus rather than attributable to either chance or idiosyncrasies of x.

The design and interpretation of every experiment presuppose a number of hypotheses, so that the data obtained by experiments are hypothesis- and often theory-dependent. These presuppositions can be grouped into *generic*, i.e., shared by all experiments, and *specific*, i.e., characteristic of every type of experiment. The latter consist in particular hypotheses or theories about the nature and behavior of the objects of experiment or the experimental means. The generic presuppositions are of two kinds: philosophical and statistical. As the statistical principles of experimentation are amply explicated in the relevant textbooks, we are focusing on the philosophical presuppositions, which are rarely if ever analyzed. These philosophical presuppositions are:

1. *Ontological realism:* The members of the experimental and control group, as well as the measuring instruments, exist really although some of the hypothesized objects may be imaginary. This assumption is necessary because, if all the things involved in an experiment were figments of our imagination, then imaginary experiments would suffice: all experiments would be Gedankenexperimente.
2. *Lawfulness:* All the objects in the experiment behave lawfully. This assumption is needed because there would be no point in performing experiments if nature were to give significantly different "answers" every time we pose the same "question", or if the instruments behaved arbitrarily.

As our main subject is the philosophy of biology, the presupposition of lawfulness deserves some elaboration. As mentioned above, we need to check the outcome of experiments for methodological reasons by repeating them with either the same object or with other objects of the same kind. The latter option presupposes that—variation notwithstanding—there actually are concrete systems *of the same kind* or, more precisely, of the same *natural kind*, i.e., systems that are *nomologically equivalent*. (More on the concept of natural kind in Sects. 7.2.1.7-8.) With regard to biology this implies that there are biological objects of the same natural kind, i.e., organisms sharing some $laws_1$. Indeed, this is what biologists tacitly presuppose when experimenting on biosystems of the same species or of some higher taxon. Moreover, only this assumption allows them to generalize their findings, if often only statistically, as in the schema "96% of the individuals of species w respond to drug x in dose y with physiological reaction z". (Still, the 4% exceptions may again have something in common, e.g., they may constitute a strain of a different genotype.) However, if the antiessentialist philosophers of biology were right in claiming that taxa, in particular species, are not classes of nomologically equivalent organisms but concrete individuals whose parts need have

nothing in common except common descent, then experiments in biology would be exercises in futility. (More on the taxa-as-individuals thesis in Sect. 7.3.)
3. *Causality:* All the things involved in the experiment satisfy some form of the causal principle, however weak, e.g., "Every event is the effect of some other event". Otherwise, no deliberate production of an effect and no effective control of variables would be possible.
4. *Randomness:* All the variables involved in the experiment are subject to some random fluctuation, both intrinsic and due to external perturbations. Otherwise, we would not be able to explain the statistical scatter of results. Note that this presupposition does not contradict the assumption of lawfulness, because random perturbations are themselves lawful.
5. *Insulation:* Objects other than the object of experiment, the experimenter, and his experimental means, can be neutralized or at least monitored for the duration of the experiment. Otherwise, no significant changes could be attributed exclusively to changes in the control variables.
6. *Artifacts:* It is always possible to correct to some extent, either empirically or theoretically, for the "artifacts", disturbances, or contaminations caused by the experimental procedures. These are not the deliberate alterations of the object but the unwanted distortions of it or of its image. (Think of color artifacts caused by optical lenses, or the staining of tissues.) If such partial corrections were impossible, we could not legitimately claim that the thing for us, i.e., as it appears to us, resembles the thing in itself, i.e., such as it is when not subjected to experiment.
7. *No psi:* The experimenter's mental processes have no direct influence on the outcome of the experiment. Alternatively, if somebody believes that, in principle, they are able to do so, we must presuppose that it is possible to shield or uncouple the experimenter from the experiment. Otherwise, the outcome of the experiment could be produced, consciously or unconsciously, by the experimenter herself. (If one assumes the existence of sundry unknown paranormal abilities, as parapsychologists do, one cannot possibly uncouple the experimenter from the experiment. Therefore, there can be no genuine parapsychological experiments.)
8. *Explicability:* It is always possible to justify (explain), at least in outline, how the experimental setup works, i.e., what it does. Otherwise, we would be unable to draw any conclusions.

This concludes our list of philosophical presuppositions of experimentation. More on experimentation in Bunge (1967b, 1983b), and on some particularities of biological experiments in Mohr (1981).

3.8 Truth and Truth Indicators

Why do scientists check almost everything they imagine or handle: data, hypotheses, theories, inferences, methods, instruments, or what have you? Apparently they are interested in knowing whether their hypotheses and theories actually "say" something about the real world, rather than being figments of their imagination, and whether their methods and instruments actually work. In other words, they are after *adequate* knowledge, i.e., they search for *true* propositions and *efficient* procedures. Disregarding the latter notion of usefulness and efficiency as belonging in technology, not basic science, we shall briefly examine the much maligned notion of truth in the following.

3.8.1 Truth

In tune with our former distinction between formal and factual sciences, i.e., sciences dealing with either constructs or facts, we follow Leibniz (1704) and distinguish two concepts of truth: *truths of reason* (formal truths) and *truths of fact* (factual truths). *Formal truth* values are assigned to propositions that either "say" nothing specific about facts, such as "Either it rains or it does not", or lack factual reference altogether, such as "7 is a prime number" and "If *p*, then *p* or *q*". This is why formal truth values are assigned and checked by purely conceptual means, such as deduction (in particular computation). By contrast, *factual truth* values are attributes of propositions making definite assertions or denials about actual or possible facts. In this case, truth consists in the (degree of) correspondence between the proposition and some factual item. Yet, as we have no direct access to such correspondence, our truth valuations of factual propositions must be checked by means of empirical operations. Consequently, whereas mathematics is self-sufficient, factual science is not: it depends on the world (as well as on mathematics).

Because formal and factual truth are predicated of propositions of radically different kinds, each of them requires its own theory of truth. That is, we need a *coherence* theory to explicate the notion of formal truth, and a *correspondence* theory to elucidate that of factual truth. Whereas model theory, i.e., the semantics of logic and mathematics, contains a coherence theory of truth, a correspondence theory is, so far, little more than a program. Moreover, the very idea of a correspondence theory of truth has been under attack by a variety of philosophers—even some who regard themselves as realists. Yet there is no room here to consider these criticisms, so that we proceed with an elucidation of the basic ideas of a correspondence theory of truth.

Let us first consider the scholastic formula *veritas est adaequatio intellectus ad rem*. How can a proposition, which is an *ens rationis*, i.e., a construct, be adequate to (or match or fit in with) a fact in the world of things? A naive-realist answer is that there is an isomorphism between a proposition (or even a theory) and the piece of reality to which it refers. However, isomorphism in the technical sense is a relation between sets. Yet neither a proposition nor the concrete thing(s) to which it refers are sets. (See also Sect. 9.3.2.2.) Furthermore, we cannot solve this problem by postulating that at least sets (or rather systems) of propositions, such as theories, can correspond to collections of facts, because theories contain infinitely many propositions, which suffices to make any theory-fact isomorphism impossible. Finally, since even a weak notion of isomorphism presupposes some structural similarity between the representation and the represented, it cannot apply here because constructs are equivalence classes of brain processes, and there just is no resemblance between those and facts in the real world (see Postulate 3.5).

To approach a solution to the problem of correspondence, let us no longer pretend that the truth relation holds between propositions and facts. Let us assume, instead, that propositions can match other propositions, and facts other facts. Thus the correspondence we seek is one between mental facts, i.e., brain processes, of a certain kind and further facts, whether mental or not. This strategy will allow us to characterize the concept of true partial knowledge of fact; once in possession of this notion we can proceed to define the concept of the truth of a proposition (see Bunge 1983b).

Consider a thing ϑ internal or external to an animal a endowed with a brain capable of learning. Call e an event occurring in thing ϑ, and e^* the corresponding perceptual or conceptual representation of e in the brain of a. Then we say that a has gained *true partial knowledge* of fact e if, and only if, e^* is identical to the perception or conception of e as a change in thing ϑ (rather than as a nonchange, or as a change in some other thing). The true (though partial) knowledge that a has acquired of event e is the neural event e^*; and the *correspondence* involved is the relation between the events e and e^*.

We have so far been talking about thoughts, i.e., concrete events, not constructs, which we have defined as equivalence classes of thoughts. To arrive at propositions, we form the equivalence class of thoughts e^* constituting true (though usually partial) knowledge of e: $[e^*]$. Note that no two members of the class $[e^*]$ are likely to be identical, for they are thoughts of a given animal at different times, or thoughts of different animals, and in either case they differ in some respect or other. However, they are all equivalent in that every one of them constitutes true partial knowledge of e; that is, for every member e^* of $[e^*]$, e^* happens if, and only if, e is (or has been or will be) the case. We identify the proposition p = "e is the case" with that equivalence class of thoughts, i.e., we set $p = [e^*]$. And we stipulate that p is true if, and only if, e happens or has happened. Thus, the correspondence relation holding between a mental fact and some other (mental or nonmental) fact carries over to propositions in relation to facts. Accordingly, truth and falsity are *primarily* properties of perceptions and conceptions (e.g., propositional thoughts), and only *secondarily* (or derivatively) attributes of those equivalence classes of thoughts we call 'propositions'.

Note that we spoke of true *partial* knowledge in the preceding (very simple) example, because facts are usually complex, so that our representations will "correspond" only approximately to some real facts. Since the correspondence between representation and represented fact is partial or approximate, our notion of factual truth can only be one of partial truth. Indeed, only a concept of approximate truth is compatible with scientific realism. (For different notions of approximate or partial truth see, e.g., Popper 1962; Bunge 1963, 1974b, 1983b; Marquis 1990; Weston 1992.) In a concept of partial truth, the truth value attributed to a proposition is a real number comprised between 0 (false) and 1 (completely true).

(Hence a half-truth is also a half-falsity.) However, since we have distinguished formal from factual truth, we do not need many-valued logic to handle the notion of partial (factual) truth. After all, logic is concerned with deduction, not factual truth: that is, logic is alethically neutral (Bunge 1974b). Furthermore, we must warn against mistaking partial truth for probability. Propositions involve no randomness, hence no probability. A probability statement is a statement about the probability of the occurrence of a certain fact, not a statement that is just plausible. For example, the truth value of the proposition "The probability of a (fair) coin landing heads up is 0.5" is not itself 0.5 but 1 (recall Sect. 1.10.2.3).

Note also that, for the sake of simplicity, we have dealt only with individual propositions, not with theories. The problem with theories is that an entire theory is unthinkable, for it contains infinitely many propositions. We can only think of a few statements of any given theory. Therefore speaking of the truth or falsity of a theory involves an inductive leap on the basis of some evidence. Note further that, if truth is an attribute of propositions, i.e., a semantic notion, then we must be aware that the term 'truth' is often used only in a metaphoric sense. For example, in the expressions 'the truth is out there', 'I discovered the truth', or 'the truth behind the appearance' the term 'truth' is equated with 'world' or 'reality', 'fact', and 'transphenomenal fact'.

Finally, we should address the main concern about any correspondence theory of truth, namely the problem that we have no direct access to any proposition-fact or, rather, mental fact-other fact correspondence. Indeed, in a correspondence theory of truth, a proposition is either (partially) true, i.e., it corresponds in certain respects to some fact, or false, i.e., it does not correspond to some fact, although we may have no knowledge of this correspondence or noncorrespondence. In other words, an individual's knowledge of a proposition *p* neither implies that *p* is true nor that he or she knows that *p* is true. We can gain knowledge of the (partial) truth of *p* only by investigating the referents of *p*. This is obvious to any scientist, who conjectures or infers a hypothesis *p* but, if rational, cannot claim any truth value for *p* before *p* has been subjected to some tests.

In sum, as we have no direct access to the correspondence involved in the notion of truth as *adaequatio intellectus ad rem*, we have to settle for truth indicators or truth symptoms, which allow us to hypothesize that there is some correspondence between our hypotheses and reality. To these indicators we turn now.

3.8.2 Truth Indicators

As truth is often not manifest, we have to rely on truth symptoms or indicators. For the scientific realist, there are two kinds of such truth indicators: empirical and conceptual. The *empirical* truth indicator of a given hypothesis is, of course, the body of empirical evidence for or against this hypothesis. Such evidence may either confirm (or support) the hypothesis, or it may disconfirm (or undermine) it. (Since the terms 'verification' and 'falsification' [or 'refutation'] suggest absolute-ness and certainty, we prefer the more cautious terms 'confirmation' and 'disconfirmation'.) The more numerous and varied the evidence for (or against) a hypothesis, the more strongly confirmed (or disconfirmed) it is. Furthermore, the weight of some empirical evidence depends on the status of the proposition to which it is relevant: it is not the same for an established hypothesis as for a new one. In other words, it takes more than a few unfavorable data to bring down a previously well-confirmed hypothesis. This holds, *a fortiori*, when the hypothesis in question is part of a theory, because, in this case, it is indirectly supported by all the positive evidence for its fellow propositions in the theory. On the other hand, any support for a new conjecture is of great value, and any conclusive negative finding is sufficient to discredit it (at least for the time being).

Whereas for empiricists empirical adequacy is all that matters (see, e.g., van Fraassen 1980), the scientific realist supposes that empirical adequacy, though necessary, is not a sufficient indicator of factual truth. This is because, in principle, two or more inequivalent hypotheses or theories can enjoy the same empirical support. Think of the Ptolemaic, Tychonian, and Copernico-Galilean models of the planetary system before Newton. In the philosophy of science this is known as the problem of the *underdetermination* of a hypothesis or theory by the evidence. Yet such underdetermination with regard to the available evidence at a certain time does not entail that further evidence at some later time never makes a difference to the empirical adequacy of the theory or hypothesis. Indeed, the three planetary theories were no longer empirically equivalent in the second half of the 17th century. Still, by increasing the number of phenomenological parameters of a theory one can improve indefinitely its fit to any set of data, though at the price of losing explanatory power.

Therefore, the scientific realist makes use not only of empirical but also of *conceptual* truth indicators. One such indicator is *internal consistency*. However, though a necessary condition for the truth of a theory, it is insufficient because it is easy to concoct consistent theories at odds with the facts. More important, therefore, is *external consistency*. By "external consistency" we mean the compatibility of a hypothesis or theory with the *bulk* of antecedent knowledge. That is, however unorthodox or even revolutionary a new conjecture may be in a certain field, there is no hope for it if it upsets the whole of science at one stroke. Furthermore, we cannot do without external consistency because we cannot pose interesting problems in a vacuum: every problem has presuppositions; and we can evaluate new ideas only in the light of some background knowledge. Finally, our background knowledge provides not only heuristic guidance but also indirect empirical support. In particular, if the hypotheses in question are logically related, then any direct confirmation (or disconfirmation) for each of them is indirect confirmation (or disconfirmation) for the other.

Lest the criterion of external consistency be suspected of being a means for enshrining dogma, stifling research, and promoting conformism, we hasten to note that it must be applied prudently. However, we need some means for distinguishing between promising new conjectures and wild ones, even though there is a gray zone between heterodox, yet sound, and wrongheaded or even pseudoscientific claims. In sum, though advocating a form of moderate conservatism, we contend that, normally, scientists evaluate hypotheses and theories according to the criterion of external consistency. (If in doubt, try to apply for a research grant in a scientific or technological field. However, your chances of getting away with "anything goes" would not be negligible if you apply for a grant in philosophy.)

A related indicator is the *unifying power* of a theory, i.e., the capacity to embrace previously separate theories. Another strong truth indicator is the *predictive power* of theories. The notions of both unifying and predictive power are related to the idea of *consilience* as proposed by Whewell: "The prediction of results, even of the same kind as those which have been observed, in new cases, is proof of real success in our inductive process. [...] But the evidence in favor of our induction [theory] is of a much higher and more forcible character when it enables us to explain and determine cases of a kind different from those which were contemplated in the formation of our hypothesis. [...] No accident could give rise to such an extraordinary coincidence" (1847, vol. 2, p. 65).

Further indicators are the *heuristic power* as well as the *stability* and the *depth* of a theory. New theories should suggest and guide research rather than just summarizing it, and they should not fail at the first unfavorable evidence. However, stability has its limits: we should not save a theory at all costs by adding further and further epicycles. As for depth, scientists clearly prefer mechanismic theories over phenomenological ones, because they are explanatory, not just descriptive. Moreover, they may have a greater predictive power.

A proverbial yet ambivalent and unreliable indicator of truth is *simplicity*. While we agree that theories should not be so complicated that any empirical tests are practically impossible (methodological simplicity), other forms of simplicity, such as logical, mathematical, and psychological simplicity, are rather dubious truth indicators. After all, the universe happens to be very complex. Moreover, scientific and technological progress has been a course of increasing complexity in all regards. Compare, for instance, Mendel's genetics with contemporary molecular genetics. (More on simplicity in Bunge 1963; and more on truth indicators in Bunge 1967b, 1983b.)

To conclude, the evaluation of scientific hypotheses and theories is performed on the strength of a whole battery of tests, some empirical, others conceptual. Depending on the outcome of such evaluation, which is usually performed tacitly rather than explicitly, we are justified in believing that the hypothesis or theory in question corresponds at least to some degree to some real fact. In other words, the degree of confirmation of a hypothesis justifies our belief that it is partially true. Usually, we do not assign precise numbers to the propositions in question, but we say that their degree of confirmation is very strong, strong, indecisive, weak, or very weak. Although we must settle in practice for such qualitative degrees of confirmation or test success on the basis of all the above-mentioned indicators, the correspondence theory of truth tells us at least what the truth of a proposition consists in—if it is true as indicated by its test success.

3.9 Upshot

We may summarize our epistemology by pointing out several of the doctrines or *isms* involved.

The first thesis is, of course, *epistemological realism*, that is, the assumption that the world can be known, if only partially (see Postulate 3.4). Obviously, our epistemological realism is not of the *naive* kind, but it is a *critical realism*. Critical realism holds that perceptual knowledge, though indispensable, is superficial, incomplete, and often wrong, so that it must be enriched with hypothetical or theoretical knowledge. This theoretical knowledge consists of constructions (e.g., propositions, theories) that may go far beyond appearances.

So another ingredient of our philosophy is *epistemological constructivism*, i.e., the thesis that concepts and their components are our own creation. Some of our constructions may represent facts in the world, however imperfectly, in a symbolic, not an iconic fashion. Others are sheer fictions without factual counterparts, e.g., mathematical and mythical objects. Note that our constructivism is epistemological, not ontological: we construct models of the world, not the world itself.

Our epistemological constructivism presupposes *epistemological naturalism*, which is the thesis that cognition is a brain process, and that there are no supernatural or paranormal, i.e., noncerebral, modes of cognition. Being functions of biological systems, cognitive abilities are subject to biological evolution. Thus, epistemological naturalism comes together with *epistemological evolutionism*. This evolutionism, however, does not imply any naive adaptationism, i.e., the thesis that we can get to know only those things to which we have been adapted. The most interesting and important cognitive abilities are processes in plastic neuronal systems, not hard-wired ones. Therefore, not all functions in plastic neuronal systems need be adaptive, while plasticity in itself may well be so. (This imposes limits on the sociobiological program.)

As far as the two great traditions of epistemology, namely rationalism and empiricism, are concerned, we must combine tenets of both. This is because both reason and experience are necessary, though not separately sufficient, to gain scientific knowledge of the world. Indeed, adopting a hypothesis without giving some reason is superstition, and adopting an even well-reasoned factual statement without some empirical support is dogmatism. Furthermore, scientific practice clearly shows the interplay between theorizing and empirical investigation: while observations, measurements, and experiments sometimes yield findings that call for theorizing, at other times theorizing precedes empirical studies. In short, science combines the sound halves of rationalism and empiricism: conceptual analysis, theorizing, and discussion come together with observation, measurement, experiment, and practice. This synthesis may be called *ratioempiricism*.

Our epistemology is furthermore *justificationist* in the sense that it requires every proposition, be it hypothesis or datum, to be ultimately justifiable either theoretically or empirically. That is to say, a scientific statement must either follow from premises in a theory, or it must be supported by controlled empirical evidence. Note, however, that, since both theories and empirical data are corrigible, justification can be only relative or conditional, not absolute. Thus, our version of justificationism is *fallibilist* (recall Rule 3.2). We further contend that scientists are also *meliorists*, because they usually hope to spot error and reduce it (see Postulate 3.6). (Incidentally, the belief that our knowledge of facts may contain errors, and must therefore be checked, presupposes the reality or autonomy of the facts in question; see Bunge 1981a, Appendix.)

Finally, we should add a version of *scientism*, which had not been mentioned before, to our epistemology. This is the thesis that anything knowable and worth knowing can be known scientifically, and that science provides the best possible factual knowledge, even though it may, and does, in fact, contain errors. This form of scientism should not be mistaken for the neopositivist unification program, according to which every discipline should ultimately be reduced to one basic science, such as physics or psychology. Neither does our version of scientism require that we should accept science in its present state. This would clearly be incompatible with fallibilism and meliorism.

The synthesis of all the preceding *isms* constitutes our version of *scientific realism*. We are now ready to apply the philosophical fundamentals expounded in this and the previous chapters to biology and propose a systematic, if only partial, philosophy of biology.

Part II

Fundamental Issues in Biophilosophy

4 Life

4.1 What Is Life? — A Philosophico-Scientific Problem

Life has long been a mystery on which mysterymongers have thrived. Indeed, as long as the question 'What is life?' was dealt with only outside science, i.e., in traditional metaphysics, a plausible answer to it was hardly to be expected. Yet, even when approached scientifically, a satisfactory answer was not to be expected as long as living beings were studied either on their own level (holistic approach) or as physical systems devoid of emergent properties (reductionism), and in either case apart from their history. Only modern biology, as a multilevel discipline between evolutionary and molecular biology, has transformed the mystery of life into the problem of life: its origin and maintenance, its evolution and extinction. The question 'What is life?' has become a *philosophico-scientific* problem.

Despite these new prospects of answering a partly scientific, partly metaphysical question, it might not be exaggerated to say that many students of living beings seem to become interested in a definition of the concept of life only during their freshman year and perhaps at the end of their career. In between, they are often discouraged from trying to elucidate that concept and, in general, from getting involved in philosophical questions. They are encouraged instead to "get on with their business", which can, of course, be done successfully by taking life for granted. Moreover, molecular biology and, particularly, theories on molecular evolution even seem to have discouraged definitions of "life" on the grounds that it would make no sense to draw any sharp distinction between living and nonliving entities: there would be only a gradual and continuous transition. Apparently only a few physicists, biochemists, and biologists dealing with such issues as the role of self-organization, synergetics, and thermodynamics in biology have tackled the concept of a living system, although their suggestions are usually (and regrettably) far too general to be of biological relevance.

Interestingly, though being a philosophico-scientific or, more precisely, an ontologico-scientific problem, the question 'What is life?' is not a hot topic in contemporary biophilosophy either. Some philosophers, mildly sympathetic to functionalism (e.g., Sober 1991, p. 763), even say that an answer to this question would not matter much anyway. (If unfamiliar with the current literature, check

Ruse 1988. Of course, there are exceptions like Wuketits 1983 and Sattler 1986, but the latter's book is somewhat mystical.) By contrast, some of the old guard did address this question (e.g., Haeckel 1866; Schrödinger 1944; Sommerhoff 1950; von Bertalanffy 1952; Hartmann 1965; Monod 1971; Rensch 1971; Mayr 1982).

As is well known, the two traditional views on life are *vitalism* and *mechanism* (or *mechanicism*). According to vitalism, living things are distinguished by special immaterial entities such as entelechies and animal spirits; or some particular properties such as goal-seeking or whole-forming; or some special forces such as the *élan vital* or the *Bildungstrieb*. Since vitalism has been bashed enough, we can safely ignore it here. Anyway, its incompatibility with a materialist metaphysics should be obvious. (Yet there are always those who still have vitalistic inclinations: see, e.g., Lenartowicz 1975; Engels 1982.)

The mechanistic answer comes in two versions, which may be called *physicochemicalism* and *machinism*, respectively. According to the former, organisms are nothing but extremely complex physical or physico-chemical systems: they have no properties or laws of their own. According to machinism, organisms are not just extremely complex physical systems but they are machine-like systems, if not machines proper. Three current examples of this heritage of Descartes and de La Mettrie spring to mind. The first is Daniel Dennett's approach to biology, according to which "biology is not just like engineering; it is engineering" (1995, p. 228). The second is the description of organisms in terms of various formal machine theories, such as the theory of self-reproducing automata. Witness the unabashed machinism of the Artificial Life project (Langton 1989, pp. 5-6), to be discussed in Section 4.4. The third is the so-called "Frankfurt School of Constructional Morphology", which aims at describing organisms exclusively as hydraulic machines and energy converters (Gutmann 1995.) We will not comment here on this approach, as it is hardly known outside Germany, and as we have criticized it elsewhere (Mahner 1995). Although mechanicism has the merit of having been extraordinarily fruitful in the past, it is inadequate if only because it flies in the face of the evidence that being alive is not quite the same as being dead.

We adopt a third option, namely *biosystemism*, which recognizes the *bios* as an emergent level rooted to the chemical one (Bunge 1979a). More precisely, biosystemism maintains that (a) living systems, though composed of physico-chemical subsystems, have emergent properties, in particular laws, which their components lack (Fig. 4.1), and (b) the units of biological science are the organism-in-its-environment, as well as its various subsystems (molecules, cells, organs) and supersystems (population, community, ecosystem).

Of course, the fact that living beings are not just aggregates of parts but integrated and coordinated wholes (i.e., systems) has been acknowledged since long, in particular among cytologists (see, e.g., the review in Woodger 1929); but it was not generally accepted due to the strong influence of mechanicism and reductionism. A systemic view of life is also known as *organicism* (von Bertalanffy 1952, 1968; Weiss 1973; Wuketits 1989; Mayr 1996). Although von Bertalanffy clearly distinguished organicism from holism, sometimes this distinction was subsequently blurred by others. For this reason, and to avoid any ambiguity, we prefer our own term 'biosystemism'.

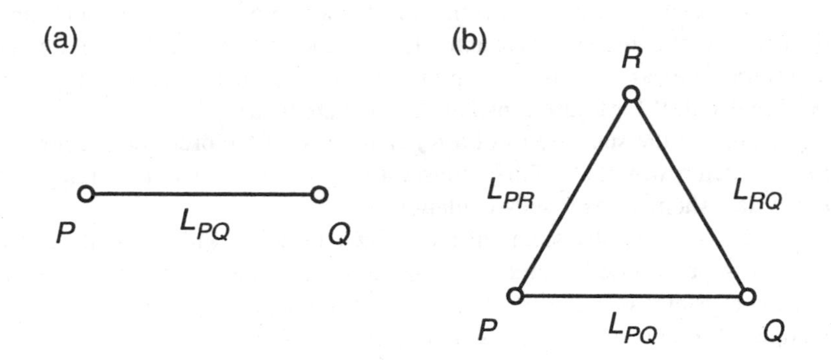

Fig.4.1 a, b. Living and nonliving things possess different properties and, hence, laws. **a** The nonliving system a possesses two properties P and Q, related by the law L_{PQ}. **b** The biosystem b possesses all of the properties and laws of the nonliving system and, in addition, the emergent property R. It is also characterized by laws of its own: L_{PR} and L_{QR}

4.2 Biosystem

Although many authors believe that a definition of "life" is an elusive task, we shall nevertheless attempt to give a list of properties which we believe jointly characterize living and only living things. Following our CES analysis of systems (Sect. 1.7.2), we assume that a living being is a material system such that

(i) its *composition* includes nucleic acids as well as proteins (both structural and functional, in particular enzymatic, the latter enabling it to exploit its habitat);

(ii) its *environment* includes some of the precursors of its components (and thus enables the system to self-assemble most if not all of its biomolecules);

(iii) its *structure* includes the abilities to metabolize, and to maintain and repair itself (within certain bounds).

We spell out this hypothesis into an axiom and a convention:

POSTULATE 4.1. There are concrete systems of a kind B such that, for every member b of B,

(i) b is composed of chemical and biochemical subsystems, in particular water, proteins, nucleic acids, carbohydrates, and lipids;

(ii) the components of b are sufficiently contiguous so as to permit continual (bio)chemical interactions amongst them;

(iii) the boundary of b involves a flexible and semi-permeable lipid membrane (biomembrane);

(iv) b incorporates some of the biomolecules it synthesizes (rather than releasing them immediately to its habitat);

(v) the possible activities of b include the assembly, rearrangement, and dismantling of components (which allow for the self-maintenance of b over a certain time) as well as the capture and storing of free energy (e.g., in ATP molecules) for future consumption (metabolism);

(vi) some of the subsystems of b regulate most of the processes occurring in b in such a way that a fairly constant *milieu intérieur* is maintained in the system (homeostasis, self-regulation);

(vii) one of the subsystems of b involved in self-regulation—its genic system—is composed of nucleic acid molecules, and its interaction with other subsystems of b (co)regulates the self-maintenance, as well as the development, if any, and the reproduction, if any, of b;

(viii) all of the control systems in b are interconnected by chemical signals (such as the diffusion of ions, atoms, or molecules, and propagating chemical reactions) and thus constitute a (chemical) signal network;

(ix) b can adjust to *some* environmental changes without jeopardizing its continued existence.

DEFINITION 4.1. The systems of kind B referred to by Postulate 4.1 are called *biosystems, living systems, living things,* or *living beings*.

It should be noted that "biosystem" is not coextensive with "organism". All organisms are biosystems but not conversely. Think of the cells, tissues, and organs constituting a multicellular organism: these are living components of organisms but not organisms themselves. For this reason, we will have to define the concept of an organism separately (see Definition 4.5).

What we have given in Postulate 4.1 would usually be called a 'definition of life'. But this would be incorrect. What we have done is to *hypothesize* the properties jointly necessary and sufficient for a material system to constitute a living system. Since we do not define things but solely concepts and signs, we can only define the concept "life". This definition may read thus:

DEFINITION 4.2 Life $=_{df}$ the collection of all living systems—past, present, and future.

In other words, "life" is the extension of the predicate "is alive". Life is neither a material nor an immaterial entity, nor a substance or force, nor a property: it is a collection, hence a conceptual object. (See also Mayr 1982.) What is a property is *being alive*, and this is a property of some complex systems with a certain composition, environment, and structure. There is, however, no concrete system composed of all living beings—present, past, and future—that would itself possess the property of being alive. This must be stated explicitly because some authors' metaphysics allows them to talk of biological taxa as being "parts of life" (Nelson

1989). Although Nelson takes the word 'life' to denote a monophyletic group consisting of all organisms, his use of the part-whole relation presupposes life or "living matter" to be a material system. As this is false, we should bear in mind that talk of 'life' instead of 'living systems' is in most cases elliptical.

The preceding, however, does not exhaust the signification of the term 'life', which has a different meaning in expressions such as 'the life of *b*'. Here, 'life' means the (partial or total) *history* of a biosystem *b*.

Since characterizations of living systems give always rise to controversy, a few remarks concerning Postulate 4.1 will be in order:

1. Clause (i) restricts the notion of a living system to the familiar biosystems on Earth. Some people, from science fiction writers to proponents of Artificial Life, are likely to object that we should allow for a broader conception of life. After all, we do not know how living systems on other planets, if any, might be composed. True, but the biosystems on our planet are the only living things we know, and thus the only ones that require scientific investigation and understanding. Everything else is so far pure speculation. For example, what is called 'exobiology' has no subject matter (yet), hence it is no scientific discipline proper. As for Artificial Life, it deserves a section of its own (Sect. 4.4). Finally, our characterization of living systems is not dogma but postulate (i.e., hypothesis), which can be either discarded or else corrected and improved in the light of future research.
2. Although every system has a more or less definite boundary separating it from its environment (Bunge 1979a, 1983c, 1992), the boundary of living systems is peculiar in that it ultimately involves a biomembrane—even if it is overlain by a cellulose wall, a horn or wax layer, a shell, or what have you. As this comparatively sharp boundary restricts the exchange of substances with the environment, biosystems are semi-open systems, although they are usually said to be open systems. In general, a *semi-open system* is a system which has a boundary that restricts the class of exchanges between the components of the system and the items in its environment. This is why biosystems interact *selectively* with environmental items.
3. Not all of the functions (properties and activities) that we attribute to a biosystem are actually carried out by it at all times during its life history. Metabolism may be temporarily reduced or perhaps entirely suspended, as is the case with spores or dormant seeds, or during anabiosis. Thus, some of the properties of biosystems are dispositions that may actualize under favorable circumstances.
4. We have no use for certain fashionable notions, such as that of "dissipative structure" borrowed from irreversible thermodynamics, because they also characterize a lot of nonliving systems and are thus far too general to distinguish living from nonliving things. (More on thermodynamics and evolution in Sect. 9.3.4.) Similarly, we had no use for the term 'self-referentiality' because we consistently restrict the usage of the notion of reference to semantics. There are more suitable terms, like 'feedback', 'self-regulation' and 'homeostasis' to denote the properties usually referred to by the term 'self-referentiality'. Finally, we refrain from using

the term 'autopoiesis' (Varela et al. 1974) because it seems to be nothing but a fancy synonym for both 'self-organization' and 'self-maintenance'.
5. Some authors believe that a characterization of living systems should include reference to their origin, their evolutionary history, or at least the property of evolvability. We disagree, because "being alive" is an intrinsic property of a biosystem, so that the latter's origin and history are irrelevant. Whether a biosystem originates through self-assembly from abiotic precursors (i.e., by neobiogenesis), whether it was synthesized *in vitro*, or whether it descends from already existing biosystems has no bearing on its status as a living system. (Needless to say, any origin through a supernatural act of creation is ruled out by a naturalistic ontology. Yet even so, such a system would be alive provided it had the necessary composition, environment, and structure.) The same holds for the disposition to evolve. Of course, if biosystems lacked this property there would be neither biologists nor objects of biological investigation today. Still, evolvability is neither a necessary nor a sufficient property of a biosystem.

Note further that the property of descent is not only irrelevant for the reason given in the preceding paragraph. It is also inapplicable, as being factually false. Indeed, a characterization of living systems by the relational property of descent, such as in the axiom "every biosystem is the descendant of a biosystem" (recall the classic principle *omne vivum e vivo*) would imply that there is an infinite sequence of biosystems. However, as the universe and thus our planet are spatiotemporally finite, we must assume the (past) existence of at least one first living system.
6. Many biologists will have expected to find the ability to self-reproduce among the properties listed in Postulate 4.1. (See also Bunge 1979a.) After all, reproduction is essential to evolution, and it is very likely that the ability to self-replicate was a fundamental characteristic of the earliest biosystems on our planet. However, there are several reasons why we think that reproducibility is not a necessary property of a system to qualify as a biosystem. The most important reason is that many biosystems are simply not capable of self-reproduction: think of most living subsystems of organisms, such as tissues and organs, including the reproductive system, if any. Not even all organisms are capable of self-reproduction, such as certain hybrids or the members of certain insect castes. Moreover, the so-called sexually reproducing organisms are not really self-reproducing: it is not the individual but the mating-pair that produces offspring; and, in so doing, it does not really self-reproduce—it does not produce another mating-pair—but merely produces one or more organisms of the same species. Finally, our characterization is ontological, not taxonomic or phylogenetic. That is, although a taxonomic definition of the monophyletic group Life (on Earth) may well have to involve the property of self-replicability, an ontological characterization of biosystems need not. (Why the former is the case will be made clear in Sect. 7.2.2.2.)

However, if these considerations were to be found wanting, nothing prevents both biologists and ontologists from adding an extra clause to Postulate 4.1, referring to the ability to self-reproduce. However, in order to be reasonably comprehensive, such a clause would have to read like this: "*b*, or some subsystem(s) of *b*, or *b* together with some other system *c* in *B* is, at some time during its history, able to produce a system of the same kind or part of a system of the same kind".

If Postulate 4.1 is adopted even with some reservations, then it must be admitted that, although living systems have emerged at the end of a long prebiotic evolutionary process, their appearance was a qualitative jump—just as much as the formation of a molecule out of atoms. (See also Weiss 1973.) In a systemic-emergentist ontology *natura facit saltus*, however small the jumps may be.

This answers the old question in biology and its philosophy, namely whether there is a borderline between living and nonliving systems at all. For instance, many a molecular biologist claims that there is no such line but only a continuum between biomolecules, viroids, viruses, cells, and multicellular organisms. So much so, that the origin of replicating molecules, in particular RNA, is often believed to mark the beginning of life (e.g., Eigen et al. 1981). Yet, according to Postulate 4.1, the property of self-replicability is not even necessary for a system to be alive. Therefore, only the combination of nucleic acid molecules with metabolizing systems marks the "beginning of life" or, more precisely, the emergence of the first biosystems on Earth. (To be consistently pedantic: not being an entity, life has neither beginning nor ending. At most, it has a first and a last member.) Thus, the question of whether replicating molecules evolved first and metabolizing cells second, or whether the two originated the other way round, is irrelevant to the problem of life. (More on this in Küppers 1979; Dyson 1985.) Consequently, there was no multiple origin of life (on our planet), even though the prebiotic components of living systems may have come into existence independently from each other. In other words, however gradual molecular evolution and biogenesis may have been, the emergence of every new or systemic property is a jump. Otherwise, one has to deny qualitative novelty completely. Therefore, our characterization entails that there *is* a dividing line, however thin, between the living and the nonliving, just as there is one between atoms and molecules.

To avoid this conclusion, only two extreme alternatives are possible. The first is *physicalism* or *radical reductionism*, according to which every thing is merely a physical thing, though perhaps a highly complex one. Radical reductionists reject the qualitative distinction between living and nonliving things only at the peril of denying their own lives. Furthermore, it is inconsistent to deny a distinction between living and nonliving things and to call oneself a *biologist* (rather than a physicist or a chemist) at the same time. For example, Keosian (1974) and Dennett (1995) deny that there is a living-nonliving boundary, but then muse over the origin of life. However, one can only speculate over the origin of something if one has an idea what this something is, which is only possible if this something is distinct from everything else.

The second alternative is *hylozoism*, which takes every thing to be alive. (Nowadays this idea is going strong only in the New Age camp. To be fair to the pre-Socratics, we should perhaps better call this version *neohylozoism*.) In attributing the property of being alive to everything, neohylozoism renders the concept of life trivial and superfluous, if not incomprehensible. Indeed, in order to draw the obvious distinction between living and nonliving things, the hylozoist has to assume that biosystems are somehow "more alive" than other things; that is, that there are "degrees of life" or "degrees of being" (e.g., à la Jeuken 1975). If we took such ideas seriously, we would have to say that the dead are not really dead but just a little less alive than the living.

4.3 Elementary Biosystem, Composite Biosystem, and Organism

Having postulated that being alive is an emergent property of certain systems with a certain composition, environment, and structure, we can now attempt to identify the smallest "unit of life" or the smallest "unit of living matter". However, as these expressions are metaphorical, we had better introduce a more exact concept. This will be the notion of an elementary biosystem, which is elucidated by:

> DEFINITION 4.3. An *elementary biosystem* is any biosystem such that none of its components is a biosystem. (More precisely, $x \in B_e =_{df} x \in B$ & $\forall y\ (y \in C(x) \Rightarrow y \notin B)$, where B_e designates the set of all elementary biosystems, B the set of all biosystems, and $C(x)$ the composition of a system x.)

So, which entities are, in fact, elementary biosystems? Let us examine several candidates. According to Postulate 4.1, viruses are not alive because they do not metabolize. In other words, they do not function at all outside some host cell—so much so that aggregates of independent viruses are often crystals. Only the host cell-virus system is alive. (See also Weiss 1973.) The same holds for certain other intracellular parasites such as chlamydiae. As for the proper subcellular components of prokaryotic cells, they clearly do not jointly possess the properties listed in Postulate 4.1.

The situation is more complicated in the case of eukaryotic cells. Mitochondria and chloroplasts, for example, possess many of the properties listed in Postulate 4.1. In particular, they have their own genetic material. The latter, however, is not autonomous in the sense that it regulates all of the system's functions. Indeed, only some functions are regulated by the mitochondrial genic system, while others are regulated by the nuclear genic system (Thorpe 1984). In sum, a couple of properties, no matter how important, do not suffice to characterize a biosystem. Only the *joint* possession of all the properties listed in Postulate 4.1 is necessary and sufficient—until further notice. It seems, then, that, since neither viruses nor any proper subcellular components are biosystems, the cell is the "smallest unit of life". We spell this out in:

> POSTULATE 4.2. All elementary biosystems are cells.

Note that, since the elementary biosystems appear to be cells (or, metaphorically, since "life begins at the cell level"), and since biology is the science of living systems, the expression 'molecular biology' is an oxymoron. In other words, since molecules are not alive, there can be no biology of molecules. Biology proper starts with cell biology. What distinguishes molecular biology from biochemistry is that the former studies molecules *qua* parts of biosystems whereas the latter studies (bio)molecules *qua* molecules. If understood in this sense, the expression 'molecular biology' is harmless.

Note further that the converse of Postulate 4.2 is not true: not all cells seem to be elementary biosystems. For example, many ciliates, which are classified as unicellular organisms, contain unicellular algal symbionts, such as zoochlorellae. Clearly, these algae are not only biosystems but also components (i.e., subsystems) of a ciliate cell. Thus, when containing zoochlorellae, a *Stentor* cell, for instance, is not an elementary biosystem. However, it becomes such as soon as it loses its symbionts. (Note that, in the light of this example, one could make Definition 4.3 more precise by adding the notion of time: a biosystem is an elementary biosystem at time t if, and only if, none of its components is a biosystem at time t.)

Besides *Stentor*, there are of course many more biosystems, such as tissues, organs, and entire organisms, which are not elementary but composed of biosystems. Therefore we have to complement the notion of an elementary biosystem with that of a composite biosystem. The definition of this concept is rather straightforward and reads thus:

> DEFINITION 4.4. A *composite biosystem* is any biosystem composed of (at least two elementary) biosystems.

Note the following points. First, whereas the expression 'composite system' would be a pleonasm, because a system is by definition a composite entity, that of a 'composite biosystem' as defined here is not. Indeed, the definition comprises only living components (subsystems) of biosystems, i.e., biosubsystems. The nonliving components are irrelevant here. Second, just as with elementary biosystems, being a composite biosystem can be time-dependent. Think of most multicellular organisms, which start as elementary biosystems (namely zygotes), but soon become composite. Third, according to this definition, tissues, organs, multicellular organisms, and both unicellular and multicellular organisms containing other organisms, such as endosymbionts or endoparasites, are all composite biosystems. Therefore, the concept of a composite biosystem is neither cointensive nor coextensive with that of an organism. We must therefore define the latter concept separately:

> DEFINITION 4.5. An *organism* is a biosystem (whether elementary or composite) which is not a proper subsystem of a biosystem. (More precisely, $x \in O =_{df} x \in B \ \& \ \neg(\exists y)(y \in B \ \& \ x \lhd y)$, where O designates the set of all organisms, and $\lhd$ the relation of being a proper subsystem.)

In other (metaphorical) words, the organism is the largest "unit of life". As this definition is not only far from obvious but also somewhat unsatisfactory, it requires some explication.

First, note the expression 'proper subsystem'. This occurs because we need to distinguish accidental or alien subsystems of biosystems which are themselves biosystems, such as parasites and symbionts, from their *proper* biosubsystems, such as their own cells and organs. For example, tapeworms, flukes, the above-mentioned zoochlorellae, and many bacteria are biosubsystems of certain organisms, but they are not their proper parts. (For the notion of subsystem recall Definition 1.8.) Hence, according to Definition 4.5, they are themselves organisms. By contrast, both algae and fungus are proper parts of the lichen supersystem, which may thus be regarded as an organism. Therefore, algae and fungi in lichens are biosystems, but not organisms. Only when living separately are they themselves organisms.

All this sounds vague, and is indeed based more on biological intuition rather than a precise concept of proper subsystem. We wish we had been able to find a satisfactory definition of "proper subsystem", so as to properly elucidate the latter before introducing Definition 4.5. So we had no choice but to presuppose the notion of a proper subsystem as a primitive concept.

(Let us briefly explain why certain candidates for a definition of "proper subsystem" fail. A definition of "proper subsystem" seems to be easy in the case of sexually reproducing organisms. Indeed, we could say that all biosubsystems developing from the zygote are proper subsystems of the multicellular organism in question. The problem, however, is that some sexually reproducing organisms sometimes also reproduce vegetatively. Take for instance the fresh-water polyp *Hydra*, which, like *Stentor,* contains unicellular symbiotic algae. Now, members of *Hydra* occasionally produce daughter individuals by budding. Thus, the daughter individuals are already "born" with symbionts and do not develop from a zygote. Examples such as these also preclude a definition of "proper subsystem" in terms of genetic identity. Moreover, parts of a multicellular organism may either undergo somatic mutations or lose genetic material but are still its proper subsystems. Another idea would be to say that everything that contributes to the normal functioning of the system is a proper part of it. Yet, for one, at this stage of our analysis we do not have a concept of function, much less of normality. For another, a spider such as a daddy-long-legs that has lost one of its legs still functions quite well, but we would regard all eight legs as proper parts of a spider. The same holds for vestigial organs, such as the human appendix, which every biologist would regard as a proper part of each member of *Homo sapiens* although we can easily live without it. In sum, it is very difficult to come up with a biologically satisfactory definition of "proper subsystem of a biosystem".)

Second, Definition 4.5 refers to an entity which Haeckel (1866) called *bion* (physiological individual) as opposed to *morphon* (morphological individual). Accordingly, a colony of physiologically interconnected morphological individuals can be regarded as an organism. Think, for instance, of a strawberry patch or certain corals. However, as soon as the *morphonta* separate from each other, we have different organisms (i.e., *bion* = *morphon*). Thus, a clone of strawberry plants is not an organism just because the individuals ultimately originated from a common zygote; neither is a clone of dandelions or aphids—*pace* Janzen (1977). After all, there is no metabolism at a distance.

Third, as the organism is by definition the largest living entity, there can be no such thing as a superorganism that is itself a living system. Hence, all supraorganismic systems, such as reproductive communities (populations) or social groups, are nonliving entities. A beehive, then, often quoted as the paragon of a superorganism, is not a living system. Furthermore, we cannot follow Wilson and Sober (1989), who regard, for instance, communities as superorganisms because they show functional organization. Functional organization is certainly a necessary but not a sufficient property of a biosystem. Therefore, communities are systems but neither biosystems nor organisms. (See also von Bertalanffy 1952.) Thus, a forest teeming with plants, fungi, animals, and bacteria is not alive. This holds, *a fortiori*, for the biosphere and the ecosphere (see Definitions 5.2 and 5.4).

As a consequence, the strong Gaia hypothesis, that is, the idea that the biosphere (actually: ecosphere) is a living (super)organism, is wrong. Moreover, it is misleading and superfluous. For one, it inspires mystics and New Agers: see Gardner (1989), Levine (1993). For another, we can have all ecological research and theory without recourse to this notion. Indeed, at present the scientific content of the Gaia hypothesis seems to boil down to the idea that the ecosphere is a self-regulating system (Kump 1996)—an idea which is plausible but still *sub judice*. (For a critical, though atomistic, view see G.C. Williams 1992b.)

4.4 Artificial Life

While many biologists are likely to agree more or less with our preceding characterization of biosystems, a certain group of physicists, chemists, engineers, and computer technologists will be deeply dissatisfied with it, namely the proponents of Artificial Life (AL). More precisely, the dissatisfaction will be expressed only by the proponents of the *strong* AL program. For, just as with Artificial Intelligence (AI), a strong version of AL should be distinguished from a weak version (Pattee 1989; Sober 1991). While *weak* AL attempts to contribute to the understanding of living systems by means of mechanical models and computer simulation, the proponents of the *strong* program are much more ambitious: they hope to *synthesize* living systems. However, strong AL is not so much concerned with the biochemical *in vitro* synthesis of biosystems. Rather, the goal of strong AL is to synthesize life-forms alternative to the "carbon-chain life" as known to biology "by attempting to capture the behavioral essence of the constituent components of a living system, and endowing a collection of artificial components with similar behavioral repertoires. If organized correctly, the aggregate of artificial parts should exhibit the same dynamic behavior as the natural system." (Langton 1989, p. 3).

We have no quarrel with the weak version of AL (and likewise AI), for modeling concrete systems, whether theoretically or practically, is a legitimate, nay necessary scientific activity—provided the models and simulations are relevant, interesting, and fruitful with regard to the solution of actual biological problems. On the other hand, we believe that strong AL is wrongheaded since it rests on ontologically flawed presuppositions. (Similar arguments could be made against strong AI.) Let us therefore examine strong AL more closely.

According to Langton (1989), one of the central concepts of AL is that of emergent behavior. Indeed, as we saw in the preceding section, being alive is an emergent property of certain material systems with a special composition, environment, and structure. The claim of AL is that the composition of the systems in question is irrelevant to the emergence of the property "alive": " ... the ontological status of a living process is *independent* of the hardware that carries it" (Rasmussen 1991, p. 770). In other words, the approach of AL, like that of strong AI, is functionalist: all that would matter is the right organization. Consequently, their notion of emergence is stuff-free: it is a concept of "emergence out of the blue". Thus, it is similar to the (purely logical, not ontological) concept of supervenience, according to which the (alleged) supervenient properties of the whole are independent of the properties of its components (see Sect. 1.7.3).

The problem with this view is that the emergent properties of a system, hence the processes it may undergo, depend lawfully on the properties of its components. For example, when an oxygen molecule combines with a carbon atom, CO_2 results, and when we combine the former with a silicon atom we obtain SiO_2. The reaction products have different emergent properties although carbon and silicon share some physical and chemical properties, so that they belong to the same chemical genus—the so-called carbon-group—in the Periodic Table. Nevertheless, at normal pressure and temperature, an aggregate of CO_2 molecules is a gas, whereas an aggregate of SiO_2 molecules is a solid body—a quartz crystal. So, the combination of these molecules into more and more complex systems will result in wholes possessing very different emergent properties. We submit that the same is true for silicone (SiO), although it shares certain properties with carbon. Thus, matter does matter after all. (At least some critics of AL share a similar view: see Emmeche 1992.) So whatever nonbiochemical materials are used in synthesizing artificial life, we are afraid that anything but a living system will emerge: mimicking one or the other emergent property of a biosystem does not suffice to create a biosystem proper.

In addition to the ontological problem of emergence, AL faces the following epistemological problem. The biosystems on our planet are the only living things we know. Hence, whatever artificial system is presumed to exhibit some property or properties of (genuine) biosystems, it can be compared only to the biosystems known to biologists. Thus, AL technologists cannot show that they have succeeded in constructing genuinely *alternative* life forms, because such artificial systems, if exhibiting emergent properties others than those known from life on Earth, could be said to be alive only *by definition*: they would not be *found* to be alive but *declared* to be alive. Hence, *pace* Langton, it is not just "extremely difficult" to distinguish universal properties of biosystems (as they could be) from those common to life on our planet: it is *de facto* impossible. Since AL's *life-as-it-could-be* is in fact *life-by-definition*, strong AL is irrelevant to theoretical biology.

Yet let us return to ontology, for we still have other metaphysical bones to pick with strong AL or, more precisely, with the computational version of strong AL. While strong AL in general is concerned with creating artificial life in any substrate, its computational version claims that not so much certain machines but certain computer processes can be alive. Witness, for instance, the following statements by Langton (1989):

> [AL attempts] ... to synthesize life-like behaviors within computers and other artificial media. (p. 1)
>
> [AL] ... views life as a property of the organization of matter, rather than a property of the matter which is so organized. Whereas biology has largely concerned itself with the *material* basis of life, Artificial Life is concerned with the *formal* basis of life. (p. 2)
>
> The claim is the following. The 'artificial' in Artificial Life refers to the component parts, not the emergent processes. If the component parts are implemented correctly, the processes they support are *genuine*—every bit as genuine as the natural processes they imitate. The *big* claim is that a properly organized set of artificial primitives carrying out the same functional roles as the biomolecules in natural living systems will support a process that will be "alive" in the same way that natural organisms are alive. Artificial Life will therefore be *genuine* life—it will simply be made of different stuff than the life that has evolved here on Earth. (p. 33)

Clearly, all this reeks of Platonism. Just as the mind-body dualists speak of the material basis of the mind (see Chap. 6), as if humans were two-storied, so does AL with respect to life. For them, life seems to be an immaterial form that is "realized" or "instantiated" or "embodied" in certain material systems. Curiously, however, Langton also speaks of the *formal* basis of life, which suggests that life can also be instantiated in form. Is life, then, a third category in addition to matter and form? In any case, the emergentist-materialist option, according to which properties are not separable from things, and life is a collection of specially structured (formed) and thus particularly changing (i.e., living) material systems, seems unacceptable to AL.

According to the second quotation from Langton, being alive is not a (substantial) property of a material system but a property of a property, namely its organization (structure). This presupposes an ontology that allows for second-order substantial properties. (Recall that we admit only second-order predicates, not properties: see Sect. 1.3.3). By contrast, according to the third quotation (as well as to the above quotation of Rasmussen), being alive is a property of a process. Now, processes involve properties but they have none: recall that processes are sequences of events, which are changes of state of things, which involve properties of things

(Sects. 1.3-5). For this reason, speaking of properties of processes is an abstraction: only changing things have properties. For example, velocity is a property of a moving thing, not a property of its movement. Similarly, being alive is not a property of a process, such as metabolism, but a property of a changing (i.e., metabolizing) system. Indeed, a (natural) organism *is* not a process, it *undergoes* processes.

Since, in our ontology, a computer process is a sequence of changes of state of a computer, only the computer as a specifically structured (programmed) whole could be the (artificial) entity possessing the property of being alive. To say, then, "x lives" means that a *thing* x—whether an organism or a computer—undergoes, by virtue of its properties and hence states, a certain process (rather than another). This process is the life (i.e., history) of x. Thus AL's claim that a process in a computer x, rather than the computer x itself, is alive amounts to saying something like 'The life of x lives', which makes of course as little sense as the sentence 'The motion of thing x moves'.

What does it mean to say that correctly implemented artificial parts will support processes that are every bit as genuine as the processes they imitate? First of all, this statement presupposes that processes are detachable from things—otherwise, the same process could not be "supported" by different systems. (Here, 'genuine' seems to mean "same" or "identical".) According to our ontology, two things cannot undergo (exactly) the same process, because a process is a sequence of changes of state of a thing. But of course, two things equivalent in certain respects, e.g., things of the same kind, may undergo equivalent processes. This is why we can abstract *process classes* from changing things that are equivalent in certain respects. Examples: metabolism, evolution, selection. Thus, the correct question is whether certain specially structured (i.e., programmed) computers are sufficiently equivalent to biosystems to be able to undergo changes that belong in the same process class (i.e., life).

Second, a serious semantical and ontological mistake is involved in the failure to distinguish an imitation (or simulation) from the thing or process imitated. How can an imitation (or simulation) be as genuine as the process (or system) it imitates? So Langton either forgets the meaning of the concept "imitation", or he must have access to a wondrous metaphysics in which there is no distinction between a model and the thing modeled. As Pattee (1989, p. 68) put it succinctly: "A simulation that becomes more and more 'life-like' does not at some degree of perfection become a realization of life." Worse, if Langton's view were true, we could perform miracles. For example, since it is possible to simulate or model nonexisting systems, such as a geocentric planetary system, we could even realize nonexistents. If we take strong AL seriously, the "realized" computational geocentric planetary system would be as genuine as the nonexistent one it imitates. The problem is of course: What is a *genuine* nonexisting thing or process? Given this genuine nonsense—at least from the perspective of scientific materialism—it comes as no surprise that AL seeks salvation in a version of objective idealism such as J.A. Wheeler's information Platonism, according to which (a) the calculus of propositions is regarded as "the basis of everything", (b) physics might be formulated in terms of information theory, and—last but not least—(c) "Matter [....] can be derived from information processing" (Rasmussen 1991, p. 771). Indeed, these basic assumptions "must be true if the larger claims are to be true" (Langton 1991, p. 20). *Sapienti sat.* (Not surprisingly, the center of AL research is the *Santa Fe* Institute: after all, "santa fe" means "holy faith". A very strong dose of the latter is exactly what is needed to believe in the strong AL program.)

To conclude, weak AL—if it aims at solving (real) biological problems by means of computer simulation or mechanical models (rather than just playing computer games)—is unobjectionable. By contrast, although we can, of course, not rule out that, on an alien planet, there may be living systems of a somewhat different composition and structure as that known to biology, strong AL seems to be flawed and confused beyond repair and thus a waste of energy, time, and money. So let us quickly return to real biology.

4.5 Biospecies and Biopopulation

It is well known that organisms may assemble to form either aggregates or systems—often called 'higher-order individuals'—such as groups, demes, populations, avatars, communities, and so on. Needless to say, there is no agreement as to the adequate characterization of these entities (see, e.g., MacMahon et al. 1978, 1981; Bunge 1979a; Damuth 1985; Eldredge 1985a; Salthe 1985). In particular, the term 'population' is highly ambiguous. It can mean either (a) a *statistical population*, i.e., a mere collection of individuals, such as in the expressions 'the population of gorillas and whales' and 'the population of HIV-infected children in 1997'; or (b) an *aggregate* (or heap) of individuals, such as in the expressions 'the bacteria population in this petri-dish' and 'the fish population in this pond'; or (c) a *system* of individuals, as is the case with reproductive communities or animal societies.

As for (a), since statistical populations are collections or classes, they have members, rather than parts. Statistical populations are thus not concrete entities, but constructed according to some research interest. As for (b), the organisms in question just happen to occupy the same locality without being coupled together by bonding relations (see Definition 1.7). They thus form an aggregate of organisms—a heap—but not a system. Mere spatial proximity of organisms does not guarantee their being bonded together by biological relations. For this reason, there is no such thing as the "spatial integration" of a population—*pace* Damuth (1985). Still, aggregates are things. In order to constitute a system, however, the organisms in question must be somehow bonded together by biological links. Only thus can their assembly lead to a cohesive and integrated entity, i.e., a system. Such bonds may, for instance, consist in mating relations, such as in reproductive communities, or in symbiotic or social relations.

Now, aggregates and systems of organisms, i.e., concrete populations, may consist of organisms of either the same or different species. In other words, there are unispecific and multispecific aggregates and systems of organisms. This being so, the concept of a species is logically prior to, and independent of, any concept of aggregate and system of organisms—in particular that of population. Therefore, an elucidation of the concept of a population-as-a-concrete-individual composed of organisms must be preceded by a definition of the concept of species. Let us then propose such a definition of the concept of biological species, or *biospecies* for short, where the prefix *bio* is due only to the fact that we deal with species of *biosystems*: it does not imply any relationship to Mayr's classical concept of biospecies, of which more in Section 7.3.1.1. Our definition of "biospecies" reads thus:

DEFINITION 4.6. A species is a *biospecies* if, and only if,
(i) it is a natural kind (rather than an arbitrary collection); and
(ii) all of its members are organisms (present, past, or future).

This definition of biospecies as classes evidently contradicts the widespread view that biological species are in some sense concrete individuals (see, e.g., Ghiselin 1974, 1981, 1984; Hull 1976, 1978, 1988; Mayr 1982, 1988; Rosenberg 1985; Sober 1993). As we shall deal with species and taxa in general in Chapter 7, it may suffice here to note that, like any other species, a biospecies is a collection, hence a conceptual object. However, as we shall also see in Chapter 7, biological species are not natural kinds in the traditional sense, so that we have to construe a definition of "kind" in tune with evolutionary biology.

In any case, we need a species-as-kind concept for logical reasons, because the notion of a species as a kind logically precedes any notions of supraorganismic aggregates or systems composed of organisms. As mentioned above, one such supraorganismic entity which is not a collection of organisms but either an aggregate or a system of such is a biological population. More precisely, we propose:

DEFINITION 4.7. A concrete aggregate, or else system, of organisms is a *biopopulation* if, and only if, it is composed of organisms of the same biospecies (i.e., iff its composition is unispecific).

To be more precise, the unispecific composition of a biopopulation may be formalized as follows: if S designates a species of organisms, and p a biopopulation, then the composition of p at the organismic level O can be written as $C_O(p) \subseteq S$. Besides unispecific populations there are also multispecific populations such as communities. We shall deal with such multispecific systems in Section 5.1.

Note the following points about Definition 4.7. First, inasmuch as biopopulations are systems, we are not concerned with the nature of the bondage among its components. Thus, our definition of "biopopulation" subsumes a vast array of more or less cohesive wholes: from the minimal system of a mating-pair through social groups and demes to the so-called most inclusive biopopulation, often wrongly termed 'species'. (For a comparison of different concepts of population in biology see Jonckers 1973.)

Second, a possible pitfall concerning the ambiguity of the term 'population' is the following. It is often useful to treat the composition of a biopopulation—a concrete system—as a statistical population—a collection, for example, when calculating average fitness values of organisms. Such averages are attributes of the statistical population, i.e., the collection of organisms in question, but they are not substantial properties of the biopopulation as a concrete whole or individual. For example, it is not the biopopulation of California Redwoods as a concrete individual that has an average height. Beware of confusing aggregates and systems—concrete things—with their composition—a collection. We submit that part of the problem of whether populations are either individuals or classes, or perhaps both (e.g., Van Valen 1976a), seems to rest on this confusion.

Having thus far dealt with biological systems in a rather static manner, we proceed to take a look at the changes biosystems undergo, that is, at the activities or functions of living systems.

4.6 Biological Function and Biological Role

Just as in ordinary language, the concept of function in biology is heavily loaded with teleological connotations. So much so, that often the term 'function' can be replaced by 'purpose' without any change in meaning. On the other hand, most authors are at pains to point out that, when using the notion of function, they invoke neither intention nor purpose. As is the case with teleology in general, we are assured that functional talk in biology is at the same time legitimately teleological but somehow not really so. Our aim, then, is to analyze the concepts of function and role, and to find out whether they have any teleological import.

We shall follow earlier suggestions to distinguish the function of a subsystem of a biosystem from its biological role (see, e.g., Woodger 1929; Bock and von Wahlert 1965; Pirlot and Bernier 1973; Bernier and Pirlot 1977; Amundson and Lauder 1994). Roughly, the *function* of such a subsystem is what it does: its *functioning* or *activity*. In other words, the function of a given organ is the set of processes occurring in that organ (Bunge 1979a, b). More precisely, we propose:

DEFINITION 4.8. Let b denote an organism, and $a \lhd b$ a subsystem of b of kind A. Further, call $\pi(a)$ the totality of processes or activities that a is undergoing during a certain period. Then

(i) any subset of $\pi(a)$ that includes any of the processes listed in Postulate 4.1 (as well as reproduction, if any), or that affects them in any way, is a *biological function*;

(ii) the *specific* biological functions π_s of a are those performed by a and its likes (i.e., by the members of A) but not by any other subsystems of b. (More precisely, $\pi_s(a) = \pi(a) - \bigcup_{x \lhd b} \pi(x)$, with $a \neq x \lhd b \;\&\; x \notin A$.)

Example 1. Let us begin with a rather unfamiliar example because it might be viewed more objectively than any of the famous examples in the literature on function and teleology. The members of the insect taxon Heteroptera, the so-called *true bugs*, are characterized by the possession of a thoracic scent gland (Schuh and Slater 1995). In a subtaxon with aquatic members, the pygmy backswimmers, this gland produces a secretion containing hydrogen peroxide (Maschwitz 1971). According to Definition 4.8, all the processes involved in the production, storage, and release of that secretion constitute the specific *function* or *activity* of this gland.

Now, the secretion of the gland is spread over the body surface once in a while by a behavior called 'secretion-grooming' (Kovac and Maschwitz 1989). The secretion has an antiseptic effect and thus protects the insects' fine hair layer from contamination by microorganisms that could interfere with its air-storing capacity, which, in turn, affects respiration. This is what the gland does in relation to the supersystems in which it is embedded. In other words, this is the *biological role* of the gland. (Thus, in aquatic bugs, it is actually not a *scent* gland proper.)

As the secretion of the thoracic scent glands among other aquatic bugs is different from that of the pygmy backswimmers, the function of the formers' glands is likely to be different, too. This is because different secretions will usually be produced by different biochemical processes. Yet the role of the glands is the same, that is, the secretion is used as an antiseptic solution. In some land bugs, on the other hand, the gland is often a genuine scent gland because its secretion may be used as a repellent. In short, an organ of a certain kind characterizing a given taxon may exert different functions and roles within the members of different subtaxa. (See also Mahner 1993b.)

Example 2. After this innocent case, let us now turn to one of the most famous examples in the literature. Most believers in teleology maintain that the function of the (mammalian) heart would be the circulation of blood but not the production of heart sounds (see the classical locus Hempel 1965; for an early dissenting view see Bernier and Pirlot 1977). According to Definition 4.8, just the opposite is the case. The function (activity) of the heart consists in the performance of rhythmic contractions, but not in the circulation of blood. The latter is the *role* the heart performs in the circulatory system as well as in the organism, i.e., in the supersystems of which it is a part. By contrast, one of the activities of the heart clearly is the production of sounds. The question, however, is whether this function plays any significant role in nature, i.e., in a nonmedical context.

Example 3. What is the function of a deer's antlers? Apparently, antlers show no significant activities or processes besides those which consist in the development and maintenance of this organ. Thus, they are (almost) functionless organs. However, they play—by virtue of their mere presence—an important role in the social life of the individual. In a similar category are, for instance, the rhinoceros's horn, the peacock's tail, the shells of mussels or tortoises, and so on.

Example 4. The human appendix is usually regarded as a functionless organ because it no longer helps break down cellulose. However, though vestigial, the appendix is not functionless: it contains lymphatic tissue and thus carries out immunological activities. What is true is that it plays only a negligible role in the body's immune system, so that its loss upon surgical removal hardly affects the person's survival or vitality. But, however small any role of an organ may be, it still is a role.

From the preceding examples we can conclude that, strictly speaking, there are no features without *any* function (activity, *Funktionieren, fonctionnement*) and biological role (*Fungieren, fonction*). However, there are features of biosystems without a *significant* (or important) function but with a significant role, and there are features which have neither a significant function nor a significant role. But the problem remains how this significance is to be characterized and measured. Furthermore, it will be apparent that what we call 'biological role' is usually termed 'function'. As this latter sense of 'function' is easily associated with 'purpose', whether literally or by analogy, it is a prime source of biophilosophical dispute. Hence, to eliminate this teleological sense of 'function' is a first step towards clarification.

The preceding examples have used the concept of biological role only intuitively by assuming that the role of a feature is what it does in a given supersystem. Accordingly, whereas the function of a feature can be understood as an intrinsic property of the system in question, its biological role is a relational property. That is, the role of a system can be understood only in relation to a supersystem of which it is part. We thus suggest:

DEFINITION 4.9. Let b denote a biosystem, $a \lhd b$ a subsystem of b of kind A, and e a supersystem or an environmental item of a or b. Then,

(i) the *biological role* of a in e, or in relation to e, is the set of bonding relations or interactions between a and e, i.e., its bonding external structure;

(ii) the *specific role* of a is the set of bonding relations or interactions between a and e that only a and its likes, i.e., the members of A, are able to hold but not any other subsystem of b.

Note the following points. Firstly, in contradistinction to the conception of role suggested by Bock and von Wahlert (1965), ours not only comprises ecological roles, but also intraorganismic roles such as mechanical and physiological roles.

Secondly, neither the concept of biofunction nor that of biorole has teleological connotations, contrary to what is implicit in the traditional notion of the "proper function" of an organ. That is, in our construal, neither the (specific) function nor the (specific) role of an organ are regarded as what the organ is "supposed" to do, or, in other words, as the purpose or the goal of the organ. Furthermore, neither the concept of specific function nor that of specific role have anything to do with normality or health, that is, with the normal function or the normal role of an organ.

Thirdly, as suggested by Cummins (1975), we could as well formulate our definitions of 'function' and 'role' in terms of capacities or dispositions rather than actual activities. This may be useful if we wish to refer to unused functions and roles. Thus, Definitions 4.8 and 4.9 do not imply an actualistic ontology: we concur with Aristotle that anything that occurs must be possible to begin with. However, defining "function" and "role" in terms of capacities makes it difficult to apply the notion of biological value to functions and roles (see below).

Fourthly, our concepts of function and role have thus far been developed independently of evolutionary considerations. In other words, they are ahistorical concepts. In this respect our view is similar to that of, e.g., Simpson (1953), Cummins (1975), Nagel (1977), Prior (1985), Bigelow and Pargetter (1987), Amundson and Lauder (1994), and Wouters (1995). This view is rejected by the so-called "etiologists", who maintain that only "functions" (i.e., functions in our sense *cum* roles) that are the result of natural selection and thus of adaptation should be regarded as genuine or proper functions. If a function or a role of a feature is not the outcome of selection, it would be a mere "effect", not a "function" proper. (See, e.g., G.C. Williams 1966; Ayala 1970; Wright 1973, 1976; Brandon 1981, 1990; Gould and Vrba 1982; Millikan 1989; Neander 1991; Griffiths 1993.)

Etiologists would consider Definition 4.9 as inadequate because, according to it, it is, for instance, a role ("function") of the human nose to support spectacles. Yet we contend that this is just the strength of the definition with regard to evolutionary biology: the etiologists either overlook or else downplay the fact that the notions of selection and adaptation logically presuppose the presence of something, namely a feature with a certain function or role or both, which can become the subject of selection and adaptation. Since, according to Definition 4.9, the role of a feature is independent of any adaptive value, there is room for the change of role, by a change either in the feature itself or in the environment, or both. Thus, a feature can more or less easily acquire a new role which can then become the subject of selection. The etiologists further neglect the importance of disciplines such as functional morphology, physiology, and ecology, which presuppose no knowledge of evolution. (See also the recent criticism of the etiological view by Amundson and Lauder 1994.)

To arrive at what the etiologists call the 'proper function' of a feature, we must supplement our analysis of the concepts of function and role with the concept of biological value. This is because the biovalue of a feature will clearly influence its fate in the selection process. And only by presupposing the ateleological and ahistorical concepts of function, role, and biovalue can we proceed to elucidate the notion of adaptation.

4.7 Biological Value

The biological functions and roles, in particular the specific functions and roles, of a subsystem of a biosystem may be valuable, disvaluable, or indifferent to the organism as a whole (Canfield 1964; Ayala 1970; Ruse 1973; Hull 1974; Woodfield 1976; Bunge 1979a, 1989). That is, they either contribute to its health, vitality, performance, or survival within the bounds of its species-specific life history; or they do not; or they are even detrimental to it. The same holds for the items in the organism's environment. We compress this idea in:

> DEFINITION 4.10. If *a* denotes a feature (organ, process, role, etc.) of an organism *b*, or some (biotic or abiotic) item in the environment of *b*, then *a* is *valuable* to *b* if, and only if, the possession of, or access to, *a* favors the ability of *b* to undergo its species-specific life history. Otherwise, *a* is either indifferent or disvaluable to *b*.

Note the following points. First, by referring to an organism's ability to undergo its species-specific life history (or life cycle), this definition makes room not only for survival but also for reproduction, if any. Indeed, a definition of "biovalue" solely in terms of survival would have to consider the reproductive system of an organism, or the specific function of reproduction respectively, as either worthless to the organism in question, because it does not contribute to the survival of the individual; or even as detrimental, as is the case with certain animals and plants in which copulation or reproduction is (naturally) followed by the death of the parental organism(s): think of mayflies and salmon. A more drastic example are the gall midges of the family Cecidomyidae (Insecta, Diptera) in which the offspring remain inside their mother and start to devour her from the inside after hatching (see Gould 1977). By referring to the species-specific life cycle in our definition of "biovalue", we can even accommodate cases like these as being valuable to the organism. At the same time, our definition precludes a sociobiological-geneticist view of biological value in terms of, say, "whatever contributes to the spread of an organism's genes".

Second, Definition 4.10 refers to individual organisms. It makes room for features that are unique to the individual in question and therefore may give it an advantage or a handicap. But it makes no reference to the biopopulation of which the organism may be part, or to the biospecies to which the individual belongs. To be sure, biologists often talk about something being valuable or disvaluable to a *species*, but this is mistaken, for species are not concrete things but collections of such, and collections are conceptual objects (see Sects. 1.2, 4.5, and 7.2), hence they are not in the survival game. Therefore the phrase '*X* is valuable to species *Y*' must be understood as short for '*X* is valuable to every member of species *Y*'. After all, species are not biosystems, hence there is no survival and no life history of species but only careless talk. (See also Cracraft 1989.) Neither is there a survival or a life cycle of a biopopulation because biopopulations, though concrete systems composed of biosystems, are not alive (see Sect. 4.3). At most, we can speak of the continued existence of a biopopulation. Indeed, it may be valuable to certain organisms to be part of a biopopulation, such as a reproductive community or a social system. Thus, the perpetuation of the biopopulation (as an environmental item) is also valuable to the organisms in question. Yet this case is covered by Definition 4.10.

Third, according to Definition 4.10, biovalues are relational properties, namely relations between organisms (as wholes) and either one of their subsystems, or else some item in their environment. Thus, there are no intrinsic or absolute biovalues, so that a statement of the form "*a* is valuable" would be ill-formed. Indeed, the concept of biological value must be construed as an at least binary predicate, such as "*a* is valuable to *b*", or "*Vab*" for short. A more precise construal would consist in formulating "biovalue" as a quaternary predicate such as in the statement schema "*a* is valuable in respect *b* to organism *c* in some circumstance *d*".

Fourth, our concept of biovalue is objective in the sense that it does not presuppose that the organism to which an item is valuable has any cognitive abilities, in particular that it is able to make value *judgments*.

Fifth, the contribution of a feature to an organism's ability to undergo its species-specific life history (comprising such properties as survival, health, vitality, performance, and reproductive capacity) comes in degrees. Thus, we could formulate a quantitative measure of biovalue (see, e.g., Bunge 1979a, 1989). Suffice it here to note that the biovalue *V* of a feature may be represented by a real number ranging between –1 (maximally disvaluable) and 1 (maximally valuable). If a feature is neutral it is assigned the value 0.

4.8 Adaptation

The notions of biofunction, biorole, and biovalue allow us now to elucidate the concepts of adaptation and adaptedness. As with the former concepts, most notions subsumed under the label 'adaptation', as well as the concept of adaptedness, do not presuppose any evolutionary concepts. In other words, all these concepts are—with one exception—logically prior to evolutionary notions. This exception is just one of the eight concepts hidden under the common label 'adaptation' that we proceed to analyze in the following.

4.8.1 Eight Senses of 'Adaptation'

The term 'adaptation' is highly ambiguous because it designates several different, though related, concepts. The resulting confusion in the literature is increased by the fact that the term 'fitness' is occasionally used synonymously with one of the senses of 'adaptation'. (See, e.g., Simpson 1953; Pittendrigh 1958; Bock and von Wahlert 1965; G.C. Williams 1966; M.B. Williams 1970; Munson 1971; Lewontin 1978; Bunge 1979a; Gould and Lewontin 1979; Bock 1980; Gould and Vrba 1982; Burian 1983, 1992; Mayr 1988; Brandon 1990; West-Eberhard 1992.) In fact, we may distinguish at least eight different senses of 'adaptation':

1. *Adaptation*$_1$ is what Simpson (1953), as well as Bock and von Wahlert (1965), call 'universal adaptation'. It refers to the fact that a living being cannot exist separate from any habitat. (Recall the distinction between "habitat" and "environment" from Sect. 1.7.) Hence, to say that an organism is adapted$_1$ amounts to saying that

it is *alive in a given habitat.* Universal adaptation, then, is a nonspecific and phenomenological concept. So much so, that it is not only a concept applicable to living things, but to *all* things, whether living or nonliving. Since every thing is either able to exist in a given habitat or not, the concept of universal adaptation has an ontological scope. Yet it is not as trivial as it appears to be, because it does not refer to existence *per se* but to existence with regard to some particular habitat. For example, a piece of iron can persist in a vacuum, but not in a bottle of hydrochloric acid. Needless to say, the range of possible environments of biosystems is rather narrow compared to that of nonliving things.

2. *Adaptation$_2$* occurs in the physiology of sense organs. It refers to the alteration in the degree of sensitivity of a sense organ depending on the intensity of the stimulus. An example is the adjustment of the eye to vision in bright or dim light. (To anticipate further notions of adaptation, the capacity of adaptation$_2$ or adaptability is an adaptation$_4$ as well as most likely an adaptation$_6$).

3. *Adaptation$_3$* refers to the physiological processes by which an organism may adjust itself to a changing habitat. The capacity to do so is also called 'adaptability'. Related terms are 'physiological adaptation', 'phenotypical adjustment', 'acclimation', or 'modification'. Adaptation$_3$ may be reversible or irreversible. For example, humans and other mammals can adapt to higher altitudes by increasing the number of erythrocytes, which compensates for the decrease in atmospheric oxygen (reversible). Some plants, such as dandelions, grow in different forms depending on the altitude (irreversible). Cuttlefish and some other animals may change colors depending on their habitat (reversible). The head shape of water-fleas depends on the temperature of the water in which they develop; accordingly, there is a seasonal variation of head form, which is called 'cyclomorphosis' (irreversible). (Further examples in Sudhaus and Rehfeld 1992.)

The degree of adaptability of an organism clearly contributes to its adaptedness (adaptation$_5$), and phenotypic plasticity itself may be a result of adaptation$_7$, hence it may be an adaptation$_6$ (Bock and von Wahlert 1965; West-Eberhard 1992).

4. *Adaptation$_4$* refers to any subsystem of a biosystem that performs a biological role with a positive biovalue with respect to some environmental item(s). Thus, fins are an adaptation$_4$ of some vertebrates with an aquatic mode of life, and the feet (tarsi) of head lice are an adaptation$_4$ enabling them to cling to human hair. Clearly, this is a concept of functional morphology regarding—to put it anthropomorphically—the "engineering adequacy of design" (Burian 1983) of an organ in relation to the items in its environment without any recourse to its history.

5. *Adaptation$_5$* is a related notion that refers to the state of adjustment of an organism to the items in its environment. Adaptation$_5$ is often—and more adequately—called *adaptedness.* (This is also one of the senses of the traditional term 'fitness'.) Adaptedness is a relational and quantitative property of an organism or, rather, a property of the organism-environment system. Accordingly, an organism's adaptedness may be altered either by a change in the organism or by a change of some item(s) in its environment, or both. Of course, environmental items may also change the organism, and the organism may change them. Thus, the notion of ad-

aptedness presupposes neither a passive organism nor a static habitat (Simpson 1953; Lewontin 1983a, b; Levins and Lewontin 1985).

When the degrees of adaptedness of two or more organisms are compared to each other, we arrive at the notion of *relative adaptedness*. In order to apply the notion of relative adaptedness, it is not necessary to propose specific values of the degrees of adaptedness of the two (or more) organisms to be compared: all that matters is the better-worse distinction with regard to some common environmental item(s). (More on adaptedness in Sect. 9.2.1.)

6. *Adaptation*$_6$ concerns organismal features whose specific role has contributed to the selective success of their bearers. In common teleological and anthropomorphic parlance, adaptations$_6$ are features that have been "designed by" natural selection "for" a given role. Although all adaptations$_6$ in a given habitat h are also adaptations$_4$ in h, not all adaptations$_4$ in h need be adaptations$_6$ in h. For example, a new feature due to a chance mutation, or which is merely epiphenomenal due to some pleiotropic correlation, may be adaptive$_4$ but it is not the result of selection. Of course, if it is adaptive$_4$ to begin with, it may soon become adaptive$_6$. At first sight, it might seem that there are also features which are adaptations$_6$ but not adaptations$_4$. For example, an adaptation$_6$ with regard to some environmental item e may no longer be an adaptation$_4$ if e changes into e'. However, it is neither an adaptation$_6$ with regard to e' because this case refers to two different environmental items. If we refer to a common environmental item e, all adaptations$_6$ are adaptations$_4$.

7. *Adaptation*$_7$ refers to the process of evolution by natural selection that produces adaptations$_6$.

8. *Adaptation*$_8$ is the operationalist concept of adaptation as fitness in the sense of Darwinian fitness or reproductive success. Clearly, adaptation$_8$ is, at best, an *indicator* of adaptation$_5$. Since this operationist concept of adaptation or fitness is the source of the famous tautology objection to the theory of natural selection, most contemporary authors acknowledge its inadequacy. We shall therefore disregard it henceforth (more on all this in Sect. 9.2).

The notions of adaptation relevant to the theory of evolution are the first, fourth, fifth, sixth, and seventh outlined above. In order to avoid indexing, it may be convenient to introduce new terms for the most relevant concepts in question. In so doing, we shall partly follow Gould and Vrba (1982), who suggested distinguishing *aptations* (adaptations$_4$, *Passungen*) from *adaptations* (adaptations$_6$, *Anpassungen*). As stated above, an aptation may, but need not, be an adaptation. For example, the ability of the (normal) human brain to recognize and distinguish hundreds of faces of conspecifics is, most likely, an adaptation of a highly social animal. By contrast, the ability to write books on the philosophy of biology is certainly not an adaptation of human brains but may, at best, be an aptation in a certain intellectual habitat.

Let us propose some more precise definitions of the various relevant notions of adaptation.

4.8.2 Aptation and Adaptation

To elucidate the concept of $adaptation_4$ or aptation, we need to recall the concept of biovalue from Definition 4.10. We can then say that, if T designates a set of time instants, then the biological *value* of features (i.e., subsystems, processes) of kind A for organisms of kind B with regard to some environmental item e of type E is representable as the function $V: A \times B \times E \times T \rightarrow \mathbb{R}$ such that, for any a in A, b in B, e in E, and t in T, $V(a, b, e, t)$ is the value of feature a for organism b in relation to some environmental item e at time t. Recalling that this value may range between –1 (maximally disvaluable) and 1 (maximally valuable), we propose:

DEFINITION 4.11. A feature a of an organism b is an *aptation* in relation to some environmental item e at time t if, and only if, $V(a, b, e, t) > 0$.

We submit that the concept of aptation ($adaptation_4$) is the most commonly used in biology, in particular in functional morphology, physiology, and ecology. In these disciplines the evolutionary origin and history of a feature are only of secondary interest. Only in evolutionary biology, in particular in evolutionary ecology, are we interested in whether aptations are also adaptations. (See also Amundson and Lauder 1994.)

Aptations perform a more or less *important* biological role. The biovalue of a feature indicates whether it performs not only *some* biological role, but perhaps a *specific* role. Compare, for instance, the biovalue of the tonsils to that of the brain. In most cases, it will be reasonable to conjecture that aptations with a specific function and a specific role are also likely to be adaptations. In other words, the specificity of functions and roles is an indicator of $adaptation_6$.

Since the value of a feature in relation to some environmental item may not only be positive, but also zero or negative, we arrive immediately at the concepts of nullaptation and malaptation:

DEFINITION 4.12. A feature a of an organism b is a *nullaptation* in relation to some environmental item e at time t iff $V(a, b, e, t) = 0$.

Whereas the biological role of an aptation is more or less significant, the biological role of a nullaptation is likely to be insignificant. Yet, the feature in question still performs *some* role because the set of bonding relations between any subsystem of a biosystem and items in its environment is not empty. Otherwise, it would be a closed system. Consequently, there are no entirely functionless or roleless features. Recall, for example, the minimal lymphatic function and role of the human appendix (Sect. 4.6).

As for the concept of malaptation, it is elucidated by:

DEFINITION 4.13. A feature a of an organism b is a *malaptation* in relation to some environmental item e at time t iff $V(a, b, e, t) < 0$.

Note that nullaptations should not be called 'nonaptations' because, logically, the term 'nonaptation' designates the complement of the class of aptations, thus comprising nullaptations as well as malaptations.

The concept of malaptation enables us to eventually define the concept of malfunction. To be consistent with our function/role distinction we should, for that matter, introduce the neologism *malrole*. However, we had better refrain from doing so, and take the term 'malfunction' to comprise both functions proper and roles. We then suggest:

DEFINITION 4.14. A subsystem (e.g., an organ) a of an organism is said to be *malfunctioning* with regard to some environmental item(s) e during some period $\tau =_{df}$ the (specific) function or the (specific) role of a, or both, is a malaptation with regard to e during τ.

Note that "malfunction" is a relative concept: what is a malfunction in one environment need not be so in another. To think otherwise presupposes the existence of some purpose beyond and above the activities and roles of biological systems. But where does such alleged purpose reside? Certainly, the physician can often disregard particular habitats and focus on the "proper function" of an organ, for there is hardly, if any, habitat in which, say, a human heart failing to pump blood might be an aptation. Yet what holds for the physician need not be true for the evolutionary biologist: he or she must make room for the fact that an organ may have different biovalues in different habitats, and that the functions and roles of organs keep changing evolutionarily.

Still, etiologists will insist that the heart has obviously been "built by" natural selection "for" the specific functions and specific roles it normally performs in the members of the given taxon. This would justify speaking of the "proper" or "normal" function of the heart, so that any serious deviation from this proper function (or physiological normality) can easily be detected as a malfunction. In other words, proper functions would be adaptations.

Although this metaphorical approach is partly correct, we should bear in mind that (a) the process of adaptation ontologically (historically) presupposes some function and role to begin with, which need not be the result of adaptation themselves; and (b) to speak of the proper function and role of an organ at a certain time presupposes that the feature in question still carries out an equivalent—though most likely improved, because adapted—function and role at that time. Otherwise, we would have to consider it to be the proper function of the human appendix to break down cellulose, and the proper function of the Kiwi's (vestigial and nonfunctional) wings to fly. (See also Prior 1985.) Consequently, the concept of adaptation logically presupposes the notions of function, role, and biovalue, and thereby that of aptation; and the detection of adaptations epistemologically presupposes some knowledge of the function and role of an organ. So we must proceed with our nonteleological, hence nonetiological, approach which has no use for the notion of proper function. Still, we need a suitable concept of adaptation. This concept is proposed in:

DEFINITION 4.15. A feature a of kind A of an organism b is an *adaptation* in relation to some environmental item e of kind E at time t if, and only if,
(i) a is an aptation of b in relation to e at t, and
(ii) the biovalue of the function(s) and role(s) of a in relation to e at t depends on (and is representable as an increasing mathematical function of) the biovalues of the functions, roles, and performances of features of kind A in the ancestors of b in relation to the environmental items of kind E at any time prior to t.

Unfortunately, the concept of dependency referred to in clause (ii) is rather vague. Although we can render it somewhat more precise by specifying that it is representable as an increasing function, this is an epistemological, not an ontological, notion. Therefore, it is in need of improvement. What we have in mind is, of course, the fact that the biovalue of a is a "result of selection". This phrase, however, only appears to be clearer, as being more familiar. Yet, as a matter of fact, the concept of selection is in need of elucidation as well (see Sect. 9.2).

In any case, it is worth noting that, if "adaptation" is defined in this way or, more explicitly, in terms of selection, we can no longer say, for example, that "to have been able to provide a scientific explanation of adaptation was perhaps the greatest triumph of the Darwinian theory of natural selection" (Mayr 1988, p. 148). Indeed, consider the subsumption

All characters that result from selection are adaptations.
c is a character that results from selection.
____
c is an adaptation.

The major premise is not a law statement but an analytic statement, and so the subsumption, though logically valid, is explanatorily empty. (Analytic statements are true either by virtue of their logical form such as "Homosexuality is innate or acquired" or, like the classical textbook example "All unmarried men are bachelors" and the first premise above, by virtue of the meaning of their concepts. Hence they are impregnable to empirical tests: see Sect. 3.7.2.) What the theory of evolution by natural selection explains, then, is (the trans-generational improvement of) aptations, not adaptations. This example illustrates once more that the concept of aptation precedes logically that of adaptation.

According to the preceding considerations, we have no use for the notion of proper function, which smacks of (crypto)teleology anyway. Suppose we discover some organism of a new kind, then all we need to practice teleology-free biology is the following: (a) we study the activities, in particular the specific functions, of its subsystems; (b) we investigate the biological roles, in particular the specific roles, of the subsystems in the organism's natural habitat; (c) we determine the biovalues of the features in question, i.e., whether the features are aptations, nullaptations, or malaptations in the natural habitat; (d) those subsystems having specific functions and specific roles which are aptations are investigated further as to whether they are adaptations. Although we might be guided in doing all that by the heuristic question 'What is this organ for?', we never give a genuinely teleological answer to it. We submit that any (sequence of) answer(s) to the preceding list of investigative steps is what most biologists and biophilosophers call 'functional explanation'. (More on teleology in Chap. 10. For the various levels of analysis subsumed under the label 'functional explanation' see Wouters 1995.)

Regrettably, but obviously, any hypothesis that assumes a feature to be an adaptation is hard to test. Thus, most such hypotheses are untested, though plausible, assumptions. Usually, the status of a feature as an adaptation is conjectured by using the indicators provided by the whole of comparative biology and systematics. For instance, the first step in the investigation of a feature's status as an adaptation is to determine the taxon-specific average of the specific functions and roles of the organisms' organs. The taxon-specificity of the specific role of a feature is usually a good indicator of its status as an adaptation. (Note that "taxon-specific" is not the same as "population-specific" or "statistically normal". Indeed, a certain organ may malfunction in an entire population, but it is unlikely to malfunction in all the members of the taxon—past, present, and future—unless the members of the taxon happen to be restricted to a single population. It goes without saying that, for practical reasons, we often have to settle for statistical normality in a small investigated population as an indicator. For the different concepts of normality see Boorse 1977 and Wachbroit 1994.)

Besides the indicator of taxon-specificity, biologists may also use adaptive evolutionary scenarios to justify their adaptationist assumptions. Yet plausible stories about adaptation can always be invented. The attitude of regarding *all* features as adaptations *a priori* as long as there is no proof to the contrary has been criticized as the "adaptationist programme" (Gould and Lewontin 1979). To be sure, adaptationism is a *heuristically* fruitful strategy, but it can turn into irrefutable dogma if one disconfirmed adaptationist scenario is replaced by another, and so on, without considering alternative, i.e., nonadaptationist, hypotheses. Such an alternative hypothesis would, for instance, consist in explaining a feature as the result of biomechanical or developmental constraints. In sum, statements regarding the alleged proper functions of traits are by no means better grounded than most other adaptationist hypotheses.

If adaptations are features of organisms whose current role in a given habitat is due to the selective success that the possession of such features conferred upon ancestral organisms in a habitat of the same kind, then there can be no pre*ad*aptations. A feature can be said to be potentially apted, i.e., *preapted*, with regard to a certain function and role in a different or future habitat, but it cannot be said to be preadapted to it. (See also Gould and Vrba 1982.) However, if there are preaptations, then we can also speak of prenullaptations and premalaptations. For example, the exocuticula of arthropods, which has originated in a marine habitat, may be regarded as a preaptation to a terrestrial habitat. On the other hand, the water vascular system of echinoderms (sea urchins, starfish, etc.) may be a premalaptation to a terrestrial habitat. This could be one of the reasons that no land-living echinoderms have ever evolved.

Just as there are no preadaptations, neither are there maladaptations or nulladaptations analogous to malaptations and nullaptations. However, there are clearly nonadaptations, namely the complement class of adaptations, i.e., the class of features which are not adaptations. For example, a feature may be due to a recent mutation or it may be pleiotropically connected to another feature. Such a feature may be useful, i.e., it may be an aptation, but its current role is not dependent on any past selective value. Gould and Vrba (1982) have suggested calling such features 'exaptations', which they (teleologically) define as "characters evolved for other usages (or for no function at all), and later 'coopted' for their current role" (p. 6). They give as an example the evolution of feathers, which may initially have been adaptations "for" thermoregulation but could be used "for" gliding (exaptation), so that selection could eventually start to adapt them "for" flight. Another example is the surplus of functions and roles of the human brain, which clearly are not all adaptations. The concept of exaptation, then, is important with regard to the conversion of functions and roles, believed to be one of the major factors in evolution (Mayr 1960). Let us, therefore, attempt to cleanse Gould and Vrba's proposal from teleological residues:

DEFINITION 4.16. Let b and b' denote organisms, where b names an ancestor of b', and let e and e' denote items of kind E in the environments of b and b' respectively. Further, let a and a' denote features of kind A, of the organisms b and b' respectively. Finally, call R the set of biological roles of a in relation to some e at some time t, and R' the set of biological roles of a' in relation to some e' at some later time t'. Then a' is an *exaptation* of b' in relation to e' at time t', where $t < t'$, if and only if
(i) feature a of b is an aptation, a nullaptation, or an adaptation in relation to e at t;
(ii) feature a' of b' is an aptation in relation to e' at t'; and
(iii) $R' \neq R$.

To sum up, the nonteleological concepts of function and role are logically prior to the notions of aptation and adaptation. Functions and roles are, moreover, ontologically (historically) prior to the process of adaptation. Hence, knowledge of functions and roles is also epistemologically prior to knowledge of adaptation, as has recently been emphasized by Amundson and Lauder (1994), for we must know what the function and role of an organ is, before we can determine whether or not it is an adaptation. Consequently, we submit that the etiologists' proposal to combine all these concepts into the single notion of *proper function* is confusing and misleading. What brings more clarity is to be aware that the ambiguous term 'function' in biology is used to refer not only to (a) the activities and (b) the roles of organs [note that both (a) and (b) are sometimes subsumed under the term 'Cummins-function', referring to Cummins's 1975 analysis, even though Bock and von Wahlert 1965 ought to have priority here], but also (c) to the activities and roles that are valuable to the organisms in question (à la Canfield 1964 and Ruse

1973); and finally, (d) to the valuable activities and roles that are moreover adaptations (that is, the etiologists' notion of function as proper function).

4.8.3 Aptedness and Adaptedness

We finally turn to the notions of aptedness and adaptedness. If we accept the aptation-adaptation distinction, it becomes obvious that the basic ontological concept of $adaptation_1$ should not be called 'universal adaptation'. A complex thing is neither able to exist in *any* habitat, i.e., universally, nor need it be *ad*apted to the habitat in question. What "universal adaptation" refers to is the basic relational property of a thing's being *apted* to a certain habitat. Therefore, this ontological concept is perhaps better called 'minimal aptedness' or 'basic aptedness'. Since the minimal aptedness of a thing will depend on different properties in different habitats, we cannot define the notion of minimal aptedness for all things. We can only formulate a criterion of minimal aptedness: A thing b is *minimally apted* in relation to environmental items of kind E during a period τ iff b is able to subsist in relation to items of kind E during τ.

Fortunately, we are in a better position with regard to living things. Here we can refer to certain properties that underlie a biosystem's subsistence in a given habitat. Moreover, we can assume that, except for the first living system(s) on this planet, there will be at least some adaptations among those properties. This allows us to use the traditional term 'adaptedness' instead of the neologism 'aptedness'. We can therefore propose:

DEFINITION 4.17. A biosystem b is *minimally (ad)apted* in relation to the items in its environment $E =_{df}$ b can, when related to the items in E, carry out all the functions listed in Postulate 4.1 to at least a minimal degree.

Employing the concept of biovalue, we could also say that a biosystem is minimally adapted to the items in its environment if, and only if, the biovalue of each of these items is ≥ 0.

Now, minimally adapted biosystems are likely to exert their biofunctions and bioroles with various degrees of intensity and efficiency. In other words, if a biosystem is able to survive at all, its performance will vary with regard to particular functions and roles (see also Bock 1980; Arnold 1983). Moreover, these particular performances will determine its overall performance. Accordingly, we define the notion of adaptedness thus:

DEFINITION 4.18. The *(overall) adaptedness* of an organism b in relation to the items in its environment E during a period $\tau =_{df}$ the degree of (overall) adjustment and performance of b in relation to the items in its environment E during τ as the result of the interplay of all its aptations, nullaptations, and malaptations.

In contrast to minimal adaptedness, (overall) adaptedness is a quantitative concept. Thus we may stipulate that the overall adaptedness value of an organism can be represented by a real number α ranging between 0 and 1 or, more precisely, $0 < \alpha \leq 1$ (M.B. Williams 1970). Since the concept of adaptedness refers to living beings, the value 0 is excluded because it amounts to nonviability or death. In nature, the value 1 will hardly be reached because it equals optimality or perfection. Moreover, an adaptedness value of 1 is extremely unlikely because any change in either the organism or the habitat or both would be immediately disadvantageous for the organism in question, thereby reducing its adaptedness. This is the reason that ecologically (over)specialized organisms often are evolutionarily doomed.

The notion of overall adaptedness has been criticized for being useless to evolutionary theory (Byerly and Michod 1991). Granted, the concept of adaptedness appears to be theoretically and practically intractable for it refers to a systemic relational property, the precise value of which seems hopelessly unmeasurable. Moreover, it is a phenomenological concept of restricted explanatory power—unless the particular underlying properties of the particular organism in question are specified (van der Steen 1991). However, the fact that we can approach this systemic property only through crude simplifications, for example, by restricting our analysis to a single or a few characters and by averaging adaptedness values over populations (Arnold 1983), does not invalidate the basic concept of overall performance of individuals (see also Lennox 1991). After all, organisms are adapted *systems*, not just heaps of isolated organs.

Some of the previously elucidated concepts, in particular that of adaptedness, will be further explored in Section 9.2, where we shall deal with the theory of natural selection. However, we have still a long way to go before we are ready for the issue of evolution. The next step on this way will consist in expanding on the issue of supraorganismic entities, whereby we are led to take a look at ecology.

5 Ecology

Like the problems of life and developmental biology, ecology has received little attention from philosophers of science (Ruse 1988). However, just like every other research field, it raises many ontological and methodological problems. One of the ontological problems of ecology is the ontological status of communities and ecosystems. In particular, the individualism-holism controversy, i.e., the controversy over whether communities are just aggregates of organisms or else systems, has haunted community ecology since its inception (Loehle 1988; Hagen 1989; Taylor 1992; Shrader-Frechette and McCoy 1993; McIntosh 1995). One of the methodological problems of ecology concerns its very status as a scientific discipline. For example, Peters (1991) has recently made the case for an instrumentally oriented ecology, focusing on prediction to solve pressing environmental problems rather than on, as he believes, insoluble issues such as explanation and the building of general theories.

Let us begin with some of the ontological problems of ecology, for this topic is a natural continuation of the preceding chapter.

5.1 Supraorganismic Entities

In Chapter 4 (Definitions 4.5 and 4.7) we defined the concepts of organism and of biopopulation. We now proceed to consider two further higher-level entities relevant to ecology, namely the community and the biosphere. And we will have to distinguish them from the concepts of ecosystem and of ecosphere. Of course, the philosopher can propose only general definitions of these concepts, for it is the task of the ecologist to investigate whether these concepts have an extension.

We begin with:

DEFINITION 5.1. A concrete system is a *biocoenosis* or *community* iff
(i) it is composed of organisms belonging to (at least two) different biospecies (i.e., iff its composition is multispecific); or
(ii) it is composed of (at least two) different biopopulations of unispecific organisms.

DEFINITION 5.2. A concrete system is a *biosphere* iff it is composed of all the communities on a planet (i.e., iff it is a planetary community).

In other words, a community is a multispecific biopopulation, which may consist of either interacting organisms belonging to different species or interacting unispecific biopopulations. (However, whether biopopulations really interact as wholes is questionable.) To be more precise, if c denotes a community composed of two biopopulations p and q, and if A and B designate two biospecies, then the organismic composition of c may be expressed by '$C_O(c) = C_O(p) \cup C_O(q)$', where $C_O(p) \subseteq A$ and $C_O(q) \subseteq B$. Alternatively, we could write '$C_O(c) \subseteq A \cup B$'. We introduce these formulas here because they will come in handy later (Sect. 9.1).

Note the following points. According to Definition 5.1, any combination of organisms or populations of different species qualifies as a community (see also Ricklefs 1990; Taylor 1992). For example, a hive of honey bees and the latters' host plants constitute a community. However, this is not a satisfactory definition for ecosystem ecology, which needs the notion of the total community within an ecosystem; more on this below.

Often, definitions of "biopopulation" and "community" include reference to a common habitat or to the co-occurrence at a certain location (Taylor 1992). This is unnecessary in our case because we have defined those entities as concrete systems: they would not be such unless there were bonds amongst their components, and such bonds are possible only if the organisms concerned are not far apart. However, as we shall see in a moment, reference to a distinct habitat will be necessary to define the concept of ecosystem.

It is controversial whether there actually is a cohesive world community, or whether local communities only cohere to form regional biota or so-called community groups. Anyway, the philosopher cannot answer this question. He or she can only point out that such a world community, if it exists, must be a cohesive system: the mere notion of spatial localization, i.e., restriction to our planet, is insufficient to establish such cohesion.

We have not counted ecosystems among the bioentities because they are usually conceived of as including the immediate environment of a community (Odum 1971; MacMahon et al. 1978; Salthe 1985; Ricklefs 1990; Taylor 1992; Shrader-Frechette and McCoy 1993). For example, a whole lake or forest is usually regarded as an ecosystem, and not only that part of it composed exclusively of living beings and biopopulations of such. This distinction carries over to our conception of "biosphere", which differs from common usage in that it is also conceived of referring solely to systems composed of biosystems or biopopulations. Often all communities together with the whole surface of the Earth are called 'biosphere'. For reasons of consistency, we suggest calling the latter system *ecosphere*, since it is identical to 'world ecosystem' or 'planetary ecosystem' in general.

However, before we can propose more precise definitions of these concepts, we need the notion of a total community. According to Definition 5.1, any system of organisms or populations of different species is a community. (We might call them 'partial communities'.) Yet when ecologists speak of ecosystems, they do not just refer to any system with a multispecific composition, such as a particular predator-prey or host-parasite community. Rather, they have in mind the total community in a comparatively distinct habitat, such as a lake, consisting of *all* the interacting organisms in that habitat. Accordingly, we propose:

DEFINITION 5.3. A concrete system is an *eubiocoenosis* or a *total community* if, and only if,
(i) it is composed of all the interacting organisms of different species in a distinct habitat; or
(ii) it is composed of all the biopopulations in a distinct habitat.

DEFINITION 5.4 A concrete system is
(i) an *ecosystem* iff it is composed of a total community and its immediate environment;
(ii) is an *ecosphere* iff it is composed of a biosphere and its immediate environment.

In other words, the ecosystem is the total community-environment system. Note that, in defining the general notions of a biosphere and an ecosphere, it is not necessary to specify the particular planet: the definition applies to any planet and it does not hold for the set of communities and ecosystems on all possible planets because such a set is not a concrete system.

5.2 The Ontological Status of Communities and Ecosystems

In a recent review, McIntosh (1995) has summarized the controversy over the ontological status of communities, and Shrader-Frechette and McCoy (1993) have collected sundry definitions of "community" proposed by ecologists during this century. (For a brief summary of this debate see Taylor 1992.) Ecologists following Frederic Clements's organismic approach regard communities and ecosystems as quasi- or superorganisms, i.e., as integrated, coordinated, and self-regulating systems (e.g., Odum 1971). Others make only the weaker claim that ecosystems are systems, yet neither well integrated nor homeostatic ones (e.g., Engelberg and Boyarsky 1979). The followers of Henry Gleason's individualistic approach, on the other hand, believe that communities are nothing but random aggregates of organisms lacking internal cohesion and coordination. (Ecologists often use the term 'assemblage' for what we call an 'aggregate' of organisms. The underlying concept, however, is the same: it means that the composite thing in question lacks a significant internal structure.) According to the definitions in the preceding section, the latter position amounts to saying that communities do not exist.

Each school provides empirical evidence for its own view. So, depending on the organisms and the habitats they study, some ecologists find structured wholes while others do not. For example, plant communities seem to be less integrated than animal communities; and among animals, insect communities seem to be less integrated than bird communities. Moreover, the degree of integration may even be seasonally variable. For instance, the macroinvertebrate fauna in the Salmon River (Idaho) was found to be in an equilibrium state in summer but not in autumn (Minshall et al. 1985). Still, it appears that "ecologists would be hard pressed to identify a site demonstrably lacking interspecific interactions to qualify as an assemblage, *sensu stricto*" (McIntosh 1995, p. 329).

So what seems to puzzle ecologists is the fact that the systems they study come in varying degrees of integration and coordination. This situation, however, is not at all unfamiliar to the ontologist, who knows that some systems are more tightly knit than others, and that there is no universal or cross-specific measure of integration or cohesion of a system. For this reason, the analogy of regarding communities as superorganisms seems misleading. Since organisms are paragons of highly integrated and coordinated systems, the analogy may lead ecologists to expect a similar degree of integration and coordination in communities, thereby predisposing them for disappointment.

Given the difficulties with the investigation of partial communities, it might prove to be more promising to study total communities in ecosystems rather than any combination of plants or animals, such as grass, bird, or insect communities. The fact that, for example, an insect community appears to be only weakly cohesive does not entail that the total community in a certain ecosystem is also weakly cohesive. Rather, the significant bonding relations might occur among the members of different taxa, such as insects and plants, plants and soil bacteria, trees and fungi, and so on (see, e.g., Jordan 1981).

The controversy over the status of communities and the best way to study them is an exemplar of the seemingly perennial philosophical debate between individualists (or atomists) and holists. Individualism rests on an atomistic ontology according to which the world is an aggregate of units of a few kinds, and on a reductionist epistemology according to which the knowledge of the composition of a whole, if any, is both necessary and sufficient to understand it. By contrast, the metaphysics of holism features organic wholes that are not decomposable into parts. The epistemological concomitant of holism is a form of intuitionism, according to which wholes must be accepted and grasped at their own level rather than analyzed.

What is often classified as holistic in community ecology appears to be actually closer to a *systemic* approach rather than to a genuinely holistic one. Indeed, many community ecologists in the Clementsian tradition are likely to treat communities as systems of interrelated and, moreover, interacting organisms, and they study them at both micro- and macrolevels. On the other hand, genuine holism is more likely to be adopted by antiscientific (or, if preferred, romantic) environmentalists than by scientific ecologists.

As mentioned above, the atomistic tradition is exemplified by Gleason's individualistic approach, which was a reaction against the holistic or quasi-systemic approach of early community ecology (McIntosh 1995). This reaction has been partly healthy, because it forces community ecologists to show whether the organisms and populations they study are actually systems rather than mere aggregates. The systemicity of the "communities" investigated is too often assumed *a priori*, in particular by invoking the allegedly omnipresent bond of competition. An example of such assumption is the idea that the morphology of organisms is affected by interspecific competition. Yet, as some individualists have shown, the distribution of certain characters, such as size ratios of morphological features, among different syntopic species in the "community" concerned is often indistinguishable from that in so-called null communities, i.e., in statistical aggregates randomly sampled from different habitats.

Results like these have spawned the call for the testing of null hypotheses in ecology before engaging in the testing of substantive hypotheses (Strong 1982; Hagen 1989). However, as Sloep (1986) has pointed out, it is not always clear whether a supposed null hypothesis actually is such. Indeed, in some cases, an alleged null hypothesis is, in fact, a substantive alternative hypothesis, and in others, it is not even a genuine alternative hypothesis, because it is complementary rather than rival to the other hypothesis. An example is the explanation of the finding that the species/genus ratio, i.e., the number of species per genus, on islands is lower than on the mainland. While the competitionists explain this finding as an effect of increased interspecific competition on islands due to their lower habitat diversity, the anticompetitionists claim that this ratio can be explained just as well as the effect of random colonization. The anticompetitionist hypothesis, however, is not a null hypothesis, because (a) random dispersal is an alternative process (or mechanism), not the absence of such, and (b) any combination of competition and partly random dispersal may produce the same species/genus ratio (Sloep 1986). In conclusion, not every hypothesis called a 'null hypothesis' is in fact one, and a hypothesis is not refuted by a rival hypothesis equally confirmed by the data at hand.

This may be the place to note that the individualism-holism controversy in ecology has parallels in ethics as well as in social philosophy and the social sciences. Here, the individualists (e.g., Hobbes, Locke, Smith, Bentham, Mill, Dilthey, the neoclassical economists, Weber, Hayek, Popper and, more recently, most sociobiologists) focus on individuals and either deny the existence of social bonds and social systems, or assert these to be reducible to individuals and their actions. By contrast, the holists or collectivists (e.g., the Romantics—in particular Hegel—, Marx, Comte, Durkheim, and the members of the Frankfurt school) hold that nature and society are "organic wholes" that cannot be understood by breaking them down into their components. (More in Bunge 1989, 1996.) We must leave it to the psychology and sociology of biology to explore the relations, if any, between the ethical and sociophilosophical views of ecologists as well as of biologists in general and their preferences for biological concepts, hypotheses, and theories that presuppose either an individualistic-atomistic or a holistic philosophical outlook.

In any case, neither individualism nor holism is tenable in ecology. (Neither is it in any other area, such as ethics and social science.) Ecological individualism fails because ecological predicates, such as "predator", "parasite", "symbiont", "pollinator", and "competitor" are at least binary, i.e., they presuppose the existence of related individuals. Yet two (or more) causally related individuals constitute a system. However, as we saw above, a moderate dose of methodological—not ontological—individualism, may be a healthy antidote to naive holism. (As for social philosophy, individualism forgets that human individuals constitute social systems; in other words, that they exist only as components of such and thus cannot be understood as isolated individuals. Individualism, by the way, has not only been traditionally adopted by liberalists but, today, also by many conservatives. In ethics, individualism is part and parcel of all varieties of moral egoism, and thus accompanied by the overemphasis of rights at the expense of duties.) Ecological holism must also fail, since it contains the wrong thesis that everything is connected to everything else, and since it is basically anti-analytical, hence heuristically barren. For this reason, genuine ecological holism is prone to attract irrationalists: think of "deep ecology" and the New Age version of the Gaia thesis. (As for social philosophy, it should be noted that holism is common to all totalitarian ideologies: "You are nothing, your x —insert: people, church, party, or what have you—is everything".)

Another noteworthy clash between the individualistic and the holistic outlook can be discerned in the rather recent field of environmental ethics—in the broad sense of the collection of approaches attempting at extending the moral sphere to nonhuman entities (Callicott 1980). Here, the animal liberation and animal rights camp appears to adopt an individualistic approach, as being concerned with the (putative) rights and interests of individual animals. On the other hand, the preservationist or ecological camp (i.e., environmental ethics in the narrow sense, originally dubbed 'land ethics') is concerned with the preservation of species and ecosystems and, ultimately, with the preservation of the ecosphere. It embraces a holistic approach, for the interests, if any, of the individual components of an ecosystem (including those of the members of *Homo sapiens*) are seen as subordinate to the higher good of the whole. Evidently, a consistent environmental ethics in the broad sense cannot be based on both an atomistic and a holistic outlook. Current work in this area, therefore, aims at the conception of a unified and consistent environmental ethics (see, e.g., Warren 1983; Callicott 1988).

Clearly, in all the aforementioned cases, the *tertium quid* is *systemism*, which admits the existence of wholes but analyzes them in terms of their composition, environment, and structure. (Recall the notion of a CES analysis from Sect. 1.7.2; for a recent defense of systemism in ecology see also Tuomivaara 1994.) Regrettably, systemism is often confused with holism, and even outspoken systemists often use the terms 'holistic' and 'holism' in characterizing their position (e.g., Wuketits 1989). So much so, that most scientific ecologists, though often classified as holists, are actually systemists rather than holists proper. We urge to distinguish systemism from holism, because the central (ontological) thesis of systemism, "Every thing is connected to *some* other thing(s)" is far weaker than the holistic thesis that everything is connected to everything else. And, epistemologically, systemism is not antianalytical as is genuine holism. In short, at a closer look, the individualism-holism controversy in ecology is actually an individualism-systemism controversy.

Even though not every old association of organisms studied by ecologists need be a system, and even though not every old system need be a self-regulating one, it can hardly be doubted that ecological systems such as biopopulations, communities, and ecosystems do exist. It is equally apparent that such systems come in sub- and supersystems, that is, in a hierarchy of nested systems. This hierarchy of systems will be tackled next.

5.3 Biolevels

Talk of levels of organization (or complexity or integration or individuality) and hierarchy is rampant in biology as well as in other fields of science. (For a quick review see Grene 1987.) Unfortunately, there is no consensus on the signification of the terms 'level' and 'hierarchy', which are used in a variety of ways. (For earlier studies on levels see, e.g., Woodger 1929; Novikoff 1945; Bunge 1959b, 1963, 1973a, 1977b, 1979a; Beckner 1974; Zylstra 1992.) Recently, Eldredge (1985a) has introduced the distinction between an *ecological* and a *genealogical* hierarchy. Whereas the ecological hierarchy is the traditional hierarchy consisting of systems such as organisms, populations, communities, and regional biota, the genealogical hierarchy is supposed to consist of "individuals" such as codons, genes, chromosomes, organisms, demes, species, and monophyletic taxa. The integrating relations of the former would be economic interactions through "matter-energy transfer"; the "glue" holding the latter together would be "information transfer" by way of replication. Furthermore, biologists speak of the Linnean hierarchy and of hierarchies of homologies. Apparently, hierarchies are everywhere, and almost everything is claimed to constitute a "level of reality" (e.g., Ghiselin 1974, 1981; Hull 1976, 1980; Vrba and Eldredge 1984; Rosenberg 1985; Salthe 1985; Collier 1988). Let us attempt to clarify this situation by analyzing the level structure of biotic systems, leaving the analysis of the systematic hierarchy for Chapter 7.

Leaving aside the prebiotic as well as the suprabiotic levels, we have thus far distinguished, more or less explicitly, six different biotic levels, namely those of elementary biosystems (or elementary cells), composite cells, multicellular organisms, biopopulations, biocoenoses (or communities), and biospheres. To take into account the various organs of multicellular organisms, we shall add another one, the organ level. (For either similar or else alternative levels see Löther 1972;

Griffiths 1974; Popper and Eccles 1977; MacMahon et al. 1978; Mayr 1982; Petersen 1983; Eldredge 1985a; Salthe 1985; Zylstra 1992.) We may display them for reference purposes in the following definitions:

B_1 = elementary cell level = the set of all elementary cells
B_2 = composite cell level = the set of all composite cells
B_3 = organ level = the set of all (multicellular) organs
B_4 = multicellular organismic level = the set of all multicellular organisms
B_5 = population level = the set of all biopopulations
B_6 = community level = the set of all communities (biocoenoses)
B_7 = biosphere level = the set of all biospheres.

Note the following points. First, the set B_7 of biospheres has so far only a single known element: the biosphere of our planet. Second, our list of biolevels is not supposed to be complete. One may interpolate further levels between any two. For example, one may easily insert another level between B_4 and B_5 comprising physiological mutualisms like lichens. Or, one may subdivide some levels into sublevels. For example, an organ in B_3, such as the mammalian brain, may contain subsystems of different complexity from, say, cortical microcolumns through the so-called nuclei to the hemispheres. And a population in B_5 can be nested in groups, demes, reproductive communities, Mendelian populations, or what have you. (Needless to say, there is no agreement as to the proper definition of any of these terms.) In general, there are as many levels as there are kinds of (sub)systems. Third, we can also merge levels if their distinction is irrelevant to our analysis. For example, we could define a cellular level B_C as the set $B_1 \cup B_2$. Fourth, our list of biolevels does not include "levels" whose elements are taxa, clades, monophyletic groups, or lineages, just because all the latter are not concrete systems but collections, hence constructs (see Sect. 7.2).

As is obvious from our construal, levels, too, are sets or classes, hence concepts, not things. However, they are not arbitrary concepts but notions that represent something real, namely systems of a certain complexity. Since a level is a set, belonging to a level is identical to set membership. For example, the sentence "b is a multicellular organism" can be shortened to "$b \in B_4$" (b belongs to the set B_4), yet not "$c \sqsubset B_4$" (b is a part of B_4). Another consequence of the construal of levels as sets is that levels cannot be said to interact: only individual members of levels can act upon one another. In particular, the higher levels can neither "command" nor "obey" the lower ones, or conversely. All talk of interlevel action is elliptical.

On the other hand, the members or elements of biolevels are concrete things and, more precisely, concrete systems. Moreover, the systems constituting those levels are related in a particular way, namely thus: any system on a given level is composed of things belonging to preceding levels. For example, the nervous system of an animal belongs to level B_3 and is composed of members of B_1 (neurons, glial cells, etc.). A biopopulation of multicellular organisms belongs to level B_5 and is composed of members of B_4, which, in turn, are composed of members of B_3, which are composed of members of $B_1 \cup B_2$, and so on. This is what it means to say that B_1 *precedes* (or *is a lower level than*) B_2, or that B_4 is a lower level than B_5.

To be more precise, the general concept of the relation of precedence between levels can be elucidated by:

DEFINITION 5.5. If B_m and B_n designate two levels, then B_m *precedes* B_n if, and only if, the L-components of members of B_n belong to B_m. That is, $B_m < B_n =_{df} \forall x(x \in B_n \Rightarrow C_L(x) \subset B_m)$.

Note the following points about this convention. For one, though motivated by biological considerations, it is not limited to biolevels. In fact, it carries over to all system levels, whether physical, chemical, social, or technical. For another, there is nothing obscure about the notion of level precedence as long as one sticks to Definition 5.5, instead of construing $B_m < B_n$ as "the B_m's are inferior to the B_n's" or something of the sort.

Let us now tackle the proposition that the entire biosphere has a hierarchical structure. Call B = $\{B_1, B_2, B_3, B_4, B_5, B_6, B_7\}$ the *set of biotic levels*. This set is ordered by the relation < of level precedence, a relation which is asymmetrical and transitive. (Note that B is not ordered by the relation $\subset$ of set inclusion, as is the systematic hierarchy, for it is not true that every cell is an organ, or that every cell is an organism, and so on. See Definition 7.2.) The set B together with the level precedence relation deserves a name of its own:

DEFINITION 5.6. The set B of biolevels together with the level precedence relation <, i.e., $\mathcal{B} = \langle B, < \rangle$, is called the *biolevel structure*.

Although a classification differs in structure from the *biolevel structure* $\mathcal{B}$, both are usually subsumed under the term 'hierarchy'. (The original concept of a hierarchy, namely that of a social hierarchy, involves a subordination or dominance relation.) However, $\mathcal{B}$ should be called neither an *ecological hierarchy* nor a *hierarchy of life* because B contains living as well as nonliving systems: whatever is alive or composed of living beings is in B. Accordingly, our next assumption is:

POSTULATE 5.1. Every biosystem and every system that, at some level, is composed exclusively of biosystems belongs in some level of the biolevel structure $\mathcal{B}$.

Consequently, neither ecosystems nor ecospheres belong in the biolevel structure. They belong to suprabiotic levels because they are also composed of systems which, at some level, are not exclusively composed of biosystems, namely components of the (abiotic) habitat. This is why $\mathcal{B}$ is not a complete ecological hierarchy but only a substructure of it.

As our version of the hierarchy of biolevels talks about levels (more precisely, concrete systems belonging to a certain level) and their order, but not about their origin and evolution, it is still static. Biologists, however, would want to say that, historically (or evolutionarily), every biolevel has emerged from one of the preceding (prebiotic or biotic) levels. Yet such a statement would be metaphorical: since levels are sets, they cannot emerge from one another. Fortunately, the notions of self-assembly and self-organization elucidated in Section 1.7.4 allow us to reformulate this idea. Thus, we obtain from Definition 1.10:

COROLLARY 5.1. Every concrete system belonging to a given level L has self-assembled or self-organized from things at preceding levels.

For example, a biopopulation is a system that has self-assembled from organisms of the same kind, which belong to the immediately preceding level. By contrast, the organs of a multicellular organism develop simultaneously with the individual or at a certain stage in its development—an instance of self-organization. This is why it is not true that every system at a given level has self-assembled from things at *the* preceding level, although it is true that it is composed of them.

An immediate consequence of this hypothesis is:

COROLLARY 5.2. Every system at a given level L_n is preceded in time by its components at *some* lower level L_{n-i}.

As stated before, some systems at a given level may not be preceded in time by their components at the preceding level because the latter self-organize during the history of the system, as is the case with the development of multicellular organisms. On the other hand, the molecular components of a higher-level system always precede the latter in time.

If a system is preceded in time by its components, the latter can be termed *precursors*. However, since not all components are precursors, level precedence and temporal precedence are neither coextensive nor cointensive (*pace* Bunge 1979a). Anyway, the hierarchy of biolevels, though consisting of static sets, is now compatible with an evolutionary outlook. Metaphorically speaking, we could say that levels may succeed each other in time and they do so by virtue of a general mechanism, namely the self-assembly or self-organization of things.

So much for the ontology of levels. An epistemological consequence of the multilevel structure of the world is that there is no such thing as *the* absolutely appropriate level of analysis, regardless of our research interests and goals. Moreover, since every unit of analysis (except for the universe as a whole) is embedded in a higher-level system, and since all systems are composed of lower-level things, we should not overlook the adjacent levels when describing our system of interest. That is, our discourse ought to refer to both the *central* (or target) referent and the *peripheral* (i.e., compositional and environmental) referent(s). (See also Salthe 1985.)

5.4 Ecological Niche

Another contentious issue in ecology is the notion of an ecological niche. We shall take a quick look at this concept here, for it relates to the concept of adaptedness, which we examined in Section 4.8.3. Furthermore, it is unclear—as with many other biological notions—what the proper referents of the niche concept are: organisms, populations, species, higher taxa, or perhaps, depending on the model, all of them together.

To begin with, we should recall that Definition 4.18 identifies the overall adaptedness of an organism with the latter's degree of (overall) adjustment and performance in relation to the items in its environment as the result of the interplay of all the organism's aptations, nullaptations, and malaptations. This definition is clearly an ecological notion, so much so that some biologists might object that Definition 4.18 does not define the concept of adaptedness, but rather that of ecological niche. However, the two concepts are different. We suggest equating the niche of an organism with the collection of the really possible relationships with its environment. In other words, we define the concept of ecological niche as follows:

> DEFINITION 5.7. Let b name an organism of kind (e.g., taxon) B with environment E. Then the *ecological niche* of b is the set of *nomologically possible*, i.e., B-specific, bonding relations of b with the items in its environment E that have a positive biovalue to b.

Accordingly, the niche of an organism is not a thing, such as its habitat. Instead, it is its potential kind-specific external structure with a positive biovalue to it. Usually, the relevant kind is some taxon or other, in particular the species; but, if necessary, any kind, systematic or extrasystematic, can be referred to. Note the following points about the above definition.

Definition 5.7 is partly analogous to Definition 4.9, which equates the biological role of an *organ* with the latter's bonding exostructure. At first sight, we could thus say that Definition 5.7 is a definition of the role of an *organism* in the environmental supersystem it is part of. Indeed, "niche" is sometimes defined as "the ecological role of a species in the community" (e.g., Ricklefs 1990, p. 817). However, there are significant differences between our two definitions. For one, we defined earlier the role of an organ as its *actual* bonding external structure at any moment. Definition 5.7, by contrast, refers to the nomologically *possible* bonding exostructure of an organism. For another, we defined the role of an organ as the latter's *total* bonding external structure, whereas Definition 5.7 refers only to the bonds with *positive biovalues* to the organism in question. Finally, our definition of "niche" restricts the external structure of the organism in question to kind-specific bonds, thereby excluding idiosyncratic features of the organism.

The reason for this construal is that ecologists want to compare the niches of organisms *qua* members of different taxa, in particular species, not *qua* individuals.

(The statements in scientific models and theories are supposed to be general in some respect, i.e., they are—as is often said—about classes of things, not individuals. The latter formulation, however, is imprecise: a scientific model does not refer to a class as an [abstract] whole, but to each and every of its members, i.e., concrete individuals sharing certain properties *qua* members of a certain class or kind.) Furthermore, we submit that the ecological niche concerns only the *needs* that an organism must meet to survive and reproduce in a given habitat. For example, the fact that an organism is the host of a parasite, or the prey of a predator, does not belong to its survival needs, hence its ecological niche. (However, the host and the prey belong to the ecological niches of the parasite and the predator, respectively.) This is why "niche" should not be defined as "the sum total of organism-environment relations", but should be restricted to the relations with a positive biovalue. Indeed, the branch of theoretical ecology called *niche theory* restricts the concept of niche as referring to the frequency of resource use by some population and represents it as a "utilization distribution" (Schoener 1989)—a concept of niche that is even narrower than ours.

Not all of the nomologically possible relations of an organism need, or indeed can, be actualized at a given time. However, in order to be and stay alive, an organism must realize or actualize a *minimal* niche at all times. (For the notion of a minimal niche see also Hurlbert 1981.) In other words, it must be minimally adapted at all times (recall Definition 4.17). Thus, whereas the notion of minimal actual niche coincides with that of minimal adaptedness, the concept of the actualized (or realized) niche of an organism is not the same as that of the (overall) adaptedness of an organism. Indeed, the former comprises only the organism's actual exostructure with a positive biovalue, whereas the latter refers to its *total* actual exostructure. Thus, the more of its potential niche "dimensions" above those essential for bare survival an organism can realize in a variable habitat, the better adapted it is likely to be.

Obviously, the referent of our definition of "niche" is the organism—the living entity which is either able or unable to survive and reproduce in a given habitat. Interestingly, only a few authors define the concept of ecological niche with explicit reference to organisms (e.g., MacMahon et al. 1981). Most authors either refer to species in their definitions, or they use the term 'niche' with reference to both species and organisms, as if there were no difference between the concept of the ecological niche of an organism and that of the ecological niche of a species. However, as we proceed to show, species do not and cannot have ecological niches.

First of all, as will be recalled from Section 4.5 and as we shall elaborate in detail in Chapter 7, we cannot do without the notion of a species as a class. However, if species are classes, i.e., conceptual objects, not concrete ones, it makes no more sense to speak of the niche of a species than to speak of the metabolism of a species; the same holds for supraspecific taxa as well as for any other class of organisms. Thus, talk of the niche of a species is as careless (and mistaken) as to say that a species is carnivorous, or that "the Turtle Frog is a burrowing species that lives on termites"—a statement found in an exhibition hall of the American

Museum of Natural History. In Section 1.2, we called such statements 'metaphysically ill-formed'. Indeed, it is not species but individual organisms that can be carnivorous or that burrow and eat termites, i.e., other organisms. This should be obvious not only if species are classes, but also if species are taken to be concrete systems, such as biopopulations (see Definition 4.7), sometimes also referred to as 'species populations'. In the latter case, one would not attribute substantial properties to constructs, but one would attribute properties possessed only by entities at a particular biolevel to entities belonging to another level. We call this mistake, which also leads to ontologically ill-formed statements, *level-mixing*. We presume that this mistake is due to the treatment of organisms as members of statistical populations, such as in plotting population (number of organisms) against the total food consumed (number of food items). It is, then, tempting to say that the population consumed a certain amount of food, even though it is in fact organisms that eat and digest, not populations as supraorganismic individuals.

To be sure, since biopopulations are concrete systems, we could, if necessary, form a concept analogous to that of the niche of an organism: the population niche. In so doing, however, we should only make use of genuine populational properties such as population density. Yet, when speaking of the niche of a "species" (actually population), most authors engage in level-mixing, as referring to organismal, not populational, properties. After all, it is questionable whether populations as supraorganismic entities do have needs concerning certain resources which are not reducible to the needs of the constituent organisms.

Even if we disregard the level-mixing in question, the concept of the ecological niche of an organism would not be coextensive with that of the ecological niche of a population (or "species-as-individual"). The reason is the following (Mahner 1993a). The ecological niche of an organism includes the latter's relations to conspecific organisms, which clearly belong to an organism's environment. By contrast, the ecological niche of a population (or "species-as-individual") cannot contain relations to conspecifics by definition but, at most, relations to other populations or to organisms of a different species. (The latter possibility presupposes that a population as a whole can interact with single organisms—a rather questionable assumption.) Thus, while the relations to conspecific organisms belong to the external structure of each organism in the population, the relations among conspecifics belong to the internal structure of the population (as a whole). The notion of an ecological niche, however, is concerned only with the external structure of the system in question.

To conclude, the concept of ecological niche refers to organisms, not to populations or species. (See also Reig 1982; Eldredge 1985b.) What is true, however, is that the niche concept is concerned only with species-specific or, more generally, kind-specific properties of organisms. As said before, ecologists are interested in the niche of individuals *qua* members of a certain species or *qua* members of any higher taxon. For example, it makes perfect sense to speak of the niches of (individual) bats *qua* bats, i.e., *qua* members of the taxon Chiroptera. Moreover, ecologists are also interested in the niches of organisms *qua* members of extrasystematic classes, such as age groups, morphs, guilds, and ecotypes (Hurlbert 1981; Ricklefs 1990). Yet again, it would be mistaken to claim that having a niche is a property of the age class, the morph, the guild, or the ecotype rather than a property of the individual organisms belonging to these classes. Of course, speaking of a niche of a species, a guild, a morph, and so on, is a convenient *façon de parler*; but it is important to realize that doing so is only an elliptic shorthand.

In the light of the preceding analysis, a statement such as "Every species has its own ecological niche" is an ellipsis, and we insist that it must be properly translated into "All organisms of a given species have the same ecological niche". This is not mere quibbling, for ecological niches have been literally ascribed to species, as in Hutchinson's (1957) famous paper on the ecological niche, or in Mayr's (1982) definition of a species (as an individual): "A species is a reproductive community of populations (reproductively isolated from others) which occupies a specific niche in nature" (p. 273). (Note, incidentally, another instance of level-mixing in Mayr's definition: it attributes the property of [sexual] reproduction to populations rather than organisms or, more precisely, mating-pairs.) Moreover, some authors have argued that species (as alleged individuals) are in some sense ecological "units" or "entities" or "wholes" because they possess an ecological niche (e.g., Van Valen 1976c; Sudhaus and Rehfeld 1992). This idea crumbles upon realizing that it is organisms, not species, that have ecological niches. Still, it is, of course, legitimate to use ecological properties of organisms to define a given taxon. A species defined predominantly by ecological features may then be classified as an *ecospecies*, just as a species defined predominantly by morphological characters is called a *morphospecies*. The term 'ecospecies', however, should not be misinterpreted as referring to an ecologically interacting concrete whole.

Our concept of "ecological niche" belongs in the family of the so-called *functional* niche concepts (Hurlbert 1981; Schoener 1989; Griesemer 1992; Colwell 1992). In a functional niche concept the ecological niche is a relational (or functional) property of an organism, so that there are no vacant or empty niches waiting to be occupied by organisms. In short, no organism, no niche. (See also Günther 1950; Hutchinson 1957; Bock and von Wahlert 1965; Odum 1971; Osche 1972; Schmitt 1987; Ricklefs 1990; Sudhaus and Rehfeld 1992.) Only a geographical space or *habitat* can be occupied or unoccupied, not a niche.

In contrast to the more recent functional niche concepts, the early concepts of niche, as introduced by Roswell Johnson, Joseph Grinnell, and Charles Elton, belong to the family of the *environmental* or *habitat niche* concepts (Schoener 1989; Colwell 1992). Here, the niche is a property not of the organism but of the environment. That is, there are niches that become vacant upon the extinction of a species, and there are empty niches waiting to be filled by either evolving or immigrating "species", just as in a firm there are vacant positions to be filled by some prospective employee.

Now, the expression 'environmental niche' has two meanings. For one, it is occasionally used simply in the sense of "habitat" (micro or macro), in which case it is redundant. For another, it refers to the potentials, possibilities, or opportunities a given habitat has on offer for prospective occupants. These potentials are sometimes also called 'roles' or 'prospective roles', as well as 'ecological licenses' (Osche 1972; Schmitt 1987). For example, in every ecosystem there are "roles" for producers, predators, parasites, and so on, which can be "played" by different organisms (species) in different regions or at different times, or both. Furthermore, there are so-called *ecological equivalents*, such as antelopes (and others) in the African savanna, horses in the Asian steppe, and kangaroos in the Australian grasslands. Another example are hyenas and arctic foxes. These examples illustrate the rather vague notion of an environmental niche. We leave it to ecologists to assess its relevance to ecological theory.

After these ontological problems in ecology, we now turn to some of its methodological problems, in particular to the question of the scientific status of ecology.

5.5 The Scientific Status of Ecology

Some ecologists and philosophers think that ecology is an "immature" or even "anomalous" science (Hagen 1989). Others claim that it is an applied science rather than a basic or pure one (e.g., Peters 1991; Shrader-Frechette and McCoy 1993). According to the latter, this (alleged) state of affairs is all right considering the fact that society is entitled to expect useful advice from ecologists for the solution of pressing environmental problems. Anyway, considering the enormous complexity and variety of ecological systems, all the ecologist could hope to achieve is to make case studies (rather than to search for general theories) and to make useful predictions based on the analysis of the correlations of environmental variables.

Before we can comment on these views, it will be helpful to have a characterization of basic (or pure) science, applied science, and technology (see Bunge 1983b). We can then apply this characterization to ecology in order to examine its status with regard to either one of the aforementioned fields.

5.5.1 Basic Science

Science in general, as well as a particular science such as ecology, can be viewed at the same time as a group of people, as an activity, and as a body of knowledge. For obvious reasons the latter is the most interesting aspect for the philosopher of science. We must be aware, however, that this view is an abstraction because there are no self-existing bodies of knowledge (recall Chaps. 3 and 6).

We start by defining basic (factual) science in general as a *family of scientific research fields* (or *disciplines*), where a family of particular disciplines is a collection every member s of which is characterizable by a 10-tuple

$$s = \langle C, S, D, G, F, B, P, K, A, M \rangle,$$

where, at any given moment,

(i) C, the *research community* of s, is part of the scientific community in general and is composed of persons who have received scientific training, hold strong information links among themselves, and initiate or continue a tradition of scientific research;

(ii) S is the *society* that hosts C and encourages or at least tolerates the activities of the components of C;

(iii) D, the *domain* or *universe of discourse* of s, is a collection of (actual or putative) concrete or real things and their changes, past, present, and future;

(iv) G, the *general outlook* or *philosophical background* of s, consists of the ontological, epistemological, axiological, and moral principles that guide the study of D. More precisely, G comprises a naturalistic ontology, a realist epistemology, and a system of internal values, an endoaxiology, which is particularly characterized by the ethos of the free search for truth. The internal value system of science includes such *logical values* as exactness, systemicity, and logical consistency; *semantical values* such as meaning definiteness, hence clarity, and maximal truth or adequacy of ideas to facts; *methodological values* such as testability and the possibility of scrutinizing and justifying the very methods employed to put ideas to the test; and, finally, *attitudinal* and *moral values* such as critical thinking, open-mindedness (but not blank-mindedness), veracity, giving credit where credit is due, and so on. The endoaxiology of science is often called 'the ethos of science' (Merton 1973; Mohr 1981). Basic science is value-free only in the sense that it makes no value judgments about its objects of study or referents—except perhaps about their suitability as objects of study with regard to a certain technique at hand. That is, basic science has no external value system or exoaxiology;

(v) F, the *formal background* of s, is a collection of up-to-date logical and mathematical theories that are (or can be) used by the components of C in studying the members of D;

(vi) B, the *specific background* of s, is a collection of up-to-date and reasonably well-confirmed knowledge items (data, hypotheses, and theories) obtained in other scientific disciplines relevant to s;

(vii) P, the *problematics* of s, consists exclusively of *cognitive* problems concerning the nature, particularly the laws, of the members of D;

(viii) K, the *fund of knowledge* of s, is the collection of up-to-date and well-confirmed knowledge items (data, hypotheses, and theories) obtained by the components of C at previous times;

(ix) A is the collection of *aims* of the components of C with regard to their study of the members of D, in particular the discovery of the laws of the members of D as well as their description, explanation, and prediction;

(x) M, the *methodics* (often misnamed 'methodology') of s, is the collection of checkable and explainable methods utilizable by the components of C in the study of the members of D;

(xi) s has strong permanent links with other scientific disciplines;

(xii) the membership of every one of the last eight coordinates of s changes, however slowly, as a result of research in s as well as in related fields.

The first three components of the 10-tuple constitute what may be called the *material framework*, and the last seven the *conceptual framework*, of a scientific discipline. (The latter resembles one of the many meanings of Kuhn's vague notion of a paradigm.) The former may be called thus because both the research community C and its host society S are concrete systems, and the domain D is a collection of material things. (The point of listing C and S explicitly is to remind ourselves that knowledge is not self-existing but an activity performed by real people in a concrete social environment.) On the other hand, the remaining seven coordinates comprise conceptual items. (Note that, even though M may involve concrete things, namely artifacts such as microscopes, microscopy techniques are conceptual items.)

Finally, we say that a discipline that satisfies only partially the above 12 conditions is a *semiscience* or *protoscience*. If a discipline is evolving towards the full compliance of them all, we may call it an *emerging* or *developing* science. On the other hand, if a discipline fails to meet the above conditions, it is a *nonscientific* discipline. While the epithet 'nonscientific' is not supposed to be pejorative, the epithet *pseudoscientific* is. We use the latter to label disciplines that, though nonscientific, are nevertheless advertised and sold as scientific. Think of "scientific" creationism. (For an analysis and evaluation of creationism according to the preceding conditions see Mahner 1986.)

5.5.2 Ecology as a Basic Science

We now invite the reader to specify this general characterization of a basic scientific discipline to biology as a whole or to any of its subdisciplines, such as ecology. If we specify the above conditions to ecology, it becomes apparent that, though ecology meets all of them in the broadest sense, it still has problems to meet some of the conditions to a full extent. For example, its fund of knowledge contains many data and low-level hypotheses but few, if any, law statements and general theories (Bell 1992). When ecological laws_2 such as "No population can increase without bound" are suggested (Loehle 1988), they are either too general, perhaps to the point of triviality, or highly controversial, such as the competitive exclusion principle. For this reason, the explanatory and predictive power of ecology is, so far, limited. However, ecology shares this shortcoming with the social sciences. And the reasons for the immaturity of both ecology and the social sciences appear to be quite similar.

We agree with McIntosh (1982) and Bell (1992) that the main reason for ecology's immaturity is mostly due to the complexity and diversity of the systems constituting the domain of ecology: "...it is just the variability from habitat to habitat or taxon to taxon which frustrates the search for ecological regularities and a satisfying theory and gives rise to continuing controversy" (McIntosh 1982, p.

34). This complexity and diversity is accompanied by conceptual vagueness and ambiguity. Indeed, many key ecological concepts, such as those of community, stability, balance, and equilibrium are notoriously vague (McIntosh 1982; Shrader-Frechette and McCoy 1993). And, of course, conceptual fuzziness, in turn, is an impediment to the theoretical progress of a discipline. In particular, certain concepts such as those of balance and competition often seem to have their place in ecology due to intuition, common sense, and perhaps ideological prejudice rather than because of scientific evidence. (The situation in the social sciences is parallel.) Witness the above-mentioned debate about null hypotheses in ecology.

Another problem that arises from the complexity and diversity of ecological systems is the difficulty of making field experiments (Carpenter et al. 1995). In particular, field studies may not be exactly replicable, and the variables and parameters are hard or not at all to control. Furthermore, as the conditions in the field are much more complex and varied than in the laboratory, it remains always contentious whether the results of laboratory experiments also hold for in certain respects equivalent yet, in fact, much more complex systems in nature (Mertz and McCauley 1982; Lawton 1995; Roush 1995). However, this problem is not peculiar to ecology: all factual sciences face it.

Finally, the complexity and diversity of ecosystems makes it hard to classify ecosystems into clear-cut types or kinds. Although some ecosystems, such as fresh-water lakes or deciduous forests, are certainly similar in certain respects, it remains questionable whether such typology is more than superficial, namely deep or theoretically significant. In other words, a general theory about ecological systems of a certain kind is only possible if ecological systems actually come in natural kinds, that is, if there are ecological systems possessing the same $laws_1$. If an ecological system is unique, it is, of course, not lawless in the ontological sense, yet we have the methodological problem of distinguishing its lawful from its idiosyncratic properties. (See also Sect. 7.3.1.4.)

A further aspect that makes ecological systems appear so unruly is the fact that they are often systems showing chaotic behavior. Since the (modern) notion of chaos nurtures an important and promising field of studies in several scientific disciplines, including ecology, and since, at the same time, the concept of chaos often involves certain misunderstandings, it deserves a brief section of its own.

5.5.3 Foray: Chaos in Ecological Systems

Scientists work on the (usually tacit) hypothesis that the objects they study are lawful or orderly, even if they appear not to be such. This article of philosophical faith impels them to look for patterns. When they find pattern, their faith in lawfulness is reinforced. When they fail to uncover pattern, they suspect that it is hidden, or they doubt their own ability rather than the principle of lawfulness (Postulate 1.4).

Now, a mess may be real or merely apparent. For example, a pile of debris or a garbage dump are really messy: there is no order in them. However, one may assume that every item in the pile landed on it following a lawful, if possibly complicated, trajectory. So, although it may be pointless to search for order in the debris, it may pay to try and reconstruct hypothetically the trajectory of every item in the debris. This is, in fact, what paleontologists do with sites where fossils of various kinds mingle. In short, such a site is the messy final resting place of items that, presumably, got there in perfectly lawful, if separate, ways.

Some patterns are apparent: think of tilings, pulse beats, circadian rhythms, or the markings on some animals' bodies. Others are imperceptible, so that they must be guessed. Among the imperceptible patterns the random and the "chaotic" ones are perhaps the most intriguing, as well as the hardest to disclose and distinguish. In fact, both random and "chaotic" patterns look alike, namely highly irregular. Thus, mere visual inspection of an irregular time series will not tell us whether it is random, "chaotic", or neither. The only way to solve this problem is to invent a mathematical model (either probabilistic or "chaotic") and confront it with the given time series, such as a population vs. time graph.

There are two major differences between random and "chaotic" patterns. Firstly, whereas the former satisfy probabilistic laws, the latter can be nonprobabilistic. Secondly, whereas randomness is hard to control, "chaos" can be controlled by varying the relevant parameter—e.g., r in the famous logistic equation "$x_{t+1} = rx_t (1 - x_t)$". Even a small change in the numerical value of r may result in qualitative changes of the population concerned, such as transitions from steady cycles to bursts or crashes.

It has been suggested quite a while ago that chaos plays an important role in ecology, particularly in population ecology (see, e.g., May 1974; Simberloff 1982). Indeed, a "chaotic" model can help describe and even predict and thus control population cycles, explosions, and crashes. However, solid empirical evidence to support this claim has been produced only very recently. We proceed with a brief description of such work (Costantino et al. 1995).

Costantino and coworkers studied populations of flour beetles of the species *Tribolium castaneum*. (The members of the genus *Tribolium* are well-known model organisms in ecology and evolutionary biology.) First, they set up a (free) mathematical model of a population of this species. (For the notion of a free model see Sect. 3.5.3.). Then they designed, performed, and analyzed an experiment in the light of that model. The model was built on the strength of time series of the larval, pupal, and adult populations of the Flour Beetle. It consists of a system of three finite difference equations that relate these three populations at any given time and a unit time later—the two week larval maturation period. (One such equation reads: $A_{t+1} = P_t\, exp(-c_{pa}A_t) + (1 - \mu_a)A_t$, where A_t and A_{t+1} denote the adult populations at times t and $t + 1$ respectively, P_t the pupae population at time t, $c_{pa}\, A_t$ the survival probability of a pupa in the presence of A_t adult cannibals, and μ_a the adult mortality rate.) The equations contain six parameters, five of which are derived from previous empirical studies. The sixth is the adult mortality rate

μ_a. It is the tuning parameter or "knob", since it is manipulated by the experimenter.

The experimental design involves four populations of each of two genetic strains of the Flour Beetle. The crux of it is that these populations are randomly assigned to each of six treatments: adult mortality rates μ_a of 0.4% (control), 4%, 27%, 50%, 73%, and 96%. The artificially induced mortalities compete with the spontaneous processes of reproduction, cannibalism, and natural death.

The results are as follows. Even a tenfold increase in adult mortality rate has no dramatic effect on the number of larvae: the population decreases but remains in stable equilibrium. But when the adult mortality rate is raised to 27%, small regular fluctuations appear in the adult population, and large and sustained oscillations in larval numbers. An increase to 50% mortality rate does not introduce any further qualitative novelties: the populations fluctuate periodically but are still in stable equilibrium. At 73% mortality rate, the fluctuations in one of the strains appear to dampen. The great qualitative leap only emerges at the 96% mortality rate, when aperiodic (apparently irregular) oscillations appear in both strains. That is, a drastic quantitative change does not result just in less of the same, but in a new demographic pattern. Moreover, the aperiodicities appear close to the boundary at which, according to the model, a bifurcation should occur—and bifurcation is, of course, the trademark of chaos. (When the mortality rate is set at 10%, the number of larvae can either drop or rise dramatically. The two possible population trends coexist until they coalesce at a mortality rate of about 60% for one of the genetic strains, and 70% for the other. When μ_a gets close to 100% the larvae population starts to fluctuate wildly.)

As stated above, both field and laboratory ecologists had observed population cycles and aperiodic fluctuations in several populations prior to the completion of this work, and even the occurrence of chaos was occasionally suggested. We repeat, however, that any given time series can be interpreted in alternative ways. In particular, it can be interpreted in either chaos-theoretical or probabilistic terms—e.g., as resulting from combining linear dynamics with "noise" (stochastic fluctuations). Only the experimental manipulation of the tuning parameter(s) occurring in a mathematical model, and the successful prediction of the resulting changes in population numbers, can offer some guarantee that a given dynamics is, indeed, at stake. The methodological moral is obvious: design ecological experiments in the light of some precise model.

This may be the place to warn against four either partially incorrect or utterly false but nevertheless popular beliefs about chaos: that chaos equals formlessness; that it is the subject of a fully-fledged theory invading all research fields; that it involves utter unpredictability; and the so-called butterfly effect. The first mistake is understandable: the choice of the word 'chaos' to designate the seeming randomness that originates in certain nonlinearities was unfortunate, because 'chaos' is an ancient word with totally different but related meanings, namely formlessness, irregularity, and lawlessness. On the other hand, the processes described by chaos "theory" are perfectly lawful. *Mock randomness* would have been a better choice, but now it may be too late to promote such rechristening. The second popular belief listed above, namely that chaos theory is a theory proper, i.e., a hypothetico-deductive system (see Sect. 3.5), is incorrect because, so far, there is only an embryonic general theory and a rapidly growing collection of examples of greater interest to mathematicians than to factual scientists.

Still, the field is being cultivated so intensively, that we must expect it to yield a rich harvest of particular interest to ecologists. In fact, important results are appearing with increasing frequency. One of the latest is the hypothesis that, when certain parameters characterizing insect populations are tuned so that the system is placed near the transition point between cyclical order and "chaos", the regular cycles are momentarily interrupted, to return after a while: they undergo an intermittent transition (Cavalieri and Koçak 1995). However, so far only a model simulation of this hypothesis has been performed: the optimistic hypothesis that there is hope of regaining simple order even after the onset of "chaos" is still awaiting experimental test.

The belief that chaos implies unpredictability is only partly correct, for there are two main sources of unpredictability. The first one is inaccuracy of the initial data. Thus, if the initial condition(s)—e.g., the initial numerosity of a biopopulation—and the value of the parameters, in particular the tuning one(s), were known exactly, then the equations of nonlinear dynamics would enable us to compute exact predictions provided they had exact solutions, not just approximate numerical solutions. (That such predictions may turn out to be falsified by the data, is another matter.) For example, in the simplest and most popular case, that of the logistic equation $x_{t+1} = rx_t (1 - x_t)$, the variable x_t eventually grows exponentially for certain values of the "knob variable" r, so that even a small error in the knowledge of the initial conditions entails an enormous error in the predicted trajectory. This, then, is a case of predictability in principle but unpredictability in practice: here unpredictability is due only to our lack of knowledge of either the initial condition(s) or the parameter(s).

On the other hand, the second type of unpredictability is inherent in certain nonlinear relations between numerical variables, such as $x_{n+1} = 2x_n^2 - 1$. Paradoxically, this does not hold for $x_{n+1} = x_n^2 - 1$. Indeed, iterating the latter for any number between 0 and 1 yields a periodic sequence of numbers between 0 and –1. On the other hand iterating $2x_n^2 - 1$ yields an aperiodic sequence of numbers between –1 and +1 that *looks* random. Moreover, a slightly different value of x_n produces at first no noticeable change in the output but, as the iteration proceeds, the new sequence is totally different from the first. Yet both sequences are manifestations of exactly the same underlying pattern. Thus, for certain values of the control parameter, a real process suddenly "branches out into two or more possible futures". The only way to find these out is by either computation or experiment. Yet, since the process is not random, there is no way of assigning a probability to every branch or possible future. In this case, the future itself is indeterminate, hence our knowledge of it uncertain.

In the case of the Flour Beetle model, the unpredictability occurs for only two values of the adult mortality rate μ_a. In particular, as this parameter reaches 10% for one of the genetic strains, the corresponding population is presented, metaphorically speaking, with two choices: either to decrease to one fourth or to multiply sixfold. This is a case of underdetermination. It is an open question whether the underdetermination might be removed by adding an equation for the rate of change of μ_a.

As for the "butterfly effect", it would consist in this: the fluttering of a butterfly's wings could cause a storm in the antipodes. This effect was first suggested by analyzing Lorenz's famous equations—and overlooking that these, though the brain-child of a meteorologist, do not describe any possible atmospheric processes, if only because they are rather simple and, moreover, not dynamical but kinematical. The energy of the shock waves generated by a butterfly's fluttering wings is soon dissipated in the surrounding air. Real storms involve huge energy transfers and build-ups beyond the power of even a gigantic swarm of Monarch butterflies migrating from Canada to Mexico. In any event, almost all the equations studied by chaos theorists are strictly kinematical: that is, they do not involve any mechanisms, let alone forces. Even the model of a Flour Beetle population discussed above involves only two causes: cannibalism and the external (experimental) regulation of adult mortality. But neither of these causes is treated in a dynamical way, i.e., representing them as forces. Indeed, cannibalism is represented only by its effect on the populations, and adult mortality by the numbers μ_a.

The morals of all the above for the experimentalist, in particular the ecologist, are these: (a) empirical tables and graphs, such as time series, do not reveal deep patterns, let alone concealed mechanisms; (b) depth is found in theories, not in figures or graphs; (c) the most conclusive experimental results are derived by combining experiment with mathematical modeling.

Having pointed out some of the potential benefits of chaos "theory" to ecology, a final warning is in order: 'chaos'—like 'fractal' and just like 'information' and 'catastrophe' earlier on (see Chap. 8)—has become a buzzword. It is sometimes used to make believe that sophisticated mathematics is involved where there is actually none, or even to lend an air of respectability to wild speculation. For this reason, we recommend asking those who speak of chaos too rashly and generously the simple question 'Where are your nonlinear equations?'.

In sum, chaos appears to inhere in the behavior of many ecological systems. However, the distinction between chaos and randomness is difficult: it can only be approached by mathematical modeling. Worse, the successful modeling of laboratory systems does not guarantee applicability to systems in nature, for what behaves chaotically in the lab need not do so in nature—and conversely (Wimsatt 1982b). So let us be patient with the young and complex science of ecology. Ecology may be an immature science but—*pace* Hagen (1989)—it is certainly not an anomalous science.

5.5.4 Applied Science and Technology

So far, we have been concerned with ecology as a basic science. That is, we have been concerned with ecology as it studies the relations of organisms to their environments as well as the kinematics and dynamics of biopopulations, communities, and ecosystems without regard to any practical application. Yet, just as with most other scientific disciplines, some branches of ecology are applied.

As the expression 'applied science' is ambiguous, we must distinguish at least two meanings of it. First, we speak of an applied science when one scientific discipline (e.g., its fund of knowledge or a certain method) is applied to another. For example, just as we can apply physics and chemistry to biology, so we may apply ecology to, say, functional morphology or evolutionary biology. Second, we speak of an applied science when scientists investigate cognitive problems with *possible* practical relevance. For example, the botanist who studies a plant because it may turn out to be useful for agricultural or pharmacological purposes engages in applied research. The same holds for the ecologist who investigates insects of a certain taxon because they are crop pests, so that knowledge of their ecology may be useful for pest control. In the following, we shall deal with applied science only in this second sense.

In so doing, we can distinguish *applied science* in general, or any applied scientific discipline in particular, from their basic or pure counterparts by the two following characteristics. First, although applied scientists try to solve cognitive problems, they do so under ultimately practical auspices. That is, they are supposed to come up not only with the finding of a certain item X, but also with the suggestion that X seems useful to produce a useful Y or prevent a noxious Z. Second, the domain D (or scope), as well as the fund of knowledge K of an applied science, are subsets of the scope and fund of knowledge of the corresponding basic science. In other words, applied science is narrower than basic science.

If the researcher takes the step from knowing to doing, that is, if he or she actually designs or produces a useful item, or else helps prevent the occurrence of an undesirable state of affairs, by means of the knowledge borrowed from basic and applied science, he or she is a *technologist*. In other words, we regard *technology* as the design, realization, operation, maintenance, or monitoring of things or processes of possible practical value to some individuals or groups with the help of knowledge gained in basic or applied research (Bunge 1983b, 1985b). If no scientific knowledge is involved, though perhaps a vast body of empirical prescientific knowledge as well as excellent craftsmanship, we speak of *technics*.

Note that our definition of "technology" is very broad: it subsumes not only the classical physical technologies, such as mechanical and electrical engineering, but also biotechnologies such as genetic engineering, medicine, and agronomy; psychotechnologies such as psychiatry and education; sociotechnologies such as law and city planning; and, finally, general technologies such as linear systems and control theory as well as computer "science". Lastly, just as there are pseudosciences, there are *pseudotechnics*, such as dowsing, and *pseudotechnologies*, such as psychoanalysis, astrology, and homeopathy.

Whereas a basic science can be characterized by a 10-tuple $\langle C, S, D, G, F, B, P, K, A, M\rangle$, as explicated in Section 5.5.1, a technology must be characterized by an 11-tuple, namely $\langle C, S, D, G, F, B, P, K, A, M, V\rangle$. The new coordinate V points out that technology has, in addition to the internal value system shared with basic science, an *external value system* or *exoaxiology*. That is, whereas basic scientists do not make value judgments other than for cognitive purposes about the referents or objects of their research, technologists must evaluate natural and artificial things as to their practical utility and efficiency. For example, while the value judgments of basic scientists about their objects of research are at most statements such as "The members of *Drosophila* are very suitable to study genetics and development", applied scientists make value judgments of the form "Knowledge of the life history of organisms of species S may be useful for controlling their population dynamics". Technologists eventually make value judgments such as "Chemical C is useful and efficient in order to control or diminish populations of pest organisms of species S".

Besides V, there are also several differences between science and technology with regard to the remaining coordinates. Suffice it to mention only a few of them (more in Bunge 1985b; Mitcham 1994). First, the technological community C is neither as open nor as international as the scientific community, because patents and industrial secrecy limit the circulation of technological knowledge. Second, within G, the ethos of technology is often not that of the free and disinterested inquiry in the service of mankind, but that of task-oriented work. Third, major differences between science and technology concern mostly the coordinates P and A, for the problematics and aims of technology are, of course, practical, i.e., action-oriented, rather than cognitive. So the technologist is not primarily interested in things in themselves but in things for us and under our control. We could also say that, whereas scientists, whether basic or applied, change things in order to know them, technologists study things in order to change them.

In sum, although there are no sharp borderlines between basic science, applied science, and technology, science and technology are quite different and should not be confused, as it is often done—in particular by the antiscience crowd both within and outside academia.

5.5.5 Ecology: Basic, Applied, or Technological?

When we study the ecology of, say, aphids for purely cognitive purposes, we do basic science. When we study the ecology of aphids in order to gain knowledge useful for finding a certain chemical or predator species which might help control aphid populations, we do applied research. And when we design or improve certain chemicals or when we experiment with different predator populations (e.g., ladybugs) in order to find the best way to control aphid populations because we regard them as pests, we engage in *ecotechnology*. We also do ecotechnology if we simply try to prevent certain ecosystems from changing, i.e., if we want to preserve them. After all, the conservation of ecosystems involves design, planning, and management. Often we must make decisions about the adequate size of the area to be preserved, about the necessity of connections (or corridors) among similar ecosystems, and about the need for artificial controls (e.g., hunting) of particular populations in the habitat to be preserved. All this is ecotechnology.

Given the pressing environmental problems of our time, it is undoubtedly admirable that many ecologists attempt to do something useful for the environment, that is, focus on applied ecology and ecotechnology. However, the philosopher of science must warn against the temptation to claim that ecology simply *is* an applied science (Shrader-Frechette and McCoy 1993, p. 150), or even that ecology should be nothing but applied and instrumental, i.e., an ecotechnology: "The division of pure and applied science is unnecessary and dangerous. If there are pressing environmental problems, then the world needs a science to manipulate and control the environment..." (Peters 1991, p. 186).

As we saw earlier, due to the complexity and variability of ecological systems, ecologists are having a hard time in coming up with general theories and law statements. Yet does this entail that they therefore should better give up basic ecology altogether and focus on ecotechnology instead, as Peters suggests? We do not think so. First, ecotechnology proper (as opposed to ecotechnics or even pseudoecotechnology) is supposed to be based on scientific knowledge. This knowledge is not provided by ecotechnology itself but by basic or applied ecology. For example, if we need a rule of the form "In order to obtain B, do A" (or, alternatively, "In order to avoid B, either refrain from doing A or prevent A") it is not sufficient that A and B be merely correlated: they must be *lawfully* related. (The corresponding law statement is "If A, then B".) This is why we call law-based rules 'nomopragmatic statements' (recall Sect. 3.5.8). However, laws_2 are sought for only in basic science, and they involve at least free models or, better, general theories. Since ecology apparently lacks both general and true theories, we must at least require a free model to show that some correlation of variables is lawful. Of course, purely empirical tampering with variables may accidentally bring about some effect, so that we can infer from the repeated confirmation of, say, "$A \Rightarrow B$" that (the referents of) A and B must apparently be lawfully related. Yet this is not doing science, and it betrays a remarkable lack of curiosity if one is not interested in what is behind this correlation and how it relates to other variables. Indeed, "[...] some of the longest-standing and most contentious issues in ecology evaporate when cause, mechanism, and explanation are ignored" (Peters 1991, p. 146). Regrettably, at the same time, ecology also evaporates as a science.

It comes as no surprise, therefore, to learn that Peters defends instrumentalism. This is the thesis that scientific theories are nothing but descriptive summaries of past observations as well as instruments for prediction. This idea was held, for instance, by Cardinal Bellarmino, by positivists like Comte and Duhem, and by logical empiricists like Ryle and Toulmin. Recently, Rosenberg (1989, 1994) has espoused instrumentalism with regard to biology, psychology and the social sciences. The problem with instrumentalism is not that it is false. After all, the scientific realist also believes that theories describe and help predict. The problem lies in the thesis that this is all that scientific theories are able or should be able to do. Instrumentalism discourages explanation, because it is interested neither in truth nor in the things behind the appearances, and it thus can explain neither the success nor the failure of scientific theories. By contrast, scientific realism, by reference to transphenomenal entities and the systemicity (or, if preferred, consilience) of scientific knowledge, encourages explanation. In short, instrumentalism is inferior to scientific realism. (And if a nonphilosophical judgment is allowed, it is just boring. More on instrumentalism in Popper 1962 and Vollmer 1990.)

Still, one might argue that, even though barren from a realistic point of view, instrumentalism is acceptable as long as it contributes to solving environmental problems. Granted. But we can have all the benefits attributed to instrumentalism in a scientifically realist ecology, too, which, as a matter of fact, comes as a basic and an applied science. So, *pace* Peters, it is not the distinction between basic and applied ecology that is dangerous but the distortion, nay, castration of science that is involved in instrumentalism.

To conclude, let us nurture ecology in three forms: basic, applied, and technological. However, we should not be surprised to find that, as basic ecology is still a developing science, the power of ecotechnology for the solution of environmental problems is limited. After all, the strongest flow of knowledge goes from basic ecology to applied ecology to ecotechnology. If there is no basic and general knowledge to borrow from, applied ecology and ecotechnology remain piecemeal enterprises of very limited power. This is why Shrader-Frechette and McCoy (1993) are able to claim that ecology is a science of case studies. Although this is, in fact, so up to a point, we ought not to be satisfied with this state of affairs.

5.5.6 Ecology: An Autonomous Science?

It is often said that biology is an autonomous science (e.g., Ayala 1968; Mohr 1981; Mayr 1982). In turn, ecology is sometimes said to be an autonomous science within the biological sciences. Yet what exactly does "autonomous" mean? (See Simon 1971 for an alternative analysis.) Etymologically, "autonomous" means "having laws of its own". Curiously, the existence of biological laws (or at least that of nontrivial or significant laws) is just what is questioned by Mayr and other autonomists. Similarly, as we saw above, the existence of laws$_2$ in ecology is also a matter of controversy.

A related connotation of "autonomous" is "independent". We say that a science *A* is *independent* of another science *B* if, and only if, all the problems in *A* can be solved without using any findings in *B*. Otherwise, *A depends* upon *B*. Thus, while physics is independent of biology, biology depends on physics. Indeed, biology (hence ecology) makes ample contact with other scientific disciplines as well, such as chemistry and the earth sciences: biology (hence ecology) is part of the tightly knit system of the natural sciences. (This becomes apparent when we examine the background knowledge of biology or ecology, respectively: see the coordinate *B* in Sect. 5.5.1.) If biology or ecology, respectively, were an independent science, it would either replace physics as the basic factual science or else be an isolated field—in which case conditions (vi) and (xi), as given in Section 5.5.1, would be violated, and we would have to classify it as a pseudoscience. Both alternatives are obviously not true.

What the "autonomists" actually mean when speaking of biology or ecology as autonomous sciences is that biology, and likewise ecology, are sciences of their own, for biology is not reducible to physics and chemistry—in particular molecular "biology", and ecology is not reducible to other biological disciplines. Now, much has been written about biology and reductionism (see Sect. 3.6.3, as well as Rosenberg 1985), and we see no need to repeat all the arguments here. Yet we wish to emphasize that the strongest argument against radical reductionism is ontological, not epistemological. If the emergentist-materialist ontology underlying biology (and, as a matter of fact, all the factual sciences) is correct, the *bios* constitutes a distinct ontic level the entities in which are characterized by emergent properties. The properties of biotic systems are then not (ontologically) reducible to the properties of their components, although we may be able to partially explain and predict them from the properties of their components. Recalling the CES analysis of systems (Sect. 1.7.2), it becomes obvious that the belief that one has reduced a system by exhibiting its composition, which is indeed nothing but physical and chemical, is insufficient: physics and chemistry do not account for the structure, in particular the organization, of biosystems and their emergent properties (see also Fig. 4.1).

So Mayr (1982) is right in using the ontological concept of emergence to argue for the autonomy of biology. (Recently, Gasper 1992 has argued for the nonreducibility of genetics to molecular biology by reference to ontology, although he regrettably used the ill-conceived notion of supervenience criticized in Sect. 1.7.3; and Rosenberg 1994 has claimed that nonreducibility entails an instrumentalist view of biology.) However, he fails to realize that the thesis that biosystems have emergent properties entails that they have $laws_1$ of their own, so that his defense of the autonomy of biology and his doubt about the existence of laws in biology are mutually inconsistent. (Recently, Mayr seems to have turned less antinomianist and to admit that biology has laws of its own: see his 1996, p. 105.) Moreover, in the light of emergentist materialism, all scientific disciplines deal with entities belonging to some level of organization and which thus (must) have $laws_1$ of their own. Therefore, all scientific disciplines are autonomous in this sense, so that a special defense of biology as an autonomous science appears to be unnecessary from this point of view.

This defense, however, is not trivial from a historical point of view. For one, the unification program of neopositivism attempted to unify all the sciences by reducing them to physics (recall Sect. 3.6.3.2). For another, although the attempt to reduce biology to physics may no longer be pursued, we now face the claim that most, if not all, of biology is reducible to molecular biology. We trust that this program will fail just as miserably as its neopositivist forerunner.

As is implicit in the preceding, what holds for biology as a whole is also true for ecology. In particular, the distinctness (autonomy) of ecology derives from its domain, the entities in which belong to several distinct biolevels. Therefore, we cannot expect population ecology to be reducible to organismal ecology (i.e., autecology), or ecosystem ecology to be reducible to population ecology. (Moderate reductionism, however, is a useful strategy: see Sect. 3.6.3.2, as well as Schoener 1986.) Rather, we must expect qualitative novelty, hence new laws$_1$, at each level. Needless to repeat, the methodological difficulties with the discovery of such laws and their proper representation in terms of law statements in ecological theories are entirely different matters.

To conclude, not only biology but also ecology is a *distinct* (autonomous$_1$) science in that it deals with entities at distinct ontic levels. However, neither biology nor ecology are independent (autonomous$_2$) sciences, for they are related to other scientific disciplines in many different ways: they are part of the system of (special) sciences. Moreover, though ontologically autonomous, the two are not methodologically autonomous: only pseudosciences are. Indeed, biology shares a common method with all the other sciences, which does not imply that there is no room for possible methodological peculiarities, such as narrative or functional (teleological) explanation. However, whether there actually are such methodological differences is contentious, so that philosophers of biology will be kept quite busy for a while with either establishing real differences or reducing alleged ones to known methodological categories.

6 Psychobiology

Having elucidated the notions of function and role in Chapter 4, we may now look at the function of the most complex, intricate, and fascinating organ known to biologists: the brain, in particular the human brain. Interestingly, the philosophical problems concerning the function of this organ, in particular the famous mind-body problem, are usually not dealt with in the philosophy of biology proper, but are left to the so-called *philosophy of mind.* Only a few biophilosophers have considered the mind-body problem, among them Rensch (1971). (Regrettably, he espoused panpsychism, which not only lacks empirical support, but also suffers from defects analogous to those of hylozoism; recall Sect. 4.2 and see Vollmer 1985, Vol. 2) However, if the philosophy of mind is to be compatible with science, it must take biology into account, particularly neuroscience, so that it becomes a proper part of the philosophy of biology. So we shall make a brief foray into this area. However, before doing so, it will be convenient to recall some of the achievements of the biological approach to behavior, affect, cognition, and volition—the four main subjects of psychology.

6.1 Successes of the Biological Approach to Psychology

Classical or prebiological psychology was conceived of as the study of the soul or mind, supposedly an immaterial and perhaps also an immortal entity. This study achieved some success in describing some features of overt behavior and subjective experience, in particular perception, learning, and memory. Yet, because it was regarded as an autonomous discipline, it learned nothing from physics, chemistry, or biology—in particular neuroscience and developmental biology. And because it ignored the central nervous system, classical psychology was incapable of explaining even the few facts it was capable of describing—since explanation, it will be recalled, involves conjecturing or revealing mechanisms, which are processes in concrete things. For the same reason, it was of no help in the task of "mapping the mind onto the brain", i.e., localizing the various mental functions—an indispensable tool for neurosurgery.

All this started to change in the 19th century with the emergence of physiological psychology, fathered by Flourens, Broca, Wernicke, and von Helmholtz, followed in the early part of the 20th century by Hess, Papez, and a few others. However, most psychologists took no notice of this new approach until mid-century, when Penfield elicited mental processes by applying weak electric currents to the wakeful brain cortex, and Hebb introduced his hypotheses of the ***Hebb synapse*** and the *cell assembly* formed when two or more neurons fire together. From then on, psychobiology made some spectacular discoveries beyond the reach of classical psychology. Let the following list suffice: effect of lithium salts on depression; localization of the pleasure center; "paradoxical" sleep; loss of recent memory following the surgical removal of certain parts of the brain; localization of memories of different languages; seasonal changes in birdsong patterns accompanying the death and birth of neurons in their song nuclei; chemical changes accompanying drug addiction; effects of emotional processes on morbidity and mortality; sprouting of dendrites and synaptic boutons under the action of radiation or hormones; localization of perception of novelty; effect of hormonal changes upon mood; loss of memory in Alzheimer patients as an effect of neuronal death; effect of experience upon the organization of the cortex; and last, but not least, the realization of the strong interactions between the nervous, endocrine, and immune systems, as well as among the various subsystems of the brain, such as the cortex, the limbic system, and the pituitary body.

The sensational success of the biological approach to the mental confirmed, and considerably refined, the materialist hypotheses of Hippocrates and Galen, that mental processes are brain processes, that all mental functions (except for memory) are localized, and that mental disorders are brain disorders. However, these are the subjects of the following sections.

6.2 The Mind-Body Problem

The mind-body problem is the system of questions about the nature of the mental and its relation to the bodily. Some of these questions are: Are mind and body two separate entities? If so, how are they held together in the organism? How do they get in touch in the beginning, how do they fly asunder at the end, and what becomes of the mind after the breakdown of the body? Do these entities interact and, if so, how? Which, if any, has the upper hand? If, on the other hand, they are not different entities, is the mind corporeal? Or is the body a form of the mind? Or is each a manifestation of a single (neutral) underlying substance?

According to the answers given to these questions, we can distinguish two families of doctrines. The first, which takes mind and body to be two separate entities, is called *psychophysical* (or *psychoneural*) *dualism*, an ingredient of all religions and idealist philosophies. Well-known dualists, for example, are Plato, Descartes, Leibniz, Freud, Wittgenstein, Eccles, and Popper. The most popular varieties of dualism are animism and interactionism. According to animism, the immaterial mind (or perhaps soul) acts on the body, but not conversely, while interactionism, as the name suggests, asserts that the (immaterial) mind and the (material) body (or brain) interact. (Needless to say, the nature of this interaction remains a mystery. Besides, such interaction would violate the law of conservation of energy.) Obviously, the notion of an immaterial mind steering human behavior is deeply entrenched in human history. Moreover, this idea is behind the teleological world view, which consists in explaining the world in terms of purposefully acting agents: God rules the cosmos, natural selection pushes and directs evolution, and the genetic program guides and directs development. All this has been superseded *de jure*, though not *de facto*: indeed, some scientists and philosophers still have a hard time giving up these primitive ideas.

The second family of doctrines, which assumes a single entity, is known as *psychophysical* (or *psychoneural*) *monism*. Monism can be idealist (Berkeley, Fichte, Hegel), neutral (Spinoza, Carnap, Feigl), or materialist. The latter, in turn, comes in several varieties: eliminative (Watson, Skinner, Ryle), reductionist or physicalist (Epicurus, de La Mettrie, Smart, Armstrong, Quine), and emergentist (Diderot, Ramón y Cajal, Hebb, Luria, Mountcastle, Bindra, Vollmer). Since the varieties of dualism and monism have been examined elsewhere (Bunge 1980; Bunge and Ardila 1987; see also Vollmer 1985, Vol. 2), and since we are dealing here with the philosophy of biology, not that of psychology, we shall restrict our musings to the sole kind of materialist monism which we claim to be consistent with biology: *emergentist materialism*.

However, before doing so, we must take a quick look at a view which is harder to classify. This is the view that everything mental is substrate-neutral algorithm or computation—a position known as *functionalism*. Since, according to this view, function is detachable from substance or "substrate", that is, from matter, functionalists believe that any given mental function can be discharged not only by chunks of nervous tissue but also by machines (particularly computers), and perhaps even disembodied spirits (see, e.g., Putnam 1975, Vol. 2; Dennett 1978; Block ed. 1980; Fodor 1981; Pylyshyn 1984; more on the heterogeneous collection of views subsumed under the label 'functionalism' in Block 1980a, b). Thus, functionalists "abstract from physical [material] detail": they are not interested in neurons, glial cells, dendrites, neurotransmitters, synaptic boutons, or even in multicellular systems, such as the striate cortex, the amygdala and the hippocampus—the very organs of behavior, emotion, and ideation. They are only interested in "functions" in themselves (in particular abstract computer programs or algorithms) regardless of the way in which they are "embodied" or materialized.

Thus, there appear to be two entities, function and "substrate"—or mind and matter—which are related to each other like software and hardware. Moreover, function (or software) is clearly the more important of the two. For this reason, we regard functionalism as a dualist doctrine—the claims of some functionalists notwithstanding that it is materialist (e.g., Dennett 1978), or else neither dualist nor materialist (e.g., Fodor 1981). (Note that the opposite of dualism is monism, not materialism; but all materialist views are monistic, whereas the converse is not true.) After all, if the mind is some sort of software it can, in principle, be detached from the brain and transmitted to, and run on, some other substrate or "carrier". Perhaps we could even run two or more minds on one brain, such as that of a deceased friend. All this is clearly impossible according to emergentist materialism.

In any case, functionalism is more popular among certain philosophers and workers in Artificial Intelligence as well as Artificial Life than among biologists and psychologists. Indeed, neuropsychology, in particular cognitive neuroscience, rejects what Kosslyn and Koenig (1995) aptly call the *dry mind approach* in contrast to the *wet mind approach*, according to which "the mind is what the brain does". (See also Beaumont et al. 1996.) Ignoring matter (except perhaps as a compositionally neutral carrier of function), functionalism discourages neuroscientific research; and, focusing on computation, it disregards the neurobiological and psychological study of nonalgorithmic mental processes, such as perception, emotion, imagination, concept formation, analogical thinking, conjecturing, and evaluating. Thus, functionalism does not suggest a heuristically fruitful research strategy for psychobiology. As a matter of fact, one of its founders has repudiated it after defending it for three decades: see Putnam (1994). We concur and turn to emergentist materialism at last.

Emergentist psychoneural monism is part and parcel of the view that the universe is material and always in flux. It boils down to the following theses (Bunge 1980):

1. All mental states, events, and processes are states of, or events and processes in, some organismal brains.
2. These states, events, and processes are emergent relative to those of the cellular components of the brain.
3. The so-called psychophysical (or psychosomatic) relations are interactions between different subsystems of the brain, or between some of them and other components of the organism, such as the muscular, digestive, endocrine, and immune system.

We submit that only by adopting this view is it possible to speak of *psychobiology* proper. From this psychobiological perspective the very expression 'mind-body problem' is a scientific anomaly. For we do not speak of the motion-body problem in mechanics, of the reaction-substance problem in chemistry, or of the respiration-lung problem in physiology. We do instead speak of the motion *of* bodies, of the respirative function *of* the lungs, and so on. We do not reify properties, states, or processes—except when it comes to the properties, states, and processes of the nervous system. Thus, we can still find anomalous, i.e., dualistic, expressions in the neuroscientific literature, such as 'the neurophysiological *basis* of the mind', 'neural *correlates* of mental functions', 'physiological *equivalents* of mental processes', or 'neural *representation* of mental processes'. By contrast, hardly a physiologist, for instance, would speak of the 'renophysiological basis of excretion' because excretion is just what the kidneys do. (Yet there are occasional lapses into process reification. Thus, in the wake of Whitehead and Woodger, Løvtrup 1974, p. 226, speaks of the "material basis of development".)

For reasons of scientific and philosophical consistency, then, it is imperative to eliminate such anomalies from neurobiology. There are no processes or activities in themselves, i.e., apart from concrete changing systems. Furthermore, if mentation is (identical with) an activity of the brain (or of some of its subsystems), then it should be obvious that there is no such thing as 'mind-brain identity'—an expression occasionally to be found in the literature. An organ is not identical to its function or activity. There is no mind-brain identity any more than there is a respiration-lung, an excretion-kidney, a walking-leg, or a facial muscle-smile identity.

Let us briefly expound how concepts such as those of mind, consciousness, and self might be conceived from a psychobiological point of view (following Hebb 1949, 1980; Bindra 1976; Bunge 1977b, 1979a, 1980, 1981a; Bunge and Ardila 1987).

6.3 Mental States and Processes

The components of nervous systems of greatest importance to psychobiology are, of course, neurons and the systems or networks into which several neurons may be connected. In Chapter 3 we have introduced several basic notions, such as the connectivity of neurons, and defined the concept of a plastic neuronal system (Definition 3.1). But before we can tackle the concepts of a mental state and process, we have to make the assumption that the plastic neuronal systems of an animal are not isolated but organized in a supersystem. That is, we assume:

POSTULATE 6.1. The plastic neuronal systems of an animal are coupled to form a supersystem, namely the *plastic neuronal supersystem (P)* of the animal.

The psychoneural identity hypothesis states that every fact experienced introspectively as mental is identical with some brain activity. However, not every brain activity is mental, and not all neuronal systems are capable of carrying out mental functions. We hypothesize that only some of the functions of plastic neuronal systems can be identical to mental functions. Hence, functions in nonplastic systems are nonmental. We thus take, for instance, hunger, thirst, and sexual urge to be nonmental functions because they are assumed to be located in committed (or "prewired") neuronal systems. What *can* be mental is the consciousness of any such processes.

DEFINITION 6.1. Let b denote an animal endowed with a plastic neuronal system P. Then

(i) b undergoes a *mental process* (or performs a *mental function*) during the time interval τ iff P has a subsystem n such that n is engaged in a specific process during τ;
(ii) every state (or stage) in a mental process of b is a *mental state* of b.

(For the notions of function and specific function see Definition 4.8.) Some of the immediate consequences of Definition 6.1 are:

COROLLARY 6.1. All and only animals endowed with plastic neuronal systems are capable of being in mental states (or undergoing mental processes).

COROLLARY 6.2 All mental disorders (dysfunctions) are neural disorders.

This corollary contradicts the standard dichotomy between organic and functional (or behavioral) disorders. This dichotomy and the accompanying division of labor into neurology and psychiatry on the one hand, and clinical psychology on the other, is inspired by psychoneural dualism. According to psychoneural monism, all mental states, whether normal or abnormal, are organic: they are all states of the CNS. The difference is not one between organic and psychological disorders but one between sickness originating at the cell level, e.g., dopamine deficiency or serotonine excess, and sickness at the system level, i.e., "wrong" (or malaptive) connections. Because of the plasticity of a large part of the human cortex, systemic or behavioral disorders can often be cured by relearning, e.g., undergoing behavior therapy, or a mere change of environment. By contrast, sick neurons call for a biochemical approach instead of logotherapy: individual cells do not listen.

COROLLARY 6.3. Mental functions (processes) cease with the death of the corresponding neuronal systems.

Though obvious to the biologist, this conclusion is seldomly stated explicitly (e.g., by Maynard Smith 1986, p. 80). And of course, this is an unpleasant corollary for those who believe in either afterlife or reincarnation. To avoid it, one must either reject emergentist materialism, or resort to miracles, e.g., by claiming that God, being omnipotent, can resurrect both body and mind on the Day of Judgment (Priestley 1776).

COROLLARY 6.4. Mental functions (processes) cannot be directly transferred (i.e., without any physical channels) from one brain to another.

If one admits that the mental is a brain function, then extrasensory perception (ESP) is out of the question. On the other hand, only technical difficulties are currently in the way of inter-brain communication with physical means other than optical, acoustical, tactual, and other conventional signals. However, telepathy via electromagnetic waves emitted, received, and decoded by brains is impossible if only because the radiation emitted by the brain is far too weak for that purpose. Not surprisingly, parapsychology has to rely on paranormal modes of communication, which is the reason why it never will become a science (Beyerstein 1987; Bunge 1987b).

6.4 Mind

6.4.1 Basic Concepts

We are now ready to tackle the concept of mind. If the mind is not regarded as a spiritual entity, the emergentist materialist has no problems with the concept of mind. Indeed, recalling Definition 4.8, which elucidates the notion of a specific function of an organ, a definition of the concept of mind may read as follows:

> DEFINITION 6.2. Let P denote a plastic neural supersystem of an animal b of species K. Then the *mind* of b during the period τ is the union of all the mental processes (specific functions π_S) that components of P, i.e., plastic neuronal systems n, engage in during τ. More precisely,
>
> $$M(b, \tau) = \bigcup_{x \lhd n} \pi_S(n, \tau).$$

As with the concept of life (Postulate 4.1, Definition 4.1), the concept of mind refers to a set of specific activities, not to some (material or immaterial) entity. (See also Mayr 1982.) It should also become clearer now why, since the members of the set called 'mind' are brain processes, it makes no sense to say that the brain is the "physical basis" of the mind, just as it makes no sense to say that the gut is the "physical basis" of digestion. Furthermore, it makes no sense to speak of the "collective mind" of humankind as if it were an entity or even a functional system. By the same token, there is no collective memory (à la Durkheim) and no collective unconscious.

Although we have already dealt with constructs in Chapters 1 and 3, it will be helpful to embed what has been called the 'objects of the mind' into the psychobiological context. We therefore propose:

> DEFINITION 6.3. Let x denote an object and b an animal endowed with a plastic neuronal system. Then
>
> (i) *x is in the mind of b* iff x is a mental state or a mental process of b;
>
> (ii) *x is in the mind* (or *is mental*) iff there is at least one animal y such that x is in the mind of y.

This definition suggests the following philosophy of mathematics. Mathematical objects are objects that can exist only in some mind. That is, mathematical objects are patterns of possible brain processes: they do not exist elsewhere, or by themselves, apart from thinking brains. Likewise, mythical objects, such as Zeus and Donald Duck, are no more than that. Further, the mind is finite, that is, we can think only finitely many mathematical objects. However, we make up for this finiteness by *pretending* that all those possibilities that are not actualized exist (formally). Thus, nobody can think of all the numbers, not even of the whole ones. Likewise, we shall never be able to derive all the infinitely many theorems of a theory, but we *feign* that they exist (formally or ideally): this alone authorizes us to speak of *the* theory. We may call this a fictionist and materialist philosophy of mathematics. (More on this in Bunge 1985a, 1997. Disappointingly, not even recent papers on the metaphysics of ideal objects and the philosophy of mathematics, e.g., Linsky and Zalta 1995 and Lowe 1995, mention, let alone consider, a psychobiological approach to constructs.)

6.4.2 Mind-Matter Interaction

The interactionist dualist faces the problem of how the immaterial mind may conceivably interact with matter, i.e., the brain. Not so the emergentist materialist: there can be no mind-matter interaction because—unlike individual mental processes and brains—mind and matter are sets, hence conceptual objects. However, it does make sense to speak of 'mental-bodily interactions' provided this expression is taken to abbreviate "interactions among plastic neuronal systems, on the one hand, and either committed neuronal systems or bodily systems that are not part of the CNS on the other". Thus, there are interactions between sensory and motor areas, between ideational neuronal systems and external receptors, between the cortical and subcortical regions of the brain, between the brain and the endocrine and immune system, and so on. Because mental events are neural events, and because the causal relation is defined for pairs of events in concrete systems (recall Sect. 1.9), we have:

> COROLLARY 6.5. Mental events can cause nonmental events in the same body, and conversely.

Consequently, disturbances of nonmental biofunctions may influence mental states and, conversely, mental events such as acts of will may influence nonmental bodily states. This is what neurochemistry, neurology, psychiatry, psychosomatic medicine, psychoneuropharmacology, education, and propaganda are all about. As well, scientific and effective psychosomatic medicine, unlike psychoanalysis, is the application of psychoneuroendocrinoimmunopharmacology to the treatment of mental disorders. (Note the extraordinary length of the name of this discipline born from the merger of previously disconnected research fields. This is one more reminder that scientific progress comes not only from reductions but also from mergers: recall Sect. 3.6.3.3.)

6.4.3 Where Is the Mind?

The Cartesian synonymy of 'body' and *res extensa*, on the one hand, and of 'mind' and *res cogitans*, on the other, epitomizes the dualistic tenet that, whereas the physical is extended, the mental is not. Since in our ontology there are no events in themselves but only events in some concrete thing or other (see Sect. 1.5), the question of the space "occupied" by an event is the question of the extension of the changing thing. Thus, not the *firing* of a neuron but the *firing neuron* is spatially extended. In other words, events occur wherever the "eventing" things may be. In particular, mental events occur in some plastic neuronal system or other. So, in principle, and often also in practice, the neurophysiologist can determine the location of a thought process. Various imaging techniques, from the almost classic electrical stimulation of precise cortical sites to positron emission tomography and magnetic resonance imaging, are now available and are producing exciting findings in quick succession.

However, the postulate that mental events occur only in brains does not entail that the mind, as the set of all mental events, is in the head. Since it is a set, hence a conceptual object, the mind is nowhere: only brains, whether minding or not, are somewhere. A related question is the one about the "location" of ideas in general. The answer to this question depends on the construal of 'idea'. If taken as ideation processes, ideas are in the brains that happen to think them up—but only there and only at the time they are being thought. On the other hand, the so-called *product* of any such process, i.e., the "idea in itself", is nowhere in spacetime because it does not exist by itself: we only feign it does for purposes of analysis. For example, although thinking of the number 3 is a brain process, the number 3 is nowhere because it is a fiction existing by convention or fiat, and this pretense does not include the property of spatiotemporality. What holds for the number 3 holds for every other conceptual object. In every case, we abstract from the neuropsychological properties of the concrete neuronal system that does the ideation, and come up with a construct that, by definition, has only conceptual or ideal properties. In other words, a *construct is* not an individual brain process but *an equivalence class of brain processes* occurring in different brains or in the same brain at different times. (More on this in Bunge 1983b.)

Although the preceding is only a programmatic hypothesis, we believe that it is basically true and, moreover, heuristically powerful. Of course, the particularities of the brain processes that are identical to thinking a given concept are likely to vary from individual to individual, from one circumstance to another, and even from moment to moment in one and the same individual. Thus it may well be that nobody undergoes exactly the same brain processes when thinking of the number 3 at different times. However, any such thoughts of the same construct must have the same general pattern, for otherwise they would not consist in thinking of the number 3.

6.5 Consciousness

Let us now tackle the highest of all brain functions, namely consciousness or self-knowledge. However, before we can do so, we must distinguish two related but different concepts which are often mixed up, namely reactivity and awareness.

All things, whether alive or not, are sensitive to some physical or chemical agents, though none responds to all: think of photosensitivity or chemical sensitivity. Thus, sensitivity to external physical or chemical stimuli does not require the presence of a nervous system. Consequently, if we were to identify consciousness with mere reactivity (or sensitivity) to external stimuli, as some people do, we would have to adopt animism or panpsychism, thus giving up the naturalist ontology of modern science. Let us then make a first distinction:

> DEFINITION 6.7. Let b denote a thing (living or nonliving) and x an action on b or on a part of b, and originating either outside b or in a part of b. Then b is *x-responsive* if, and only if, b reacts to x (i.e., if x causes or triggers a change of state of b), either always or with a certain probability.

The concept of awareness, on the other hand, applies only to living beings of certain species, namely those who possess a CNS of a certain complexity. Awareness, then, is one of the specific functions of a sufficiently complex CNS:

> DEFINITION 6.8. If b denotes an animal, b is *aware* of change x (internal or external to b) iff b feels or perceives x —otherwise, b is unaware of x.

For example, in the case of blindsight, a person is able to sense, and react to, external objects in spite of not being aware of them. Thus, awareness requires not only sensory organs (in the case of external stimuli) but also perceptual neuronal systems. (In principle, the latter are sufficient for awareness, as is the case with hallucinations.) Hence, plants and animals lacking such organs cannot be aware of anything. *A fortiori*, machines cannot attain awareness, although, if equipped with suitable "sensors" such as photocells, they can react to certain stimuli.

Note that an animal may be aware of its surroundings but not of what it is feeling or doing itself. To cover the latter possibility, we need also:

> DEFINITION 6.9. If b denotes an animal, b is *self-aware* (or has self-awareness) iff b is aware of some of its inner changes and actions.

To be self-aware is to be aware of oneself as something different from everything else. A self-aware animal notices, however dimly, that it is the subject of its own feelings and doings. Yet self-awareness does not require *thinking* about one's own perceptions or conceptions. Satisfaction of this additional condition qualifies as consciousness:

> DEFINITION 6.10. If b is an animal, b is *conscious* of brain process x (e.g., perception or thought x) in b itself iff b thinks of x —otherwise, b is not conscious of x.

According to this convention, an animal can only be conscious of some of its own higher mental processes: not just feeling, sensing, and doing, but also thinking of what it perceives or thinks. An animal conscious of mental process x (in itself) possibly undergoes (either in parallel or in quick succession) *two* different mental processes: x —the object mental process or content of its consciousness, and thinking about x —i.e., being conscious about x. We thus postulate that a conscious event is a brain activity consisting in monitoring (recording, analyzing, controlling, or keeping track of) some other activity in the same brain. Although scientists do not know yet for sure in which systems these activities take place, we formulate:

> POSTULATE 6.3. Let P denote a subsystem of the CNS of an animal b engaged in a mental process p. Then the CNS of b contains a neuronal system Q, other than P but connected with P, whose activity q equals b's being conscious of p.

A remark on the difference between awareness and consciousness may be in order. The animals of some species can become aware of certain stimuli, and some are capable of attention, but they cannot be conscious of anything unless they can think, that is, form concepts and propositions. Conversely, a person lost in daydreaming or in deep, productive thought may be unaware of her surroundings. Consequently, the two concepts are mutually independent, and should therefore not be confused. Moreover, the neurophysiological processes involved in awareness might well be different from those of consciousness: awareness might consist in the oscillatory synchronicity of the neurons of a given neuronal system rather than in the monitoring of one neuronal system by another (see, e.g., Zeki 1993).

All consciousness is consciousness of something. This something is called the *content* or *object* of consciousness. Consciousness without a content, as certain forms of meditation attempt to attain, is no consciousness at all, but merely a state of mindlessness, similar to deep sleep.

Being conscious of a mental process in oneself is to be in a certain mental state —which is the same as the brain being in a certain state. For this reason, we adopt:

> DEFINITION 6.11. The *consciousness* of an animal b is the set of all the states (or, rather, processes) of the brain of b in which b is conscious of some perception or thought in b itself.

Since consciousness is not an entity, it is incorrect to speak of *The Unconscious* as if it were an entity. Instead, there are simply some mental processes that remain nonconscious or preconscious, even though, on occasion, they can be manifested behaviorally. Furthermore, since being conscious of something is a *state* of the brain (or, rather, a sequence of states, i.e., a process), there can be no such things as "states of consciousness" nor, *a fortiori*, "altered states of consciousness". Such are instances of reification. (But of course, there may be altered brain states, e.g., due to drugs or injuries.) Neither can there be a "collective consciousness", because there are no collective brains. However, there can be degrees of being conscious because such degrees are nothing but the intensities of the activities (functions) of the corresponding neuronal systems.

Just as self-awareness is one rung higher than awareness, so self-consciousness is one step higher than consciousness. A subject is self-conscious only if she is conscious of her own perceptions and thoughts as occurring in herself. At first sight, the term 'self-consciousness' is a pleonasm. However, there is solid clinical evidence that subjects in certain pathological conditions are confused about the source of some of their own mental experiences and even actions. Therefore, we need:

DEFINITION 6.12. An animal is *self-conscious,* or has a *self* at a given time if, and only if, it knows who and what it is.

Again, the self is not an entity, but a state of an entity, namely a sufficiently complex brain. Therefore, to say that "the self has a brain" (Popper and Eccles 1977) amounts in our view to saying that certain brain states have a brain, that sneezes have noses, or that wheels have rotations.

Now, in order for one to know who and what one is, one must have some recollection of one's past. On the other hand, the animal need not be able to extrapolate its own life into the future, that is, it may not be capable of imagining or planning its next move (e.g., because it lacks sufficiently developed frontal lobes). Accordingly, we lay down:

DEFINITION 6.13. A self-conscious individual is
(i) *antero-self-conscious* iff it recalls correctly some of its past;
(ii) *pro-self-conscious* iff it can imagine (even wrongly) some of its own future;
(iii) *fully self-conscious* iff it is both antero- and pro-self-conscious.

All these abilities come in degrees. Thus, an individual with Alzheimer's may recall only some episodes in the distant past, and a lobotomized individual is unable to imagine more than the immediate future.

6.6 Intention

Our definitions of "awareness" and "consciousness" did not involve the concept of intention. Indeed, only *some* conscious behavior is intentional. For example, what began as a voluntary act may, if learned, become automatic, i.e., still goal-striving but no longer intentional. And conscious behavior may be aimless, as in daydreaming. Acts of will, however, are conscious and intentional. We thus propose:

DEFINITION 6.14. An animal act is *voluntary* (or *intentional*) if, and only if, it is a conscious purposeful act. Otherwise, it is *involuntary.*

It goes without saying that the will is not an entity but a neural activity: x wills y iff x forms consciously the purpose of doing y. Nor is it a mysterious faculty of an immaterial mind, but a capacity of a highly evolved CNS. More precisely, it seems to be a specific function of the frontal lobes.

Voluntary acts can be free or compelled. The general who decides to launch an attack may act freely, but those of his soldiers who go unwillingly to battle act voluntarily, though under compulsion. Thus, free will is volition with a free choice of goal, with or without foresight of possible outcome. We make then:

DEFINITION 6.15. An animal acts of its own *free will* if, and only if,
(i) its action is voluntary; and
(ii) it has free choice of its goal(s)—i.e., is under no programmed or external compulsion to attain the chosen goal.

Vulgar materialists, as well as positivists and behaviorists, regard free will as illusory, hence the very notion as nonscientific. The former deny that it can possibly be exerted by a physico-chemical system, and the latter reject it as being unobservable. Most idealists (or spiritualists) accept free will but deny it a scientific status because they regard the free voluntary act as lawless or spontaneous, hence unpredictable. None of these characteristics is implied by our definition.

The philosophical literature is littered with confusions regarding free will. Two of them are the alleged identities "determinism = predictability" and "free will = indeterminacy". However, the concept of determinacy is an ontological category, whereas that of predictability is an epistemological one. Hence, in principle, we can have the one without the other. For example, even though a process may be perfectly determinate (i.e., lawful and subject to constraints and antecedent conditions), we may know it only imperfectly, and therefore may not be in a position to predict it. Most physical processes are of this kind. Likewise, the concept of free will is an ontological category, so that predictability does not count against it, and unpredictability cannot be taken as a test or criterion of free will. Since free volition is assumed to be a neural process, it must be lawful. Yet it is not causal, because no spontaneous process is. But causality is just one mode of lawfulness (see Bunge 1959a). Being lawful, free volition must be capable of repetition (*ceteris paribus*) and predictable. For example, if we know a person reasonably well, we may be able to predict that he or she will choose of his or her own free will to perform a certain action of kind *A*, whenever confronted with a problem of kind *B*.

To define the notion of free will is nice but it does not tell us whether there actually is such thing as free will. Fortunately, we need not postulate that free will does, in fact, exist but can derive the hypothesis of its existence from Postulate 6.3 and Definition 6.15:

THEOREM 6.1. All animals capable of being in conscious states are able to perform free voluntary acts.

The concepts of self-awareness, self-consciousness, self, and intention are relevant to the elucidation of those of person and moral agent, which play central roles in bioethics. However, we cannot pursue this subject matter here.

To conclude, sensing and feeling, thinking and evaluating, planning and deciding are biological processes. Hence, they should be approached biologically rather than as immaterial entities or processes. However, since humans are social animals, their mental (or subjective) life is strongly influenced by their social environment —to which, in turn, they contribute to modifying. Therefore an adequate understanding of the mental calls for the combined effort of psychologists, neuroscientists, and sociologists.

7 Systematics

Things are similar: this makes science possible. Things are different: this makes science necessary. (Levins and Lewontin 1985, p. 141)

Now that we have a rough knowledge of what an organism is (Sect. 4.3), we can proceed to explore how organisms are classified. Without such knowledge no *biological* classification would be possible, because in order to classify living systems rather than nonliving ones we must be able to tell the difference between the two. Telling the difference between two objects or classes of objects, however, is itself an act of classification. Therefore, we shall first examine the general principles of classing before we turn to the peculiarities of biological classification.

Interestingly, though one of the oldest practices in biology, systematics is still found among the main topics in contemporary philosophy of biology because it continues to be haunted by spirited controversies. A sample of the most important problems is the following: What is the proper way to do systematics, that is, by which methodological principles and by which methods should the systematist abide? What is the status of taxa, especially of the species taxon? Are they sets, classes, or kinds, hence conceptual objects, or are they real systems? Are classifications conventions or else theories? What is the relation, if any, between systematics and evolutionary theory? Should they merge or be kept separate? What is the difference, if any, between classification, systematics, and taxonomy?

7.1 Philosophies of Taxonomy

We submit that the major philosophies underlying biological taxonomies are idealism, nominalism, and conceptualism. Every one of these schools has a time-honored pedigree: idealism goes back to Plato, nominalism to Ockham, and conceptualism to Aristotle. And all but idealism, which has remained stagnant since the days of Romantic *Naturphilosophie*, have branched out into rival schools.

Idealism or *typologism* holds that (a) all the members of a given taxon share certain essential properties that serve to define the taxon—whence the name *essentialism* often given it; and (b) species are ideas and, moreover, archetypes or ideal forms that individual organisms resemble only imperfectly. As idealist essentialism has been extensively criticized elsewhere (notably by Mayr 1982) and no longer plays any significant role in taxonomy, we shall disregard it henceforth—all the more since it is incompatible with the materialist ontology of modern science.

Nominalism is, of course, a reaction against idealism. There are two variants of it: traditional and contemporary. Traditional nominalism, found in Buffon, Lamarck, and Darwin, held that there are only individuals (organisms), species being just conventional names adopted for purely practical reasons. (See Mayr 1982. According to Gayon 1996, there are passages in Buffon's work suggesting that he even regarded species as individuals.) Contemporary nominalism or *neonominalism* comes in two versions: weak and strong. Weak neonominalism claims that *species* are neither names nor concepts but concrete individuals, that is, material entities composed of organisms (e.g., Mayr 1963, 1982, 1988; Ghiselin 1966, 1974, 1981; Löther 1972; Hull 1976; Sober 1980, 1993; Reig 1982; Ax 1984; Rosenberg 1985; M.B. Williams 1985; Willmann 1985). Strong neonominalism takes not only species to be concrete individuals but holds that all taxa are so-called "historical entities", if not concrete composite wholes (e.g., Bock 1974; Ghiselin 1974, 1981; Griffiths 1974; Hull 1976, 1978, 1980, 1988, 1989; Wiley 1978, 1980, 1981, 1989; de Queiroz 1988, 1994; de Queiroz and Donoghue 1988; Mayr and Ashlock 1991).

Neonominalism seems to be restricted to biology, so that it is more aptly called *bionominalism*. Bionominalists usually do not doubt that there are classes or, more precisely, natural kinds of nonliving things, such as the chemical elements (e.g., Sober 1980; Hull 1989). However, they believe that evolution precludes the conception of taxa as classes or kinds. Since bionominalism has gained wide acceptance among taxonomists and philosophers, nay appears to be the predominant view today, we shall examine it in detail in Section 7.3.

Conceptualism is a sort of compromise between idealism and nominalism. It holds that, in the systematic hierarchy, (a) only individual organisms are real, i.e., are concrete individuals and exist independently of the knowing subject (other material systems composed of organisms, such as biopopulations, communities, and social systems are real, too, but are disregarded here because they are not relevant to systematics); (b) species and the other taxa are concepts, though not arbitrary and useless ones, for they represent objective commonalities among organisms: they are natural classes or, ideally, natural kinds. To paraphrase Christian Wolff: *taxa non existunt, nisi in individuis* (1740, §56). However, we shall see below that, in order to take evolution into account, the standard notion of a natural kind needs to be modified into the weaker notion of a biological kind. (See also Ruse 1987.)

7.2 Conceptualism

As indicated in Part I, as well as in Section 4.5, we take conceptualism to be the sole viable methodology and philosophy of taxonomy. We proceed to outline our conceptualist taxonomy and philosophy of taxonomy, before we end up by examining the currently dominant philosophy of taxonomy, namely bionominalism.

7.2.1 Concept Formation

7.2.1.1 Discrimination

The most basic relation between any two items, whether factual or conceptual, is that of equality or its dual—difference. Let us start with the latter. The perception or conception of difference is called *discrimination* (or *analysis*). Although our perceptual apparatus has a restricted discriminatory power, we can enhance it with the help of observation instruments (e.g., microscopes). Likewise, we can sharpen conceptual discrimination with the help of logic and mathematics.

Note that we are talking about *perceptual* and *conceptual* operations. That is, distinguishing, discriminating, or analyzing is not the same as detaching or dismantling, which are (concrete) actions. For example, the components of a system, though distinct and therefore distinguishable (at least in principle), are not separable: if they are materially separated, the system breaks down. We emphasize this, because holists usually reject analysis, confusing it with dismantling.

The important step in discrimination is to find out in what the distinct objects differ; and this involves taking cognizance of some of their properties. Such cognitive operation is no mere perception: it is a conceptual operation, for it consists in attributing properties. Indeed, in order to be able to attribute a property to an object, we must form some concept of such property, i.e., an attribute or predicate (recall Sect. 1.3.2).

7.2.1.2 Commonalities: Equivalence and Similarity

Perceiving or conceiving commonalities is the counterpart of realizing differences. If individuals a and b share a certain property (or property cluster) P, we say that they are *equivalent* (or *equal*) with respect to P. We can also say that a and b are P-equivalent, and write: $a \sim_P b$; and we can further say that all such individuals constitute an *equivalence class* under P. (The P-equivalence class of a, which is identical to that of b, is designated '$[a]_P$'.) This definition entails that there are as many equivalence relations as there are properties. It also entails that every equivalence relation $\sim$ is reflexive, symmetrical, and transitive. That is, for any objects a, b, and c in a given set: $a \sim a$ (reflexivity); if $a \sim b$, then $b \sim a$ (symmetry); and if $a \sim b$ and $b \sim c$, then $a \sim c$ (transitivity).

If the transitivity condition is relaxed, we obtain the weaker relation of *similarity* or *resemblance*: if a is similar to b, and b similar to c, a may or may not resemble c. That is, a and b may share some properties which are not those that b and c share: see Fig. 7.1. A qualitative measure of similarity is the intersection of the sets of properties of the similar objects. That is, calling $P(a)$ and $P(b)$ the sets of properties of things a and b, respectively, we stipulate that the similarity s between a and b, i.e., $s(a, b)$, equals $P(a) \cap P(b)$. By counting the number of shared properties and assigning them the same weight, we obtain a quantitative measure of similarity. Thus, *pace* Ghiselin (1966), it makes perfect sense to quantify similarity. (For such quantitative measures of similarity see Sokal and Sneath 1963; Sneath and Sokal 1973; Bunge 1977a, 1983a.)

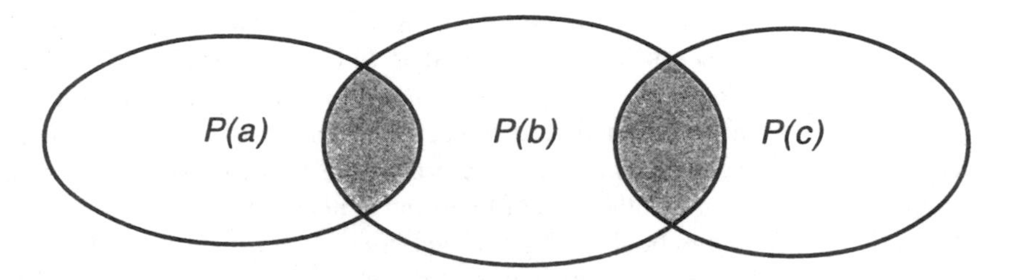

Fig.7.1. The sets of properties of three things a, b, and c: $P(a)$, $P(b)$, and $P(c)$. Since a shares properties with b, a is similar to b; likewise with b and c. However, although a is similar to b, and b similar to c, a does not resemble c, because a and c do not share any properties, i.e., $P(a) \cap P(c) = \emptyset$

7.2.1.3 Grouping

The operations of grouping and collecting and their dual, that of distinguishing and separating, seem to be basic animal behaviors. In fact, many animals gather things of some kind or other, and some higher vertebrates are also capable of collecting ideas, that is, of grouping images, concepts, propositions, and the like. Material things can be put together spatiotemporally, and all objects, regardless of their nature, can be put together in thought. In the former case the outcome is a material thing: either a conglomerate (aggregate) or a system; in the latter case the outcome is a concept: a set or collection. Since in systematics we are not concerned with the building of material aggregates, we shall consider only the collecting of items of any nature whatever (i.e., whether factual or conceptual) to form conceptual collections, i.e., classes.

If two items are distinct, yet equivalent in some respect, then they can be put into a single class. (Not so if they are merely similar.) Note the expression 'can be put' instead of 'exist'. Since differences among factual items are objective, they exist whether or not a given subject knows it. On the other hand, grouping items together in a class is a conceptual operation: classes are concepts, not real (concrete, material) entities. But of course while some groupings are arbitrary or artificial, others are natural or objective. Thus, by putting together all the persons called 'Mike' we obtain an artificial class, whereas by grouping all the persons sharing some common ancestor we obtain a natural class, that is, a collection whose members are objectively related. Consequently, while some classes are realistic (not arbitrary), no class is real.

When collecting material things in thought, i.e., conceptually, we must recall that each of them is changeable, to the point that, whereas some of them may have already vanished, others may not yet have come into being. We must therefore distinguish between a collection the membership of which varies over time, and a set proper (or collection regarded *sub specie aeternitatis*) the membership of which is constant, that is, all the elements are treated as though they were timeless entities.

The difference between the two collecting operations can be elucidated with the help of the notion of time, which, of course, is both a scientific and an ontological concept, not one in pure mathematics. Call F the attribute of interest and write 'Fxt' to indicate that individual x has a certain property (represented by the predicate) F at a time t. Then the collection S of all F's at time t will be $S_t = \{x \mid Fxt\}$. By taking the union of all the variable collections of F's for all time, i.e., $\cup_{t \in \mathbb{R}} S_t$, we obtain the (timeless) set S of F's, called the *extension* of the predicate F (recall Sect. 2.2). Thus, to say, for example, that Charles Darwin *is* human, or belongs to the human species S (with invariable membership), amounts to saying that Darwin belong*ed* to the collection of humans who *were* alive in, say, 1859.

The notion of a variable collection should allay the fears of those taxonomists who have argued that taxa, and in particular species, could not be conceived of as sets because the latter have a fixed membership. (Note that the expression 'variable collection' must not be understood as implying that collections are changeable, hence real, objects. All it means is that, at different times, the collections *formed by us* differ in the number of their members.)

7.2.1.4 Set

So far, we have used the terms 'set' and 'class' without explicating them, for it was sufficient to understand that they are conceptual collections of objects of any nature whatsoever. It is appropriate now to elucidate these concepts, as well as those of kind and natural kind, if only because many a student of systematics has been bewildered by these different notions (e.g., Mayr 1988, 1989).

A set is any collection of items which need not share any property in common with the exception of their shared membership in the set to which they belong. For instance, the set

$\mathcal{A}$ = {Charles Darwin, Tokyo, Mickey Mouse, 10555} is a collection of objects sharing no obvious common property. Their only commonality consists in their being members of the set $\mathcal{A}$. Thus, the set $\mathcal{A}$ can only be defined *extensionally*, that is, by listing its members.

7.2.1.5 Class

If the objects of interest do share another common property besides membership in the given set, this set is a special type of set, namely a class. For example, in the class

$\mathcal{B} = \{\text{Mozart, Beethoven, Bruckner}\}$

all three members of $\mathcal{B}$ share the property of being a composer. As some authors believe that classes are merely arbitrary collections, it should be noted that the three elements of $\mathcal{B}$ are grouped together by an objective property. What is still arbitrary, though, is our classing of just these three composers. Clearly, we can construe a more significant set by forming the class of *all* composers. This class would be identical with the scope of the property of being a composer; and it would be identical with the extension of the predicate "is a composer". Thus, unlike the above set $\mathcal{A}$, the class $\mathcal{B}$ can also be defined *intensionally*, namely as the set of all individuals x that possess the property of being a composer; in obvious symbols, $\mathcal{B} = \{x \mid Cx\}$. We submit that all biological classes are defined intensionally (see Sect. 7.2.2.2).

However, we can arrive at something stronger than a class, which requires the possession of only one property in common. This is the notion of a kind.

7.2.1.6 Kind

A kind can be conceived of as a class whose members share more than a single common property. More precisely, it can be construed as the intersection of the scopes of several properties. If we start with the scope of n properties we obtain n classes. The intersection of these classes, if nonempty, yields a kind: see Fig. 7.2. For example, consider three properties P, Q, and R. The scope $\mathcal{S}$ of P, i.e., the set of all objects p_i possessing P, is the finite class $\mathcal{S}(P) = \{p_1, p_2, \dots, p_i\}$; the scope of Q is $\mathcal{S}(Q) = \{q_1, q_2, \dots, q_j\}$; and the scope of R is $\mathcal{S}(R) = \{r_1, r_2, \dots, r_k\}$. Then the *kind* defined by the three properties P, Q, and R is the set $K = \mathcal{S}(P) \cap \mathcal{S}(Q) \cap \mathcal{S}(R)$.

However, the properties delimiting a kind need not be lawfully related: though objective, they may still be arbitrarily chosen. If we want to class things according to lawfully related properties, we have to specialize the notion of a kind to that of a *natural* kind.

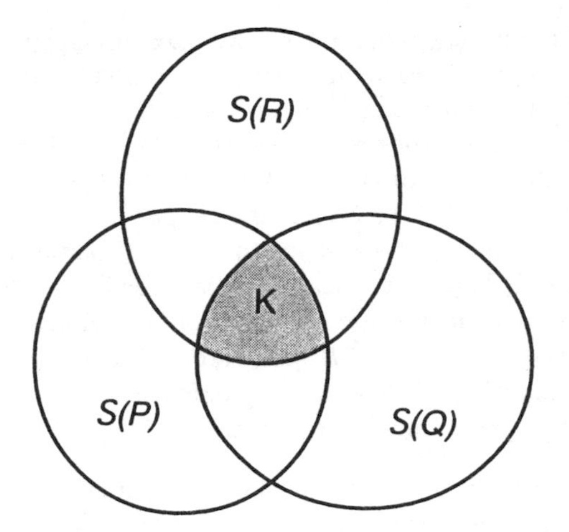

Fig.7.2. The scopes S of three properties P, Q and R. The intersection of these classes yields the *kind* K

7.2.1.7 Natural Kind *sensu lato*

The concept of natural kind has been widely discussed in the philosophical and biophilosophical literature. (See, e.g., Mill 1875; Putnam 1975; Bunge 1977a; Kitts and Kitts 1979; Dupré 1981; Ruse 1987; Splitter 1988; Suppe 1989; Wiley 1989; Leroux 1993; Wilkerson 1993. For a review see van Brakel 1992, and for a brief history of natural kind concepts see Hacking 1991.) Despite this ongoing interest in the matter, the concept has also been widely misunderstood. For example, Ghiselin (1981) thinks that natural kinds would refer to Platonic essences and would therefore be inadequate for modern biology. Others try to elucidate the notion of a natural kind with the help of possible worlds metaphysics. And some authors include properties, processes, and functions, such as diseases and behavior patterns, among natural kinds. We contend that either approach is unsatisfactory or even mistaken. In our emergentist-materialist ontology we have, of course, no use for Platonic essences. The same holds for possible worlds, which, in our view, are pure fictions unsuitable to tell us anything about the real world. Finally, neither diseases nor behavior patterns can form natural kinds, because they do not exist in themselves. As Aristotle knew, only sick organisms or behaving animals have real existence. (See also Woodger 1952, p. 325). Of course, when classifying diseases, for instance, we can *abstract* from the organisms being sick: we can *feign* that states of things or processes in things exist in themselves; but this is an instance of methodological abstraction. Consequently, we have to rely on a different explication of natural kinds. In so doing, it will be necessary to distinguish two kinds of natural kinds, namely natural kinds *in a broad sense* and natural kinds *in a strict sense*.

To elucidate the concept of natural kind, we make use again of the notion of the scope of a property. Since we want to arrive at *natural* kinds, i.e., kinds of real objects we find in nature, we recall that the notion of the scope of a property is restricted to substantial properties, i.e., properties of material things (see Definition 1.2). Comparing the scopes of any two substantial properties, P and Q, we may face four possible outcomes. First case: the scopes of P and Q are disjoint, i.e., $\mathcal{S}(P) \cap \mathcal{S}(Q) = \varnothing$. Second case: the scopes of P and Q partially overlap. Thus, we may obtain a (simple) kind, as explicated in the preceding section. Third case: the scopes of P and Q stand in a relation of inclusion, i.e., either $\mathcal{S}(Q) \subset \mathcal{S}(P)$, or $\mathcal{S}(P) \subset \mathcal{S}(Q)$. Fourth case: the scopes of P and Q are coextensive, i.e., $\mathcal{S}(P) = \mathcal{S}(Q)$. As we need the two latter notions to eventually elucidate the concept of natural kind, they deserve to be defined. Thus we lay down:

DEFINITION 7.1. If P and Q are any two (substantial) properties, then
(i) P and Q are *concomitant* if, and only if, they have the same scope, i.e., iff $\mathcal{S}(P) = \mathcal{S}(Q)$;
(ii) P *precedes* Q if, and only if, P is more common than Q, i.e., iff $\mathcal{S}(Q) \subset \mathcal{S}(P)$.

The concomitance of properties is what Hume (1739/40) called 'constant conjunction of properties'. In our ontology, however, the concomitance of properties is not coincidental but lawful. The same holds for the precedence of properties. Thus, we stipulate that two properties P and Q are *lawfully related* if, and only if, either the scopes of the properties are concomitant, or one property precedes the other, i.e., iff either $\mathcal{S}(P) \subseteq \mathcal{S}(Q)$ or $\mathcal{S}(Q) \subseteq \mathcal{S}(P)$. (Recall Definition 1.3 and Postulate 1.2.)

A natural kind *sensu lato*, then, is obtained when material objects (i.e., things or concrete systems—not conceptual objects, properties, events, processes, functions, or other nonthings) are grouped on the basis of a cluster of lawfully related properties. In other words, all the objects in a natural kind are nomologically equivalent with regard to certain properties, i.e., they share the same nomological state space with regard to *some* lawfully related properties. It should be pointed out again that the properties in question are supposed to be substantial, real, or objective properties of the things being grouped. Hence, neither negative nor disjunctive predicates qualify for generating natural kinds. (Recall Sect. 1.3.2. The differences between properties and predicates as well as between the algebra of kinds and the algebra of sets are the reasons that we analyze both the concept of a natural kind and that of a biological taxon in terms of scopes of properties rather than in terms of extensions of predicates; for details see Bunge 1977a.) Thus, when we pick three properties P, Q, and R, the intersection of their corresponding scopes, i.e., $\mathcal{S}(P) \cap \mathcal{S}(Q) \cap \mathcal{S}(R)$, yields the natural kind N whose members are P- *and* Q- *and* R-equivalent. The extension of N, however, must equal that of at least one of the scopes of the properties P, Q, or R. For instance, it must at least equal that of $\mathcal{S}(R)$ if $\mathcal{S}(P) \supseteq \mathcal{S}(Q) \supseteq \mathcal{S}(R)$.

Two examples may help illustrate this point. Let P represent the property "possessing a chorda dorsalis", Q "possessing an amniote egg", and R "possessing three ear ossicles". Then we can say that it is *necessary* for an organism possessing three ear ossicles to be at the same time an amniote and a chordate, that is, the corresponding properties are lawfully related. (This necessity is nomological, not logical.) Hence, the class Mammalia is a natural kind *sensu lato.* (We shall qualify this conception in Sect. 7.2.2.2 because we have to take evolution, i.e., descent with modification, into account.)

It might appear that the following is a counterexample to our construal of taxa as natural kinds. Indeed, what holds for the possession of three ear ossicles among mammals seems to hold also if the property R is, for instance, taken to be that of being warm-blooded. It seems to be necessary at first sight to be an amniote for being warm-blooded. Accordingly, a class Haematothermia (comprising mammals and birds) also appears to be a natural kind *sensu lato*. Since most systematists will reject a taxon Haematothermia as being polyphyletic (see, however, Gardiner 1993), we hasten to emphasize that this example does not render the concept of natural kind *sensu lato* useless. It only shows that naturalness comes in degrees, and that we must distinguish between predicates and (substantial) properties, for further analysis may show that the predicate 'warm-blooded' does not refer to one but in fact two (substantial) properties (in familiar parlance: convergences.) For instance, the average temperature in birds is somewhat higher than in mammals, and the physiological mechanisms bringing about homoiothermy as a global property are likely to be different as well.

This notion of a natural kind in the broad sense suffices to characterize the taxa of scientific classifications *in general*. Thus, taxa are natural kinds, natural kinds are kinds, kinds are classes, and classes are collections, but not conversely. Since taxa are natural kinds, and since species are taxa, species, too, are natural kinds. Moreover, since natural kinds are conceptual objects, not material ones, species also are constructs, yet not arbitrary or idle ones. In fact, species are those kinds of things we arrive at when we consider *all* their lawfully related properties. In other words, the broad conception of natural kinds has to be narrowed down to arrive at natural kinds *sensu stricto* or *species* (in the ontological sense).

7.2.1.8 Species or Natural Kinds *sensu stricto*

A characterization of natural kinds *sensu stricto* can be obtained in terms of the concept of nomological state space (recall Sect. 1.4.3). Whereas any subset (or cluster) of nomologically related properties of a thing induces a natural kind *sensu lato*, i.e., a nomological state space with regard to some law(s)$_1$, the set of *all* laws of a thing determines its full nomological state space, hence the natural kind *sensu stricto*, or (ontological) *species*, it belongs to: see Fig. 7.3. In other words, whereas full nomological equivalence generates a species, nomological equivalence with respect to *some* properties yields only a natural kind in the broad sense. Thus, halogens constitute a natural kind in the broad sense, while fluorine, chlorine, and iodine are natural kinds in the strict sense, i.e., atomic species. (Note that we must distinguish, say, the atomic species *chlorine* from the natural kind *chlorine gas*, which is an aggregate composed of diatomic chlorine molecules and thus has emergent properties that its components lack.)

The situation is analogous in organisms, but it is much less obvious what the natural kinds *sensu stricto* are. For example, all mammals belong to a natural kind *sensu lato*, and the same is true for all primates and all hominids. But what about all humans? Even though all humans belong to a natural kind too, this may still not be a natural kind *sensu stricto*. After all, there are not only different morphs, such as males and females, but also a number of varieties (e.g., races), which are nomologically different. Thus, the natural kind *sensu stricto* is apparently not to be found at the level of *Homo sapiens*, but more likely at the level of either the morph or the variety; and this seems to hold for most species of organisms.

This example shows that a species in the sense of a natural kind *sensu stricto* is a purely ontological concept, not a taxonomic one. So much so, that in the context of systematics such a species may be ranked as a variety, a subspecies, a subgenus, or even a genus, or that it may have no taxonomic status at all. For this reason, we must clearly distinguish the notion of an ontological species or natural kind *sensu stricto* from that of a taxonomic species. The latter is a natural kind in either the broad or the strict sense as it occurs in a scientific classification. (Unless a classification consists of natural kinds of whatever scope, it cannot be said to be a natural one.) The former deserves a definition of its own, because we will need it for the analysis of the concept of speciation:

> DEFINITION 7.2. Any class of fully nomologically equivalent entities is called a *natural kind sensu stricto* or (*ontological*) *species*.

(For the notion of a species of organisms see Definition 4.6; for an earlier attempt at using the notion of law in defining species see Ruse 1969.)

Finally, we must distinguish both the ontological and taxonomic concepts of species from the logical notion of species. A logical species is any kind of objects, whether natural or artificial, whether material or conceptual, that is subset of a more comprehensive kind, which is called *genus*. For example, if the (logical) genus is Vertebrata, then the *differentia specifica* "four-legged" gives the (logical) species Tetrapoda. Thus, these terms merely correspond to the notions of the generic (or general) and the specific. (See also Mayr 1982.)

In sum, we must distinguish a logical, an ontological, and a taxonomic concept of species. However, whether or not a taxonomic species coincides with an ontological one, a biological classification is scientific and natural only if it contains taxa that are classes of nomologically equivalent organisms. For this reason, it is useful to distinguish different taxa by means of the notion of a nomological state space: see Fig. 7.3.

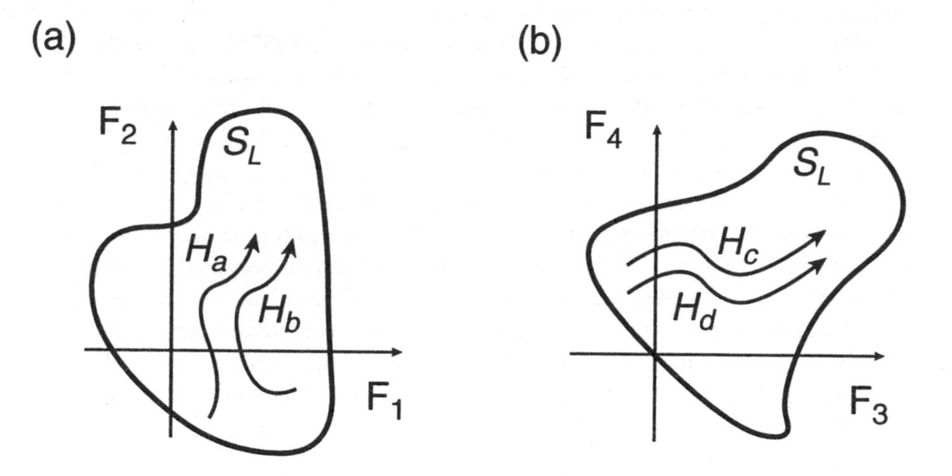

Fig.7.3 a,b. Organisms of different species "have" different nomological state spaces. **a** The state space of organisms of species *A* is spanned by the state functions F_1 and F_2. The trajectories H_a and H_b represent the histories of two individual organisms, *a* and *b*, of species *A*. **b** The same for species *B*, with the two individuals *c* and *d*, characterized by the properties F_3 and F_4

7.2.1.9 Foray: The Psychobiology of Classing

Most organisms seem to be capable of distinguishing between organisms belonging to their own kind from those belonging to some other. In so doing, organisms lacking nervous systems are likely to rely on (bio)chemical properties. Organisms possessing plastic nervous systems may be said not only to *react* specifically to other organisms but to *recognize* organisms belonging to the same or to a different kind. This holds, for instance, for the often very specific relations between mating partners, parasites and hosts, predator and prey, and so on. Thus, a koala does not know anything about common descent when it recognizes plants of the kind botanists call *Eucalyptus*.

What holds for the abilities of animals to distinguish organisms of some kind by means of (objective) properties, holds *a fortiori* for humans. Comparative anthropologists have observed that traditional classifications of plants and animals are—to a large degree—culturally invariant. In other words, the prescientific classifications of traditional peoples usually come up with kinds of plants and animals similar to those formed by scientific taxonomists. Interestingly, the best match between traditional kinds and scientific taxa occurs at the genus level. This congruence decreases with higher or lower taxonomic rank. Despite some cultural differences—e.g., males and females may be differently familiar with certain plants and animals, and foraging societies distinguish fewer taxa than agricultural societies do—prescientific common sense classifications seem to be much less influenced by a utilitarian point of view (e.g., matters of edibility, religious or symbolic significance) than previously assumed (Boster 1987; Berlin 1992).

Biologists, in particular psychobiologists, and realist epistemologists will not be too surprised about these findings. After all, there is ample evidence that the process of forming kinds, such as the class of cats, is a function of special neuronal systems: many neurons and neuronal systems react specifically and selectively to certain properties of the things perceived. In other words, one of the basic functions of the sensory modules of the visual cortex is the *categorization* of the incoming stimuli (Zeki 1993). Assuming that such single property units (or modules) may be combined and interconnected to form discriminating supersystems we may further conjecture that forming a concept of the "concrete" kind, i.e., a class of real things or events, consists in responding uniformly to any and only members of a given class of objects (Bunge 1980). We thus suggest the following hypothesis:

> POSTULATE 7.1, Let C represent a set of (simultaneous or successive) things or events. Then there are animals equipped with plastic neural systems whose activity is triggered, directly or indirectly, by any member of C, and is independent of what particular member activates them.

However, this basic ability is in need of considerable refinement before a scientific classification can be achieved. Since Postulate 7.1 accounts first of all for everyday classing, it helps explain why deep (i.e., scientific) classifications are often felt to be counterintuitive, while common sense classifications appear more natural to us. For example, grouping birds and crocodiles together is, at first sight, counterintuitive, whereas forming a class of reptiles is not. This shows how ambiguous the notion of a *natural* classification is.

7.2.2 Classification

7.2.2.1 Classification by Partitioning

Just as we can analyze individual things and individual ideas, so we can analyze collections of either. The most basic mode of analysis of a collection is its *partition* into homogeneous subcollections. The simplest such partition is the *dichotomy*, which is so simple that it almost always occurs as a first stage in analysis. For example, we could start classifying organisms by partitioning dichotomously the collection of living beings into edible and inedible. A more refined classification of organisms could be a tetrachotomous one, as when we classify them, say, according to their mode of life, which may result in four classes: aquatic, amphibian, terrestrial, and aerial organisms.

The key to such partition is the concept of sameness in some respect, i.e., of equivalence, examined above. One says that an equivalence relation *induces the partition* of a collection into a family (or collection) of *equivalence classes*, and that all the members of each equivalence class are equivalent in the given respect. The operation resulting in the formation of such family is called the *quotient* of the given set S by the given equivalence relation $\sim$ (written '$S/\sim$ '). For example, T-shirts usually come in (at least) three sizes: small, medium, and large. In this case, the equivalence relation "the same size as" ($\sim$) partitions the set of T-shirts T into a family of three equivalence classes: S, M, and L. That is, $T/\sim = \{S, M, L\}$, where $S \cap M = \varnothing$, $S \cap L = \varnothing$, $M \cap L = \varnothing$, and $T = S \cup M \cup L$.

Once a first partition $P_1 = S/\sim_1$ of an original collection has been carried out, we have completed the first rank of the classification, which consists of *species*. (Note that 'species' here refers to taxonomic species, which means being a member of any first rank category, not necessarily being a natural kind *sensu stricto* as outlined above.) These species may be grouped in turn by using a second equivalence relation $\sim_2$, defined on the family P_1, i.e., $P_2 = P_1/\sim_2 = \{G_1, G_2, ..., G_m\}$. The members of this partition may be called *genera*. A third equivalence relation, defined on P_2, will induce the third rank of the classification, for example that composed of *families*. One may also invert the process, that is, start by partitioning the original collection into higher-rank classes, and proceed to distinguish subsets within them. As one descends from the higher to the lower-rank classes one does so with the help of finer and finer equivalence relations. Every classification, then, has two dimensions: the horizontal one attached to the relations of membership ($\in$) and equivalence ($\sim$), and the vertical one linked by the inclusion relation ($\subset$). Thus, every classification by partitioning is a trivial example or model of elementary set theory.

It may be helpful to exhibit a few more examples before we proceed. An example from chemistry consists in the partition of all atoms possessing the same number of protons. By this partition we obtain more than a hundred equivalence classes: the atomic species or chemical elements. These, in turn, may be further partitioned by the equivalence relation "having the same valence" and other properties related to valence. The result is the Periodic Table of Elements. A biopopulation can be analyzed into collections of individuals of the same morph (e.g., sex or caste), the same age, or what have you. The set of all amniotes equivalent with regard to their overall skin covering can be partitioned into reptiles (scales), birds (feathers), and mammals (hair). (Cladists should not stop reading here; they will soon be put at ease.)

To summarize, classes are formed by grouping together individuals that share certain properties, even if they differ in all other respects. A single attribute A and its complement not-A allow one to make black or white statements such as "c is an A" and "c is a non-A". A pair of attributes, A and B, allow us to form four different propositions: "c is an A and a B", "c is an A and a non-B", "c is a non-A and a B", and "c is a non-A and a non-B"—i.e., the components of a 2×2 contingency table. In general, for n attributes we may construct 2^n propositions for any given individual. Each such set may be called a *Boolean partition*.

We are now ready to formulate some general principles of classification (cf. Bunge 1983a).

7.2.2.2 General Principles of Classification

A *classification* of a given collection of individuals—be they concepts, concrete things, events, or what have you—is a conceptual operation with the following characteristics:

1. Each member of the original collection is assigned to some *class*.

In traditional logic the original collection of a classification was called *genus summum*. In biological classification the broadest original collection is the set of all organisms. However, it is possible to choose any collection as the basic one. For instance, we may be interested only in classifying plants, insects, or birds. Any class an organism is assigned to in a biological classification is called a *taxon* (plural: *taxa*).

2. There are two types of class: simple (or basic) and composite, the latter being the union of two or more simple classes.

Each basic class is usually called a *species* or, more precisely, a *species taxon*. Note that this taxonomic notion of species derives from the *species infimae* of traditional logic and need not coincide with ontological species, as explicated in Section 7.2.1.8. Every higher taxon is the union of two or more species taxa. Consequently, what are called *monotypic* taxa in the Linnean classification, i.e., taxa containing only a single species, do not comply with this condition because they are identical to a simple class, a species. The set of all species taxa is called the *species category* or the *species rank*. Categories are thus classes of classes (taxa). For this reason, categories are not natural kinds.

3. Each basic (simple) class is composed by some of the *members* of the original collection, and no basic class is composed of subclasses.

Since the original collection in biological classification is the collection of all organisms, and since the basic classes are species taxa, the latter are sets of organisms, not sets of populations. This is important to note because it contradicts the tenet of the "New Systematics" that populations or species (as individuals) are the units of classification (e.g., Mayr and Ashlock 1991). As will be seen below, if populations were the units of classification, it would be impossible to formulate propositions such as "Aristotle is a human being" or "Aristotle belongs to *Homo sapiens*".

Traditional Linnean classification allows for classes below the species level, such as subspecies, varieties, and forms. Although this procedure shows that systematists intuitively distinguish ontological from taxonomic species, it presupposes that species are not the basic classes of the classification. And since not all species are thus subdivided, it violates condition (10) below.

4. Each class is a *set* whose membership is determined by a predicate or a conjunction of predicates.

This is to say that taxa are defined intensionally, not extensionally. For example, a taxon $\mathcal{T}$ is defined as $\mathcal{T} = \{x \in O \mid Px\}$, where O names the set of organisms, and P is either a predicate or a conjunction of predicates. Moreover, intension precedes extension, because one can determine the extension of a class only if one can distinguish members from nonmembers. (By contrast, what can be defined only extensionally are sets proper, not classes: recall Sect. 7.2.1.4.)

Since classes are defined by predicates, what are defined are *taxa*, not taxon *names* as the bionominalists contend (e.g., Hull 1965; Buck and Hull 1966; de Queiroz and Gauthier 1990). If taxa are concrete individuals, as is claimed by the bionominalists, one can at best *assign* proper names to them, but one cannot define taxon names. This is because nominal definitions are identities, that is, the symbol '$=_{df}$' is to be read as "identical by definition", and a sign or name cannot be identical to its nominatum (recall Sect. 2.3. and Sect. 3.5.7). In our view, naming a taxon amounts to attaching a sign to a previously defined class, which is a purely conventional operation.

The requirement that the taxa in a classification be defined by a set of necessary and sufficient properties poses no problems in nonbiological disciplines. The objects of biological classification, however, are different from nonliving things in that they (may) undergo developmental and evolutionary changes. Therefore, any developed or evolved feature may subsequently be either lost or subject to further qualitative change. This is at the basis of the antiessentialist argument against the view that biological taxa are natural kinds defined by a set of necessary and sufficient properties. The fact of evolution, so the argument goes, has refuted essentialism definitively, thereby rendering taxonomic conditions such as (4) obsolete. Let us analyze this problem more closely.

In defining, for example, the insect taxon Pterygota by the predicate "possessing a pair of wings on both the mesothorax and the metathorax", we face the following problems. First, pterygotes possess (functional) wings only as adults. How, then, are we supposed to class a caterpillar, for instance? One solution consists in looking at the whole life history of the given individual. If we observe that the caterpillar eventually turns into a butterfly possessing two pairs of wings, we are justified in classifying it as a pterygote insect. Yet what if the caterpillar dies before metamorphosis, which is not an unlikely event in view of, say, predation by birds, and other imponderables of life? Second, some pterygotes possess only one pair of wings, such as dipterans (flies, mosquitoes, etc.), and worse, some possess no wings at all, such as fleas and lice, although they undoubtedly belong to the taxon Pterygota, as is shown by other features.

Thus, it comes as no surprise that any definition of taxa by a conjunction of predicates, i.e., a list of necessary and sufficient properties, has been regarded as impossible in biology. Before the advent of bionominalism, the so-called *disjunctive* or *polytypic* (or *polythetic*) definition had been proposed as a solution to this problem (e.g., Beckner 1959; Hull 1965). Accordingly, a taxon $\mathcal{T}$ would not be defined as, say, $\mathcal{T} = \{x \in O \mid Px \;\&\; Qx \;\&\; Rx\}$, where $\mathcal{T}$ equals the intersection of the scopes of the properties P, Q, and R, i.e., $\mathcal{T} = \mathcal{S}(P) \cap \mathcal{S}(Q) \cap \mathcal{S}(R)$. Rather, it would be defined as $\mathcal{T} = \{x \in O \mid Px \vee Qx \vee Rx\}$. In this case, $\mathcal{T}$ does not equal the intersection but the union of the scopes of the properties in question, i.e., $\mathcal{T} = \mathcal{S}(P) \cup \mathcal{S}(Q) \cup \mathcal{S}(R)$. Yet, to obtain a natural kind it is necessary that the scopes of the properties in question be either included in each other or coextensive. Consequently, in the preceding example a natural kind is obtained only if $\mathcal{S}(P) \cap \mathcal{S}(Q) \cap \mathcal{S}(R) = \mathcal{S}(P) \cup \mathcal{S}(Q) \cup \mathcal{S}(R)$, a condition fulfilled only when the three scopes are identical. That is, there is no way to circumvent the definition of a natural kind taxon by necessary and sufficient properties. Any list of properties that are only sufficient does not define the whole class $\mathcal{T}$ but only a subclass of $\mathcal{T}$. A polythetic definition, then, is at best a list of indicators which allow us to *recognize* any organisms belonging to the given taxon. But of course indicators do not define a taxon.

Although strong neonominalists do not seem to realize this consequence of polythetic definitions, they are more consistent in going even further by claiming that taxa, if monophyletic, are not defined by (predicates referring to) organismal properties at all (e.g., de Queiroz 1988, 1992, 1994; Wiley 1989; de Queiroz and Gauthier 1990; Sober 1993). For instance, no matter what properties a mammal may possess, what makes an organism a mammal is its descent from a common ancestor. That is, the only necessary and sufficient property for belonging in any taxon is "stemming from the same common ancestor". Although there is a grain of truth in this proposal, it fails as a solution to the problem. The grain of truth lies in the fact that it is possible to define a class of objects by the relational property of descent. But this relation *is* a property of the objects in question. For example, "descending from a common ancestor" is a (relational) predicate by means of which we can define a taxon $\mathcal{T} = \{x \in O \mid Dxb\}$, where '$Dxb$' stands for "$x$ descends from b". This definition alone, however, is not very illuminating, since we have not the slightest idea what x and b look like. Every thing is individuated by *all* its properties, not by any single one, even if possibly essential. As Ruse (1987) has noted aptly, descent seems to have become the new essence of the anti-essentialists.

The strong bionominalists' claim amounts to the assertion that it is sufficient to individuate, for example, the individual Wolfgang Amadeus Mozart by the single relational property "son of Leopold and Anna Maria Mozart". Although this relational property may not be unimportant, it provides only a rather uninformative description. In other words, it may be sufficient to recognize Mozart, provided he was their only son, but it does not individuate him. Another problem with that proposal is that the relation "stemming from the same common ancestor" defines only the progeny of a given common ancestor. In phylogenetic systematics, however, the common ancestor is usually included in a monophyletic taxon (more on this below).

Fortunately, it turns out to be possible to take the serious concerns of polytheticists and bionominalists into account while avoiding their ontologically untenable conclusions. A first step is to make use of the notion of time in the definition of a taxon, as we did in Section 7.2.1.3 when introducing the concept of a variable collection. Now, however, we do not define a collection of organisms at a certain time, but a class of organisms that may possess a given property only at a certain time in their life history. (See, e.g., Hennig's notion of a *semaphoront*.) Let us assume again that a taxon $\mathcal{T}$ is defined by the three predicates P, Q, and R. Then, including the notion of time, we define the taxon $\mathcal{T}$ as $\mathcal{T} = \{x \in O \mid (\exists t)\ Pxt \ \&\ Qxt \ \&\ Rxt\}$. (Read: "For all organisms x, x possesses property P at some time t, x possesses Q at some time t, and x possesses R at some time t".)

Such a definition still omits organisms that have lost certain properties, i.e., that no longer possess a given property at any moment of their life history. For example, fleas and lice are classified as pterygotes not because they have wings, which they do not, but because they (a) possess other properties, such as undergoing a metamorphosis during some period in their development, that make them members of Holometabola, which are a proper subset of Pterygota, and (b) because they (must thus be assumed to) descend from winged ancestors. Thus, we can say that for any organism x to be a pterygote it must either possess two pairs of wings at any time in its life history or descend from some ancestor y that possessed two pairs of wings during some time in its life history.

To be more precise, let O designate the set of all organisms, D the relation of descent, and P, Q, and R predicates that represent other substantial properties of organisms. We now define the taxon $\mathcal{T}$ by means of the *conjunction* of three complex predicates: $\mathcal{T} = \{x \in O \mid (\exists t)(\exists y)\ (Pxt \vee [Dxy \ \&\ y \in O \ \&\ Pyt]) \ \&\ (Qxt \vee [Dxy \ \&\ y \in O \ \&\ Qyt]) \ \&\ (Rxt \vee [Dxy \ \&\ y \in O \ \&\ Ryt])\}$. (Read: "For all organisms x: x possesses property P at some time t, or x descends from y, and y is an organism, and y possessed property P at some time t, and x possesses Q at some time t, or...".) To be even more precise, we should in addition take the environment into account in which the organisms under consideration develop or have developed the character in question. After all, "...the characters of the organisms are really characters of the organism *and* its environment" (Woodger 1929, p. 346; italics in the original). However, since we shall deal with questions of development, in particular with the organism-environment relation during development, only in the next chapter, we may ignore this complication here.

Although this definition makes room for all eventualities, a simpler time-dependent definition such as $\mathcal{T} = \{x \in O \mid (\exists t)Pxt\}$ will do in many cases. For example, many developing organisms go through a so-called *phylotypic stage*, which is common to all the members of a given taxon, however different their developmental pathways may be before or after this stage (Hall 1992; Slack et al. 1993). Thus, vertebrates pass the invariant developmental bottleneck of the *pharyngula* stage, which is the period when gill arches are developed (Gilbert 1994); amniotes share a so-called *primitive streak* in their development; and a phylotypic period of arthropods is the *germband* stage (Patel 1994). Hence, though denied by antiessentialists, there apparently are candidates for invariant features in certain taxa.

Still, if we take evolution seriously, we must admit that, in involving the notion of descent, definitions of natural kinds of organisms may but need not coincide with definitions of natural kinds of nonliving things. To take this difference into account, we may use the expression *biological kind* rather than 'natural kind' when dealing with organisms, even though biological kinds are, of course, natural as well. What is more important, however, is that biological kinds are still classes, not concrete individuals.

5. Each class is *definite*, i.e., there are no borderline cases.

In practice, of course, the systematist may not be sure whether to put a certain organism into a given taxon. Yet, eventually he or she will have to make a decision, if only a provisional one. For example, the so-called mammal-like reptiles have to be grouped either with the reptiles or with the mammals, or with neither —in which case they have to be put into a separate class.

6. Any two classes are either mutually *disjoint* (i.e., have no members in common) or one of them is *included* (contained) in the other: if the former, they are said to belong to the same *rank*, otherwise, to different ranks.

In biology, we must add another requirement: each class is *nonempty*. In contrast to the Periodic Table of chemical elements for instance, which contained empty classes the members of which had still to be discovered, there are no vacant positions in a classification of organisms. The reason is that biological classifications are not constructed by proper partitioning, as explicated in Section 7.2.2.1 (see also Sect. 7.2.2.3).

7. Only two (logical) relations are involved: the *membership* relation ($\in$)—an irreflexive, asymmetric, and intransitive relation holding between the individuals of the original collection and the first rank classes—and the *inclusion* relation ($\subset$)—an irreflexive, asymmetric, and transitive relation, which relates classes of different ranks.

In a biological classification every organism is a member of some species taxon, and a species taxon may be included in some higher-order taxa. Recall that the membership relation is intransitive: if $A = \{x\}$ and $B = \{A\}$, then $x \in A$ and $A \in B$, but $x \notin B$. Hence, only organisms are the elements or units of classification, and no taxon is a member ($\in$) of another taxon. The relation between taxa is the relation of set inclusion ($\subset$). For instance, primates are a subset of mammals (Primates $\subset$ Mammalia). Since only the relations of set membership and set inclusion are involved, it is possible to make statements such as "Aristotle is a human being" (or, in obvious symbols, "$a \in$ *Homo sapiens*"), "Aristotle is a primate" (or "$a \in$ Primates"), "$a \in$ Mammalia", and "$a \in$ Vertebrata", and so on.

In traditional systematics—where ranking, i.e., the attaching of categories according to the Linnean hierarchy, is mandatory—so-called *monotypic* taxa occur (Gregg 1954, 1968; Sklar 1964; Buck and Hull 1966; Ruse 1973). A monotypic taxon is any taxon above the species level containing only a single species as the basic class; that is, it is not strictly composite. A famous example is that of the African aardvarks, belonging to the single species *Orycteropus afer*. The Linnean hierarchy now demands the formation of higher-order taxa, although one cannot distinguish among genus, tribe, or family properties. Thus, there is the genus *Orycteropus*, the subfamily Orycteropodinae, the family Orycteropodidae, and the order Tubulidentata, which all are coextensive with the basic class *Orycteropus afer*. Clearly, this is a purely conventionalist game contradicting condition (2). The formation of such monotypic higher-order classes is not due to any new equivalence relations. This is why only the relation of being a proper subset ($\subset$) should be involved in biological classification, not the inclusion relation $\subseteq$ that also allows for set identity. (See also Hill and Crane 1982.)

To distinguish this taxonomic or systematic hierarchy from what has been called the *ecological hierarchy* (Eldredge 1985a), i.e., the hierarchy consisting of cells, multicellular organisms, biopopulations, communities, and the biosphere (as elucidated in Sect. 5.3), we lay down:

DEFINITION 7.3. The set T of all taxa $\mathcal{T}$ together with the relation $\subset$ of set inclusion, i.e., $\mathrm{T} = \langle \mathcal{T}, \subset \rangle$, is called the *systematic hierarchy*.

Recall that the so-called ecological hierarchy has been defined by us as the biolevel structure $\mathcal{B} = \langle \mathrm{B}, < \rangle$, where B designates the set of biolevels and $<$ the relation of level precedence (Definition 5.8). In short, the two hierarchies, though superficially similar, have completely different structures. Thus, the very term 'hierarchy' designates completely different concepts.

8. Every composite class, i.e., every class of a rank higher than the first, equals the *union* of all its subclasses of the immediately preceding rank.

For instance, any given order equals the union of its families, and any given genus equals the union of its species. An example from evolutionary systematics would be: Vertebrata = Agnatha (jawless fishes) $\cup$ Chondrichthyes (cartilaginous fishes) $\cup$ Osteichthyes (bony fishes) $\cup$ Amphibia $\cup$ Reptilia $\cup$ Aves (birds) $\cup$ Mammalia. By contrast, in a cladistic classification Vertebrata = Agnatha (if monophyletic) $\cup$ Gnathostomata (vertebrates with jaws).

9. All the classes of a given rank are *pairwise disjoint* (do not intersect), so that no item in the original collection belongs to more than one class of the same rank.

To use the preceding cladistic example: Agnatha $\cap$ Gnathostomata = $\emptyset$. If this condition is not met, we may speak of a *typology*, but not of a classification.

10. Every partition of a given rank is *exhaustive*, i.e., the union of all the classes in a given rank equals the original collection.

Example: Obviously, all taxa of family rank taken together result in the original collection, namely the set O of all organisms. Thus, O = Hominidae $\cup$ Pongidae $\cup$ Apidae $\cup$ Lumbricidae $\cup$ Magnoliaceae $\cup$ Asteraceae, and so on.

7.2.2.3 Basics of a Natural Biological Classification

As systematists will have already anticipated, in biology a classification by partitioning, as outlined in Section 2.2.1, may result in anything but a phylogenetically significant classification. (See, e.g., Haeckel's 1866 geometry based classification, in which he attempted to classify organisms according to laws. Unluckily, though, he chose mathematical laws, not biological ones. See also the questionable attempts of building a "rational taxonomy" based on laws of form: Sect. 8.2.4.1.) Some authors, notably Ernst Mayr (1982; Mayr and Ashlock 1991), even maintain that the principles of classification applying to the grouping of living beings would be entirely different from the ones applying to nonliving things. This contention contains a grain of truth, but it holds only for consistently phylogenetic classifications. The difference between a phylogenetic and a nonphylogenetic as well as an extrabiological classification is only slight and, interestingly, it does not apply to Mayr's own evolutionary taxonomy (see below).

To begin with, from an ontological point of view, there is an objective and significant partition of the set of organisms into homogeneous subcollections. This is the partition into ontological species, i.e., natural kinds *sensu stricto*, where the equivalence relation is "possessing the same set of laws". Thus, all organisms sharing the same full nomological state space belong to a particular ontological species. Yet, as we noted before, these ontological species would not necessarily coincide with taxonomic species. Among sexually reproducing organisms we could apply the traditional criterion of the ability to mate and successfully reproduce with each other, i.e., "having the same fertilization system" (Paterson 1985). This equivalence relation would yield natural kinds as well. Yet again, they would not necessarily coincide with taxonomic species.

However, *pace* Dupré (1981), this state of affairs does not prove that there is no objective and significant partition of organisms into species. It only shows that biological systematists do not care much for ontological species, but usually assign natural kinds of a broader scope the status of taxonomic species. Therefore, taxonomic species (and actually all biological taxa) are indeed not defined by partitioning the set of organisms into homogenous subcollections, although one could, in principle, do so by acknowledging only ontological species as taxonomic species. Still, any further partition of the collection of ontological species thus obtained into collections of higher-order taxa seems impossible.

The reason is that, since Darwin, a natural classification in biology is usually regarded as one reflecting the genealogical (phylogenetic) relationships among organisms, i.e., descent with modification. But descent with modification results in a phylogenetic tree with irregular branches. In these branches qualitative novelties, by which the organisms belonging to those lineages are characterized, evolve irregularly and independently of each other. Another result of this irregular branching is that neither the branches nor the end-points of the phylogeny are equidistant from the beginning, i.e., the highest taxon or original collection. Therefore, a successive partition of organisms into homogenous subcollections by means of a set of equivalence relations of different power will result in anything but a phylogenetically significant classification. (This is not to say that such partition and classification, if possible at all, would be illegitimate. For instance, we might be interested in a functional or ecological classification.)

As a consequence, no systematist will expect that all the taxa of a given rank, such as the families Hominidae, Apidae (Bees), Lumbricidae (Earthworms), Magnoliaceae, etc., are equivalent in any biologically significant respect. In fact, the only attribute these taxa have in common is that they all are assigned the rank of family, and this attribute certainly does not correspond to any substantial property according to which the organisms involved would be equivalent. It has thus been admitted long ago that the (members of the) taxa in any category of the Linnean hierarchy have nothing in common (Mayr 1982). For this reason, the Linnean categories are to be regarded as mere formalisms, and, therefore, as ultimately dispensable (e.g., Griffiths 1974; Ax 1984, 1988, 1995; de Queiroz and Gauthier 1992; Mahner 1993b; Ereshefsky 1994. Needless to say, especially museum systematists show no inclination to abandon the Linnean hierarchy: see, e.g., Andersen 1995.)

Consequently, in biological classification condition (10) can be upheld as purely formal only if the classification is not consistently phylogenetic. If the classification is to reflect natural classes defined by properties the distribution of which is due to common descent, condition (10) must be given up. Thus, indeed, common descent results in a formal difference between classifications of living versus nonliving things. Ironically, evolutionary taxonomy à la Mayr does *not* show any formal difference in the classification of living and nonliving things because it retains condition (10).

If we want to classify organisms consistently according to their possession of qualitative novelties due to descent with modification, there is only one way to arrive at a phylogenetic classification in biology. This consists in a modified version of a dichotomous partition, or in what Mayr (1982) calls "downward classification". Whereas a single dichotomous partition satisfies all the ten principles of classification given above, the modified version of successive dichotomies outlined in the following only satisfies conditions (1) to (9).

We can start with any predicate P and obtain its extension $\mathcal{E}(P)$, i.e., the class of P-equivalent organisms, which we may abbreviate for the sake of simplicity as $\mathcal{P}$. Note that we work with equivalence relations, not similarities, because the similarity relation is intransitive. It should also be noted that we work with any properties, whether morphological, physiological, genetic, developmental, behavioral, or what have you. So there is no reason to restrict the term 'character' to morphological features. Finally, we warn against the view held by Eldredge (1979) and others, that we group characters rather than organisms. This is a clear instance of Platonism, because properties are not separable from the things that possess them.

By partitioning the set of all organisms by the attribute P (putatively referring to a substantial property), we simultaneously obtain the complement $\overline{\mathcal{P}}$ of that equivalence class, i.e., the set of all organisms not possessing P. Since there are no negative properties, the predicate not-P does not refer to any real common property of the organisms in the complement set. (See also Nelson and Platnick 1981.) Accordingly, we discard complements, that is, they may not be assigned taxonomic status, although of course we need complement classes for comparative purposes. What cladists call an *out-group*, for instance, is nothing but the complement of a given taxon. The same holds for paraphyletic groups. (Incidentally, if we consistently retained complementary classes, condition (10) above would be met.)

For example, we can use the predicate "possessing a skull", and partition the set of all animals into the class Vertebrata (in the sense of Craniata) and its complement Invertebrata. Since the latter class is defined by a negative attribute, i.e., by a predicate that does not correspond to any substantial property, Invertebrata may not be assigned a taxonomic status.

The next step is to look for another equivalence relation Q to form the class $\mathcal{Q}$ of Q-equivalent organisms. In doing so, there are several possible outcomes, as explained in Section 7.2.1.7, if we disregard all complements again. Since we strive for natural kinds, i.e., kinds corresponding to lawfully related properties, we disregard the cases in which $\mathcal{P}$ and $\mathcal{Q}$ are disjoint and in which $\mathcal{P}$ and $\mathcal{Q}$ overlap only partially. Accordingly, condition (6) allows only for pairwise disjoint classes, or classes included in each other. That is, only the cases in which $\mathcal{Q} \subseteq \mathcal{P}$ or $\mathcal{P} \subseteq \mathcal{Q}$ yield natural kinds, and only natural kinds qualify for taxonomic status. Having obtained natural kinds (or rather approximations to natural kinds), we can form law-like generalizations, such as "For all x, if x has property Q, it also has property P", or "For all x, if x has property P, it also has property Q", or "For all x, x has property P if, and only if, it has property Q". These law-like generalizations acquire the status of law statements if they become part of a phylogenetic theory. The fact that those law statements may be largely statistical due to the loss of features within a taxon—that is, having the form "For *most* x: if x possesses P, then x possesses Q"—does not undermine their nomological status: recall that we have introduced the (relational) property of descent from a common ancestor possessing the (intrinsic) property in question in Section 7.2.2.2, condition (4).

For example, if we choose as a second property "possessing a chitinous exoskeleton", the corresponding class Arthropoda will not overlap with Vertebrata. On the other hand, the class defined by the property "living in freshwater" will partially overlap with Vertebrata. If we choose "possessing feathers", the corresponding class Aves will turn out to be a proper subset of Vertebrata. Now, P and Q are lawfully related. We could also say that it is (nomically, not logically) *necessary* for an organism possessing feathers, i.e., a bird, to be a vertebrate.

Proceeding in this way one should—at least in principle, though certainly not in practice—arrive at a hierarchy of nested classes or, rather, natural kinds *sensu lato*. In this hierarchy the most comprehensive class, the logical *genus summum*, is equal to the collection of all organisms, i.e., Life. On the other hand, the smallest (or basic or terminal) classes, the logical *species infimae*, may or may not be ontological species or natural kinds *sensu stricto*.

To be sure, the procedure is not as simple and straightforward as we would wish. The ideal of obtaining a complete hierarchy of nested classes is obstructed by incongruences, i.e., overlaps, among the extensions of some predicates. In many such cases, a closer analysis may reveal that what has originally been conjectured to be a single property may turn out to be, in fact, two different properties. We shall discuss several possible origins of resolvable incongruences in a moment.

However, there is at least one case of incongruence that cannot be dissolved as a matter of principle. To exemplify this case we consider four classes $\mathcal{P}$, $\mathcal{Q}$, $\mathcal{R}$, and $\mathcal{S}$, where $\mathcal{Q}$, $\mathcal{R}$, and $\mathcal{S}$ are proper subsets of $\mathcal{P}$. Then the following incongruence may occur: $\mathcal{Q} \cap \mathcal{R} \neq \emptyset$. Assuming further $\mathcal{S} \subset \mathcal{Q}$ & $\mathcal{S} \subset \mathcal{R}$, we obtain $\mathcal{S} = \mathcal{Q} \cap \mathcal{R}$. Clearly, this pattern may be explained by a hybridization event; that is, a case where speciation is of the type $A + B \rightarrow C$ rather than $A \rightarrow B$ or $A \rightarrow B + C$. To avoid violating conditions (6) and (9) the classes $\mathcal{Q}$, $\mathcal{R}$ and $\mathcal{S}$ may enter the classification as three taxa of equal rank (see Fig. 7.7).

Having arrived at a hierarchy of nested classes or, rather, natural kinds, the latter may be attached names. However, this christening is only an accidental aspect of biological classification, although some authors seem to restrict the term 'classification' to just this formal procedure (e.g., Simpson 1961; Mayr and Ashlock 1991). The important scientific action in classifying is to find appropriate equivalence relations that allow one to define natural classes. "Defining" the *names* of taxa, instead of the taxa themselves, as the bionominalists are fond of doing, is, of course, a useful but scientifically rather unspectacular activity.

7.2.2.4 Systematics and Evolutionary Theory

The most important theory that may help refine a biological classification is, of course, the theory of evolution. However, thus far we have been concerned only with the definition of a hierarchy of nested classes and their static inclusion relations. Moreover, our viewing taxa as kinds seems to be incompatible with an evolutionary outlook, because classes are conceptual objects to which the category of change does not apply. This is one of the main reasons brought forward by bionominalists in favor of the view that species (or taxa in general) could not be conceived of as natural kinds (e.g., Rosenberg 1985). Therefore, we must show first of all that our conceptualist philosophy of taxonomy is compatible with evolution.

Graphical Displays of Nested Classes. In order to facilitate the visualization of the procedure, it is convenient to introduce two forms of graphic representation of a hierarchy of nested sets. One useful way of displaying graphically such a hierarchy is the *Venn diagram* familiar from set theory (Fig. 7.4a); another is the *tree* familiar from graph theory (Fig. 7.4b). An older name for the latter is *Hasse diagram.* The two are alternative visual representations of one and the same network of logical relations; that is, they are logically equivalent. It should be pointed out that they visualize *purely logical* relations among sets. In particular, a Venn diagram does not represent part-whole relations among things, and a tree in this case represents neither a phylogenetic tree nor a lineage. Moreover, these diagrams involve no time axes. Yet if conceptualism is supposed to be compatible with evolution, particularly with the hypothesis of descent with modification, such a logical tree must somehow be the basis for hypothesizing a phylogenetic tree. Indeed, that a classificatory tree or *cladogram* is not the same as a phylogenetic tree has been emphasized by Nelson and Platnick (1981, p. 17). However, they incorrectly state that a cladogram involves the "element of time" (p. 36), and that a cladogram "denotes" a set of (phylogenetic) trees (p. 171). The first proposition is incorrect because a cladogram depicts nothing but nested sets; and the second is incorrect because a classification refers to organisms, not to phylogenetic trees (recall the notion of denotation or reference from Chap. 2). What is true is that a cladogram is *compatible* with possibly more than one phylogenetic hypothesis or tree.

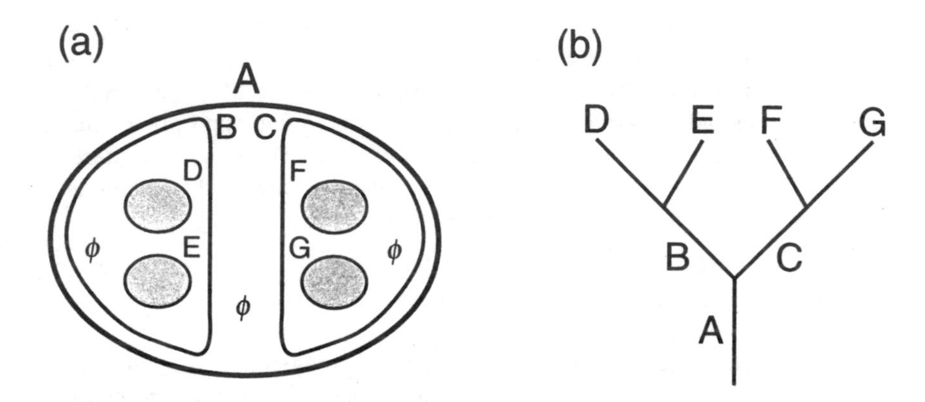

Fig.7.4 a, b. a Venn diagram of nested sets. Note that, if a Venn diagram is used to represent a biological classification, then the free space between any two boundaries is empty, i.e., it is not itself a nonempty subset of the given set. That is, in the example depicted $A = B \cup C$, $B = D \cup E$, and $C = F \cup G$. To make clear that the free space between any two boundaries is empty, the symbol $\varnothing$ should be added as shown. (Alternatively, one could use a pie diagram, which does not contain empty spaces.) **b** *Hasse* or *tree* diagram of nested sets. This representation is logically equivalent to the one shown in **a**. In biological systematics the sets involved are called *taxa*, and any representation of this type is known as a *cladogram*. Unlike a phylogenetic tree, the lines do not represent stem species (or perhaps entire stem lines) but set inclusion

From Property Precedence to Common Descent. We shall attempt to solve the problem of reconciling the static levels of our systematic hierarchy with an evolutionary outlook in a similar way as we did in the case of biolevels (Sect. 5.3). Just as we can "read" the notion of level precedence in terms of temporal precedence, we can interpret the notion of property precedence introduced in Definition 7.1 in terms of temporal precedence. (However, this procedure transcends classification.)

That is, if a property P is more common than a property Q, i.e., if P precedes Q, or taxon $\mathcal{Q}$ is included in taxon $\mathcal{P}$, then we can say that the things possessing *only* property P, i.e., the individuals in the set $\mathcal{P} - \mathcal{Q}$, have come into being before the things possessing *also* property Q, i.e., those in $\mathcal{Q}$. Still, mere temporal precedence does not imply descent. Therefore, we have to stipulate in addition that the Q-things descend from P-things, which at the same time (subsumptively) explains the lawful relationship of the properties P and Q. Furthermore, if the Q-organisms descend from P-organisms, we can say that property Q is (likely to be) an *evolutionary novelty* characterizing the class of Q-organisms, i.e., the taxon $\mathcal{Q}$. Consequently, all the properties that may be concomitant with Q will constitute the set of evolutionary novelties of the Q-organisms; that is, they define the taxon $\mathcal{Q}$. Each (nested) natural kind in the classification is thus defined by a set of qualitative novelties (*derived characters* or *apomorphies* in cladistic terminology), although the organisms in any given taxon are also characterized by the features they share with earlier ancestors (*primitive characters* or *plesiomorphies* in cladistic terminology).

May we now after all say that taxon $\mathcal{Q}$ descends from taxon $\mathcal{P}$? No; or, more precisely, not if $\mathcal{Q}$ and $\mathcal{P}$ are taxa related by the inclusion relation. Since taxon $\mathcal{Q}$ is included in taxon $\mathcal{P}$, all Q's *are* P's, which is not the same as saying that all Q's descend from P's. For example, since Mammalia $\subset$ Vertebrata, we cannot say that mammals descend from vertebrates, because all mammals *are* vertebrates. Neither can we say that *Homo sapiens* descends from Hominidae, because we *are* hominids. We could, however, say that we "descend" from other hominid species, such as *Homo erectus* and *Australopithecus afarensis*. (That there are no supraspecific ancestors has been emphasized by some authors, although for reasons different from ours, namely reasons inspired by a neonominalist outlook: e.g., Wiley 1981; Ax 1984, 1988.)

When talking of the "descent of species", we still face two problems. First, we do not find ancestral species in our hierarchy of nested classes. Second, we still have to solve the problem that biospecies conceived as natural kinds (hence, as conceptual objects) cannot descend from each other in a literal sense: only organisms (being material objects) can. Therefore, we have to remove this obstacle by firstly elucidating the notions of ancestry, progeny, lineage, and evolutionary lineage.

Ancestry, Lineage, Evolution. In contradistinction to the bionominalists, who take species to be concrete individuals and are thus able to get by with a single concept of descent and of lineage, we must elucidate two different concepts of descent and lineage: one for organisms and, if necessary, for biopopulations (real individuals), the other for kinds (constructs). Here is the first, which could also be stated in terms of biopopulations rather than organisms:

DEFINITION 7.4. Let O designate a collection of organisms of some species. Then for any individuals x and y in O,

(i) x is an *immediate ancestor* of y (or y descends immediately from x) iff x is a parent of y;

(ii) x is a *mediate ancestor* of y (or y descends mediately from x) iff there is a z in O such that x is an immediate ancestor of z, and z an immediate ancestor of y;

(iii) x is an *ancestor* of y (or y descends from x) iff x is either an immediate or a mediate ancestor of y (symbol: $x < y$);

(iv) the *ancestry* of x is the collection of ancestors of x: $A(x) = \{y \in O \mid y < x\}$;

(v) the *progeny* of x is the collection of organisms of which x is an ancestor: $P(x) = \{y \in O \mid x < y\}$;

(vi) the *lineage* of x is the union of the ancestry and the progeny of x: $L(x) = \{y \in O \mid y < x \vee x < y\}$.

As the ontological status of lineages is by no means clear in the biophilosophical literature, we emphasize that the ancestry, progeny, and lineage of a system are neither things nor "historical entities" but sets, where the ancestry relation (or, if preferred, the relation of descent) $<$ is a strict partial order in such a set. What has been called the 'genealogical nexus' (e.g., by Hull 1978, 1987; Ghiselin 1981) is not a bonding or causal relation and thus cannot be regarded as a coupling among parts of a system (recall Definition 1.7). Hence, lineages are not real entities and they neither change nor evolve.

Moreover, there are also logical reasons why lineages cannot be real entities. First, the concepts of ancestry, progeny, and lineage are relational: there are only ancestries, progenies, and lineages *of* things. Real existents, by contrast, are absolute. (Thus, *pace* Wilson 1995, a single entity such as an organism cannot be a lineage. Besides, the relations involved, namely those of ancestry and descent, are irreflexive.) Second, time is already implicit in the notion of lineage, so that the change of a lineage could only be a timeless change—a *contradictio in adjecto*. What do change are the members of the lineage from one instant of time to the next, namely organisms or biopopulations giving rise to further organisms or populations, respectively.

Room for evolutionary considerations can be made in the following way:

DEFINITION 7.5 Let $L(x)$ name the lineage of an organism x of species S. Then $L(x)$ is an *evolutionary lineage* if, and only if, at least one ancestor or one descendant of x belongs to a species S' different from S.

According to this general notion of evolution, *evolution amounts to speciation*, that is, to the emergence of qualitatively novel organisms. Since natural kinds are collections, it suffices for the formation of a new species that there be only a single new individual. (More on evolution and speciation in Sect. 9.1.)

Although the ontological categories of change and descent do not apply to conceptual objects such as species, the notions of descent, ancestry, progeny, and lineage can nevertheless be carried over to species of organisms by means of:

DEFINITION 7.6. Let $\mathcal{S}$ designate a family of species of organisms: $\mathcal{S} = \{\mathcal{S}_i$ names a species of organisms $\mid 1 \leq i \leq n\}$. Then for all $\mathcal{S}_j$ and $\mathcal{S}_k$ in $\mathcal{S}$,

(i) $\mathcal{S}_k$ *descends* from $\mathcal{S}_j$ iff every member of $\mathcal{S}_k$ descends (mediately or immediately) from some members of $\mathcal{S}_j$: $\mathcal{S}_j < \mathcal{S}_k$;

(ii) the *ancestry* of $\mathcal{S}_j$ is the collection of species from which $\mathcal{S}_j$ descends: $A(\mathcal{S}_j) = \{X \in \mathcal{S} \mid X < \mathcal{S}_j\}$;

(iii) the *progeny* of $\mathcal{S}_j$ is the collection of species descending from $\mathcal{S}_j$: $\Pi(\mathcal{S}_j) = \{X \in \mathcal{S} \mid \mathcal{S}_j < X\}$;

(iv) the *lineage* of $\mathcal{S}_j$ is the collection of species descending from $\mathcal{S}_j$ or from which $\mathcal{S}_j$ descends: $\Lambda(\mathcal{S}_j) = \{X \in \mathcal{S} \mid X < \mathcal{S}_j \vee \mathcal{S}_j < X\}$.

In Definition 7.6 the word 'species' could as well be replaced by 'taxon'. We would then be able to say that a taxon A descends from some other taxon B. For example, we could say that Aves descend from Reptilia. However, this is only possible if the taxa in question are mutually disjoint, as Aves and Reptilia are in an evolutionary classification à la Simpson and Mayr. Moreover, one of the taxa must not be defined by unique evolutionary novelties. Since in a cladistic classification all taxa are consistently defined by concomitant properties, i.e., evolutionary novelties, two mutually disjoint taxa cannot be ancestral to each other *per definitionem*. Thus, ancestors must be sought elsewhere.

Taxa and Ancestors. Neither the systematist nor the paleontologist can observe the descent of past or even of most present organisms and biopopulations: see Fig. 7.5. What they do observe, if lucky, is remains or traces of past organisms and, in most cases, present organisms of a certain kind or species characterized by a set of properties, whether morphological, physiological, biochemical, behavioral, or what have you. Thus, the only hope for systematists to reconstruct descent with modification is to trace the descent of species.

The most convenient way to do so is by means of the visual representation of a classification in form of a tree as introduced above (Fig. 7.4b). In so doing, it should be recalled that such a tree represents only a hierarchy of nested classes. The first step towards the reconstruction of common descent or phylogeny is to add a time axis to a logical tree, where time is usually plotted on the ordinate. The time axis is not quantitative, however, since all we need is to interpret set inclusion, and thereby property precedence, in terms of temporal precedence. The next step consists in taking each line between two nodes to represent the stem species of the set of species contained in the subordinated taxon. If we consider a taxon $\mathcal{P}$ containing two subsets $\mathcal{Q}$ and $\mathcal{R}$ of equal rank, then there would be a nonempty set of organisms $\mathcal{S}_P = \mathcal{P} - (\mathcal{Q} \cup \mathcal{R})$ identical to the *stem species* of all the species in taxon $\mathcal{P}$. Clearly, this assumption goes beyond classification because, according to condition (8) of the principles of classification, $\mathcal{P} = \mathcal{Q} \cup \mathcal{R}$, hence $\mathcal{P} - (\mathcal{Q} \cup \mathcal{R}) = \varnothing$. This is the reason why a classification (or a cladogram *sensu* Platnick 1977 and Nelson and Platnick 1981) does not contain ancestors and why it is *not equivalent* to a phylogenetic tree. As Patricia Williams (1992) has shown, ever since Hennig

(1966) cladists confuse classifications—logical hierarchies of nested classes—with phylogenetic trees—hypotheses about descending species—thereby giving rise to a number of problems, such as that of the "reality" and "survival" of stem species.

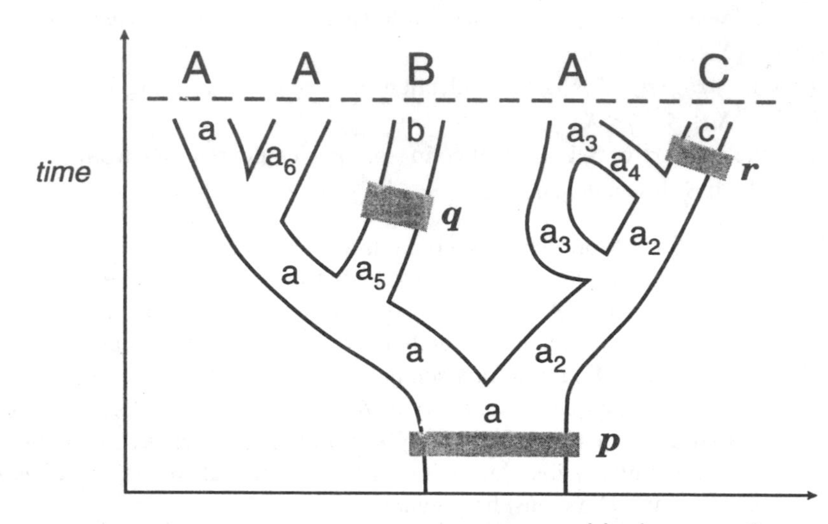

Fig.7.5. History of a biopopulation *a* composed of organisms of species *A* characterized by the qualitative novelty (or autapomorphy) *p*. The population *a* splits into several daughter populations: a_2, a_3, a_4 (which subsequently fuses again with a_3), a_5, and a_6. The organisms in a_2 as well as those in a_5 undergo qualitative changes, i.e., speciations, by acquiring the evolutionary novelties *q* and *r* respectively. Upon fixation of *q* and *r* in the respective populations, there are now two new biopopulations *b* and *c*, since they consist of organisms belonging to the new species *B* and *C*, respectively. (A *gray box* indicates the period between both the emergence and the fixation of some novelty or novelties in the population.) Note that, at the time marked by the *dashed line*, the species *A* is not extinct, for the populations *a*, a_3 and a_6 still exist after the speciations of *b* and *c*

When we assume the existence of stem species, all branches *diverging* from a given node in a phylogenetic tree depict the progeny of a single species (Fig. 7.6). If, at any given time, a stem species still has living members, then we may represent this case by continuing the line representing the stem species to the time horizon under consideration (Fig. 7.6a). The lines reaching the time under consideration are called *terminal* and may represent either single species or higher taxa. The line between two nodes may represent not only a single species but also the ancestry of the stem species within a given taxon, i.e., the *stem line* (Ax 1985). That is, there is room for interpolating new subsets within a given set, if further organisms, or species respectively, whether past or present, are discovered. Any stem species together with its progeny is called a *monophyletic group*, or *monophylum* for short (Ax 1984).

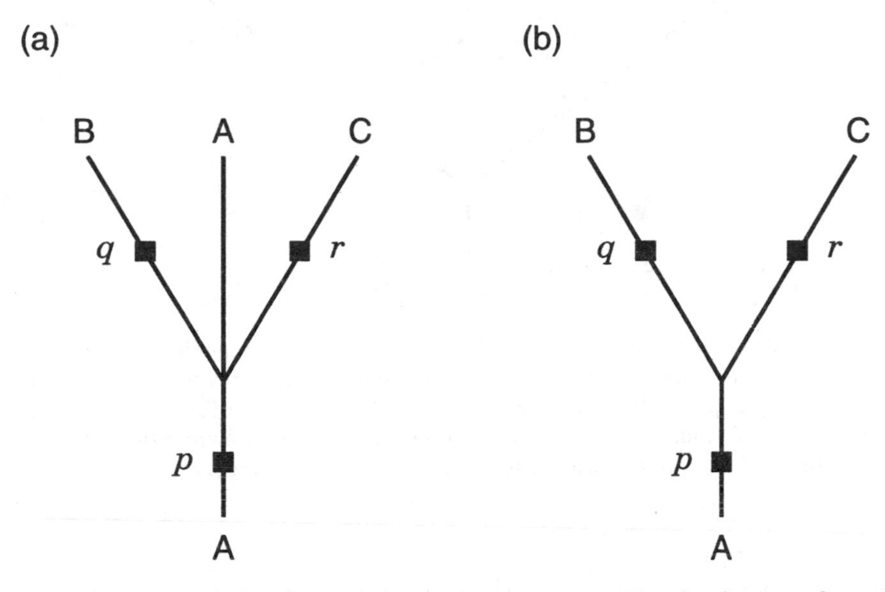

Fig.7.6 a, b. a Trichotomous phylogenetic tree representing the descent of species *B* and *C* from ancestral species *A*, which still has living members. This tree is all we can reconstruct from the actual history of the biopopulations depicted in Fig. 7.5. Unlike the gray boxes in Fig. 7.5, the *black squares* do not indicate a period of speciation but merely the presence of the defining properties, i.e., the qualitative novelty or novelties, of the given species. **b** As soon as species *A* becomes extinct, the trichotomous tree reduces to a dichotomous one, and so the character *p* will be regarded as a synapomorphy of the species *B* and *C*, whose status as sister species is now apparent. We submit that the popular cladistic belief that ancestral or stem species necessarily become extinct upon speciation rests on the confusion of the methodological principle according to which phylogenies should be resolved into dichotomous branches—a principle which underlies the Hennigian method of phylogeny reconstruction by means of synapomorphies—with the ontological assumption that speciation is, in fact, always or at least predominantly dichotomous. The situation illustrated here and in Fig. 7.5 also explains why phylogenetic analyses at the genus and species level often do not result in the desired neatly dichotomous tree but are prone to contain polytomies

Note the following points. First, even if we call such a tree a *phylogenetic tree* now, because it contains ancestors and not only nested taxa, it may not be mistaken for a history graph (see also Bunge 1987a; P. Williams 1992). A phylogenetic tree merely represents the temporal precedence of species or, in other words, a sequence of successively descending species. In particular, the lines representing stem species should not be mistaken for the histories of biopopulations (Fig.7.5). Each such line merely represents a class of organisms.

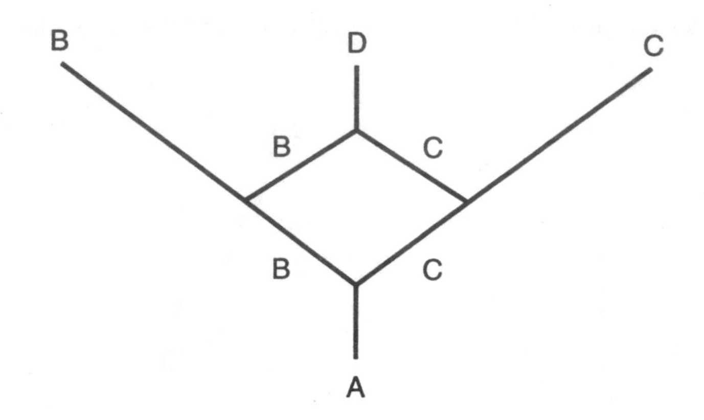

Fig.7.7. Phylogenetic tree representing a case of speciation by hybridization: $B + C \rightarrow D$; explanation in the text. Note that this representation assumes the simplest case where the hybrid species D has retained all the derived properties of its parent species. As a matter of fact, many speciations by hybridization are much more difficult to discern, since a hybrid species need not retain all of the defining properties of its parent species

Second, contrary to popular cladistic belief, it is not required that the branching of a node be dichotomous. (See also Nelson and Platnick 1981.) A species may subsequently give rise to many daughter species and yet have Recent members (Fig. 7.6a). However, since a polytomous branching cannot be proved to actually represent such but may be due to ignorance, that is, a lack of knowledge of characters allowing for a finer analysis, the striving for a dichotomous or binary branching remains a *heuristic* rule of systematics. Already Hennig (1966) called it a *methodological principle*.

Third, the abscissa of a phylogenetic tree does not represent the degree of (character) divergence—morphological or otherwise—between species or taxa: it only indicates speciation. However, since speciation involves the emergence of new characters, the number of speciations between any two species in the tree is likely to indicate also character divergence, in particular genetic and morphological divergence.

Fourth, a special problem arises in the case of speciation by hybridization. The latter (e.g., $B + C \rightarrow D$) cannot be depicted by a trichotomy such as the one shown in Fig. 7.6a, as is suggested by the classification which may contain three taxa of equal rank. The reason is that such a trichotomy signifies that some stem species with living members gave rise to two daughter species. In the case of hybridization, however, two species B and C, themselves likely daughter species of a stem species A, gave rise to a common daughter species D. That is, the daughter species D must be depicted as branching off from $B + C$. Thus two branches, B and C, will initially diverge from a common node, A, but then converge again into a single node and branch, D, indicating that two ancestral species gave rise to a single daughter species. Graphically, we thus get two bent lines forming a diamond: see Fig. 7.7. (More on character incongruences and reticulation in cladograms in Nelson 1983.) Thus, not every node in a phylogenetic tree represents a speciation.

We submit that the taxonomic procedure outlined so far allows for the closest possible approximation to the real phylogeny in nature. There is no way to arrive at a genuine history graph that depicts lineages of organisms or biopopulations: again, compare Figs. 7.5 and 7.6. These figures also illustrate why biopopulations (material systems) should not be confused with species (classes).

Evolutionary Assumptions. So far, only the hypothesis of descent with modification—the assumption that all organisms go back to a single ancestral species—was needed to guide our "translation" of a classification into a phylogenetic tree. (See also Wiley 1975; Gaffney 1979.) In turn, descent with modification implies that (most of) the characters of the ancestors are somehow stably re-constructed in the descendants. (More on this in Chap. 8.) However, no theories concerning evolutionary mechanisms or genetics were involved. Hence the use of theories in classification is thus far pretty limited. We submit that a refinement of classification should be possible when we additionally make use of *all* available relevant knowledge in biology.

To substantiate this claim, we should first of all bear in mind that the basic procedure of forming hierarchies of nested natural kinds is not as easy and straightforward in practice as its previously outlined logic and methodology may suggest. This is because, in practice, we face several problems in finding suitable properties or equivalence relations for delimiting natural classes.

One problem is that we have to deal with character transformations. For example, it is important to realize that the limbs of tetrapods are modified fins. Otherwise, we might be tempted to form two nonoverlapping classes within vertebrates, namely Pisces (fishes) and Tetrapoda. However, since fins evolved into limbs, limbs are a qualitative novelty characterizing Tetrapoda. Fins, on the contrary, are a qualitative novelty that does not characterize fishes but vertebrates (see also Platnick 1979). Similar considerations are in order in the case of complete reductions of characters such as the loss of limbs within tetrapods, e.g., among snakes and caecilians. This poses the problem of how to decide whether a character transformation is an instance of increasing complexity or an instance of reduction. Another problem consists in the fact that certain characters may have evolved two or more times independently from each other. In this case we obtain either overlapping sets or two or more alternative, i.e., incongruent, nested classes.

Before making use of biological theories, a great many problems of this sort can be eliminated by comparing the congruence, or else incongruence, of different character distributions (Patterson 1982). The computer has given rise to a whole industry supplying methods and algorithms for analyzing incongruent nested sets of properties and for finding the most congruent or most parsimonious hierarchy of such nested sets. (For an overview including literature see Zandee and Geesink

1987; Mayr and Ashlock 1991; as well as the journals *Systematic Biology*, formerly *Systematic Zoology*, and *Cladistics*.)

Clearly, such analyses can be carried out largely numerically without choosing and weighting properties according to biological knowledge. Hence it should come as no surprise that the results are of limited interest. For example, a recent computer analysis of the phylogenetic relationships of hammerhead sharks resulted in more than 30000 equally parsimonious trees (Naylor 1992). Although the number of trees could be further restricted by means of character weighting and different algorithms, the lesson from cases like this is obvious: parsimony can only be a heuristic assumption to begin with, because there is no reason to believe that nature (and particularly evolution) is actually parsimonious (Bunge 1963; Nelson and Platnick 1981). The heuristic rules of parsimony in systematics can be formulated thus: "Start with the assumption that a character has evolved only once in an ancestral species" or "Favor—until further notice—the tree which assumes the smallest number of evolutionary events". (For some of the problems of computer systematics see Wägele 1994.)

Any departure from these rules must be justified by further knowledge. Where such extrasystematic knowledge, that is, the whole of biological theory, may help refine a raw (i.e., pretheoretical) classification is in the analysis of characters. Any knowledge of evolutionary, adaptational, ecological, genetic, and developmental processes and mechanisms should be welcome to assist the analysis of characters. Theoretical knowledge may thus not only help find more suitable features for classification, but also help choose among alternative phylogenetic hypotheses, i.e., trees. For example, it may turn out that a less parsimonious tree is truer than a more parsimonious one because it is backed by a more plausible adaptational scenario, which, in turn, is in tune with ecological knowledge.

Regrettably, an assessment of some such scenarios indicates that their usefulness for phylogenetic analysis has not yet been effectively demonstrated. This is because many such scenarios just propose more or less plausible historical stories but are not based on knowledge, in particular well-confirmed theories, of the underlying evolutionary processes (Cracraft 1981). Furthermore, though adaptionist assumptions remain heuristically fruitful, not all features and evolutionary changes need actually be adaptive (Simpson 1953; Gould and Lewontin 1979).

We conclude that, whereas systematics does not—*pace* Bock (1981)—logically presuppose any knowledge of evolutionary mechanisms or knowledge from any other biological theory, such knowledge may and should contribute to the refinement of classification. Deep knowledge can only be achieved by theorizing. To use Hennig's felicitous phrase, classification and theory gain from each other by *reciprocal illumination*. (See also Hull 1979.) However, we emphasize that any extrasystematic theory does not enter the classification to become part of it. Theories only help us find, analyze, and evaluate suitable characters for classification. Yet, if not a theory, what exactly is a classification?

7.2.2.5 The Logical and Methodological Status of Classifications

When the principles of classification, as outlined in the preceding, are applied properly, one arrives at a nested system of class or, more precisely, taxon *definitions*. The following (cladistic) example illustrates the basic structure of a classification. (Recall also the remarks to condition (4) in Sect. 7.2.2.2.)

Let $\mathcal{A}$ designate the class of Amniota, M a conjunction of predicates referring to qualitative novelties, such as "possessing hairs", "possessing mammary glands", "possessing a squamoso-dentary joint", "possessing a synapsid skull", and so on, t an instant of time, and D the relation of descent. Then the taxon (Recent) Mammalia, or $\mathcal{M}$ for short, is defined as $\mathcal{M} = \{x \in \mathcal{A} \mid (\exists t)(\exists y)\ (Mxt \vee [Dxy \;\&\; y \in \mathcal{A} \;\&\; Myt])\}$. Within the class $\mathcal{M}$ we can now define the taxon Theria, i.e., the viviparous mammals, as $\Theta = \{x \in \mathcal{M} \mid (\exists t)(\exists y)\ (Vxt \vee [Dxy \;\&\; y \in \mathcal{M} \;\&\; Vyt])\}$, where V designates the conjunction of the relevant predicates, such as "viviparous". Further, we define the class Monotremata or Prototheria (i.e., the so-called egg-laying mammals) as $\mathcal{P} = \{x \in \mathcal{M} \mid (\exists t)(\exists y)\ (Ext \vee [Dxy \;\&\; y \in \mathcal{M} \;\&\; Eyt])\}$, where E designates the conjunction of relevant predicates, such as "possessing electrical sense organs" and "being a male and possessing a poison gland in the hind legs". (Note that the feature "egg-laying" is one of the defining characters of the class Amniota, so that it cannot be used to define the taxon $\mathcal{P}$. For morphological and systematic details see Ax 1995.) According to the principles of classification, $\mathcal{M} = \Theta \cup \mathcal{P}$ and $\Theta \cap \mathcal{P} = \emptyset$. Furthermore, by definition, we have $\Theta \subset \mathcal{M}$, i.e., Theria $\subset$ Mammalia, and $\mathcal{P} \subset \mathcal{M}$, i.e., Monotremata $\subset$ Mammalia.

The preceding example illustrates the thesis that a classification is a system of nested taxa definitions. Since definitions are conventions (Sect. 3.5.7.1), classifications are conventions too, hence neither true nor false. However, with a few exceptions like Ruse (1973), most authors have contended that classifications are either theories or at least quasi-theories (e.g., Løvtrup 1973, 1974; Bock 1974; Brady 1979; Nelson and Platnick 1981; Bunge 1983a; Suppe 1989; Mayr and Ashlock 1991). This latter view is usually defended by arguments like the following. Scientific classifications have a factual content. They involve empirical operations such as observation and sometimes measurement. Moreover, on the strength of new observations and hypotheses, classifications are often "corrected" and thereby "improved". Last, but not least, classifications are said to have either predictive or explanatory power, or both.

Indeed, unlike formal classifications such as those of numbers (e.g., $\mathbb{N} \subset \mathbb{Z} \subset \mathbb{Q} \subset \mathbb{R} \subset \mathbb{C}$), scientific classifications deal with material objects and thus involve empirical operations. Yet this has no bearing on the formal status of classifications. Although concrete things come in natural kinds, kinds are classes, not things. Therefore kinds can only be defined, not described. Only their individual members can be described.

Furthermore, since the definition of natural (rather than artificial) kinds involves lawfully related properties, and since the proper representation of laws involves theories (Definition 3.9), it is true that scientific classifying involves theorizing.

In particular, it involves knowledge of comparative morphology and homology hypotheses. However, the fact that scientific classifications may be altered on the basis of a scientific theory, and of observations made in the light of such theory, does not entail that the classification itself—as a formal object—is a theory. It only entails that a scientific classification is backed by a scientific theory, whereas an ordinary or nonscientific classification is not based on such theory. So we could say that while a *natural* or *nonarbitrary* or *scientific* classification is backed by a reasonably true scientific theory, an *artificial* or *arbitrary* or *nonscientific* classification is either based on a false theory or is simply without theoretical foundation. (See also Löther 1972.)

It will be obvious from the preceding how easy it is to conflate classification and theory. This temptation is nurtured by the fact that inferences as to the inclusion of certain taxa, which in the classification hold by definition, can easily be transferred to a theory. For example, although in classification viviparous mammals are mammals *by definition*, the statement "All therians are mammals" may function as a *hypothesis* in a theory about the phylogeny of mammals. (Recall from Sect. 3.5.7.1 that what functions as a definition in one context may function as a hypothesis in another, and vice versa.) This holds in particular for cladistic classifications, which are supposed to be "translatable" or "convertible" into a phylogenetic tree. A phylogenetic tree, however, is clearly a theory, not a system of taxa definitions. For example, the statement "All mammals go back to a single ancestral species" is a hypothesis that does not occur in any classification.

The confusion of a classification with the theories "interacting" with the classification also underlies the assumption that classifications have explanatory power. For example, the statement that fishes, amphibians, reptiles, birds, and mammals are vertebrates appears to explain (phenomenologically) the similarity among these animals, but only because some of their common properties were used to construct the taxon Vertebrata in the first place. So this seeming case of subsumption, not explanation, is tautological (recall Sect. 3.6.1.2). A genuine subsumption of the similarities among vertebrates can only be provided by a phylogenetic theory; and a genuine, i.e., mechanismic, explanation of the similarities among vertebrates can only be provided by evolutionary theory (including a theory of development), not by any classification.

For the same reason, a classification proper has no interesting predictive power. The biologist may believe that classifications have relevant predictive power because they sometimes allow him or her to infer "unknown" data about character distribution. For example, when the biologist identifies a newly found female arachnid as a spider because it has spinnerets, he or she can infer that its yet unknown male has pedipalps modified for sperm transfer (Platnick 1994, in litt.). However, from a methodological point of view, this is not an interesting case of prediction (recall Sect. 3.6.2), because what can be inferred in such cases is only that which is already contained in the definition of the taxon in question. In other words, if you find a spider of a new species you can only "predict" what you already know about spiders in the given taxon. You cannot predict really new features. Thus, neither the reduction, nor the complete loss of any taxon-specific character, nor the presence of any yet unknown derived character of this taxon can be predicted.

Examples of predictions due to the underlying theories of evolution and phylogeny, not the classification proper, are the hindcasts of missing links or "transitional taxa". For example, phylogeneticists and paleontologists have often inferred that, since evolution is (more or less) gradual, a gap in the fossil record of a certain taxon can be expected to be bridged by some missing link or transitional taxon. Yet all the premises of this retrodiction belong in evolutionary, phylogenetic, and paleontological theories, not in classification. A classification becomes imperfect or unsatisfactory only in the light of some theory, and only in combination with some theory does it acquire genuine predictive and explanatory power.

To conclude, classifications are systems of taxa definitions, hence conventions. They should not be confused with the theories employed in constructing and refining the classification. Scientific classifications, however, must be backed by one or more reasonably true scientific theories. Indeed, classing and theorizing are mutually complementary activities. Still, it is important to note that categorizing logically precedes theorizing if only because every theory is about some category of objects. In turn, theory allows one to refine pretheoretical classifications. For example, a biological classification logically precedes a phylogenetic hypothesis (e.g., a phylogenetic tree), but the classification may be refined in the light of the phylogenetic theory if there is evidence in favor of the phylogenetic tree that is not already contained in the classification. Thus, the claim that a biological classification must be "based on phylogeny" is wrong if understood as a claim as to the logical priority of a phylogenetic hypothesis over a classification; but it is correct if it is understood in the weaker sense that a biological classification may and should be refined in the light of phylogenetic knowledge.

7.2.2.6 Taxonomy, Classification, Systematics

We turn now to the examination of the differences, if any, between the meanings of the terms 'taxonomy', 'classification', and 'systematics'. Whereas many authors use these terms interchangeably, others have attempted to distinguish some or all of them (e.g., Simpson 1961; Griffiths 1974; Solbrig and Solbrig 1979; Wiley 1981; Ax 1984, 1988, 1995; Bunge 1985b; de Queiroz 1988; Mayr and Ashlock 1991). A first useful distinction is certainly that these terms may denote a scientific discipline, the activity of the scientists in that discipline, and the outcome of this activity (de Queiroz 1988). From an epistemological and historical point of view, the basic and most ancient *activity* is *classifying* objects, i.e., classification. Indeed, this activity does not principally presuppose any scientific knowledge. If it does not, the classification—the activity as well as its outcome—may be largely arbitrary, artificial, anthropocentric, or superficial, rather than objective, realistic, and deep. (See, however, Berlin 1992.) A deep and realistic classification is obtained only with the help of scientific knowledge. In biology it involves the whole of comparative biology (Nelson 1970). Thus, there is a long way from the four ancient elements to the elements of modern chemistry. The same holds for biological classification from ancient humans through Aristotle and Linnaeus to modern phylogenetic classification.

Since the outcome of a scientific classification is a *conceptual system*, i.e., a system of nested definitions with referential unity, it is possible to call the corresponding scientific discipline *systematics*, and the underlying research activity *systematization*. However, a modern classification or system based on scientific knowledge is still a classification. At the same time, it is a result of a classification process. In other words, every systematization is a classification but not every classification is systematic. In proposing this usage we reject the reason given by neonominalists why biological classification should be termed 'systematics', namely because taxa would be concrete systems rather than classes (Griffiths 1974; de Queiroz 1988; de Queiroz and Donoghue 1988). Taxa *are* classes, though certainly not arbitrary but natural ones. We also reject a narrow formalist conception of classification, that is, naming taxa, assigning them a categorical rank, and publishing this arrangement in written form. This is only the last step in a classification.

The term 'taxonomy' has mostly been used to denote the scientific discipline of classification as well as the outcome of the activity of classification. Since we already have the terms 'classification', 'system', and 'systematics' at our disposal for these purposes, the term 'taxonomy' is redundant. We therefore adopt Simpson's (1961) proposal, which has, however, not gained wide usage, to regard *taxonomy* as the theoretical study of the (logical and methodological) principles and rules of classification and systematics, i.e., as *metaclassification*. (Interestingly, Simpson did not consistently use the term 'taxonomy' as defined by himself.) This coincides with what Gregg (1954) called 'methodological taxonomy'. Because of its theoretical nature (methodological) taxonomy or metaclassification is sometimes called 'theory of classification' or 'theory of systematics'. We reject this usage because we reserve the term 'theory' to designate a hypothetico-deductive system, which is neither a discipline nor a set of rules for producing a classification.

The *philosophy of taxonomy* or *metataxonomy*, then, is the conceptual system or discipline providing the ontological, epistemological, semantical, and logical background for taxonomy. For example, the questions "What kind of entity is a taxon: class or individual?" or "What is the logical and methodological status of a classification: convention or theory?" belong in the philosophy of taxonomy.

7.2.2.7 Three Taxonomies: Cladistic, Evolutionary, and Phenetic Taxonomy

Cladistic Taxonomy. Systematists will have noticed that the taxonomy outlined in the preceding bears some resemblance to what has been called *transformed* or *pattern cladism* (see, e.g., Platnick 1979, 1985; Nelson and Platnick 1981; Patterson 1982). However, our taxonomy has been developed independently since it flows naturally from our ontology. Moreover, it is based on logical and ontological reasons, whereas traditional pattern cladism appears to be inspired by empiricism and falsificationism. In any case, our version is yet another "transformation" of cladism, even a transformation of transformed cladism. Still, it is an essentially cladistic taxonomy, that is, one that aims at producing consistently nested classes of organisms which are defined exclusively by evolutionary novelties. Since the taxa of cladistic classification are defined by lawfully related properties, they are—*pace* Wiley 1989—examples of biological kinds *sensu lato*.

As for the label 'pattern cladism' (Beatty 1982), it is due to its adherents' tenet that the task of systematics consists in revealing a pattern of characters in nature from which phylogeny can be reconstructed or inferred—if one is inclined to do so. No prior evolutionary considerations, i.e., assumptions about the processes resulting in that pattern, should enter the systematist's activity, because the procedure then would risk becoming circular. Only if an observed pattern, i.e., a classification, is obtained independently of the theory of evolution as well as from paleontological data, would it be possible to test evolutionary hypotheses by means of that pattern. This proposal, which seems to be inspired by an empiricist outlook and by Popperian falsificationism, ensued in what is known as the *pattern-process controversy* (e.g., Brady 1985; Rieppel 1988.)

Since our view concerning the relation between classification and theory has already been expounded in the preceding, we will add only a few comments in the following. First of all, it is interesting to note that the tenets of the pattern cladists have been widely misunderstood or misrepresented (e.g., Beatty 1982; Charig 1982; Leroux 1993). For example, if we understand the writings of the pattern cladists correctly, it is denied neither that cladistics presupposes that natural groupings are possible nor that knowledge of stratigraphic and biogeographic data as well as of theories about evolutionary mechanisms may refine systematics (Platnick 1985). What is stressed is the logical and methodological independence of systematics from particular *mechanismic* theories. (See also Ax 1988.) However, the practicing systematist usually does not care much about logic or methodology, but proceeds according to certain rules and comes up with a classification by simultaneously applying pattern analysis as well as evolutionary and adaptational assumptions. He or she clearly starts classifying *sub specie evolutionis*—an entirely legitimate procedure. (See also Hull 1979). On the other hand, the neglect of the logic of classification may lead to such naive contentions that one would first analyze phylogenetic relationships and then convert a phylogenetic tree into a classification or, worse, that the two would be equivalent representations of the same conceptual system.

The concerns about the test of evolutionary hypotheses by means of systematic patterns are unwarranted, because we no longer have to test the hypothesis that evolution has in fact occurred, as was necessary when Darwin tried to establish his hypothesis of descent with modification in the first place. Today, we have to demand that classifying and theorizing go hand in hand, because science is characterized by a feedback process of successive approximations to the facts. This feedback process is not an instance of circularity but a hallmark of science. Finally, as shown in the analysis of the principles of classification, the preference for cladistic principles of classification over the standard rules of classification by partitioning can be understood only by presupposing the hypothesis of (common) descent with modification and the evolution of qualitative novelties. In other words, contrary to the empiricist outlook of some pattern cladists, the observed systematic pattern is not theory-free, as it is produced by cladistic (rather than alternative) rules of classification. (As usual, empiricism proves to be either naive or wrong.)

To conclude, despite certain taxonomic and metataxonomic differences between pattern cladism and traditional Hennigian cladistics, we submit that the two can be reconciled if reconstructed in the light of our philosophy of taxonomy as outlined above. In any case, pattern cladism and phylogenetic systematics yield virtually the same results in practice. (For "traditional" cladistics see Hennig 1966; Wiley 1981; Ax 1984, 1988, 1995; Sudhaus and Rehfeld 1992. For a history of pre- and quasi-cladistic ideas see Craw 1992. For the development of cladistics in relation to the other schools of taxonomy, including plenty of entertaining gossip, see Hull 1988.)

Evolutionary Taxonomy. Certainly, the most famous representatives of traditional or evolutionary taxonomy are George Gaylord Simpson and Ernst Mayr (Simpson 1961; Mayr 1974, 1982, 1995; Mayr and Ashlock 1991; see also Bock 1974). The main difference between cladistic and evolutionary taxonomists lies in the latters' emphasis on anagenetic difference, i.e., in the acknowledgment of so-called *grades*. For example, although gorillas and chimpanzees are genealogically more closely related to humans than to orang-utans, the big anagenetic, i.e., morphological, behavioral and, particularly, intellectual gap between apes and humans would justify placing the former into the family Pongidae and the latter into the separate family Hominidae. Thus, overall morphological, ecological, and supposedly genetic (or genotypic) similarity—constituting a grade—takes precedence over phylogenetic relationship—constituting a clade. In other words, in evolutionary taxonomy cousins can be more closely related than sisters if the former are more similar to each other than the two sisters. (This criticism is still not properly understood by some evolutionary taxonomists: see the exchange between Mayr 1994 and Mahner 1994b.)

Evolutionary taxonomists thus want to establish a difference between phylogenetic or cladistic *analysis* on the one hand, and final ordering or *classification* on the other (Mayr 1974, 1995). According to them, the former consists in an analysis of the relevant characters of a given set of organisms or species revealing their phylogenetic relationships, while the latter consists in the ordering of the given species in a system or classification. This idea implies that the results of the previous character analysis may or may not enter the final classification. Thus, the chopping of the phylogenetic tree into classes and ranks is up to the systematist. Hence, it should come as no surprise that different systematists are likely to come up with different classifications.

To exemplify the arbitrariness of this procedure, we use another notorious example of taxonomic dispute: the classification of reptiles and birds. In this case, the cladist may classify as follows (Ax 1984): Sauropsida = (Squamata [snakes, lizards, amphisbaenians] ∪ Rhynchocephalia) ∪ (Chelonia [turtles] ∪ (Crocodylia ∪ Aves)). However, after having agreed that this classification yields a plausible hypothesis of phylogenetic relationships if converted into a phylogenetic tree, the evolutionary systematist is likely to group thus: Sauropsida = Reptilia ∪ Aves. The reason for this classification is that reptiles are much more similar to each other than to birds and that they form a grade, i.e., a conspicuous "morphological level" occupying a particular adaptive zone (Mayr and Ashlock 1991).

This proposal is open to the following objections. First of all, cladistic analysis *is* classification because the properties (i.e., evolutionary novelties) used in a cladistic analysis function as equivalence relations that define classes. For example, exhibiting the shared evolutionary novelties of crocodiles and birds is *defining* a class Crocodylia ∪ Aves, whether or not this class is named, say, 'Archosauria'. After all, it should be clear that only after this classing has been performed is it possible to hypothesize a common stem species. Consequently, to disregard this classification in favor of another amounts to classing according to completely different principles and different equivalence relations. To claim that cladistic analysis and classification are different and separable enterprises is patently inconsistent. There can be no compromise between a classification corresponding to a consistent hierarchy of nested kinds and a classification that cuts across such nested kinds wherever it seems appropriate for subjective reasons. (For further criticisms see Wiley 1981.)

Thus, the class Reptilia is nothing but the complement of the class Aves within the class Sauropsida: it is not characterized by any qualitative novelty but only by overall similarity. (Cladists call such complement classes *paraphyletic groups* because they do not contain all the descendants of a given stem species.) On the other hand, the Linnean class Pisces (fishes), i.e., the complement of Tetrapoda within Vertebrata, is usually rejected by evolutionary systematists who acknowledge at least two classes of fishes, namely Chondrichthyes (cartilaginous fishes) and Osteichthyes (bony fishes). Why? Because it is somewhat more "natural" to do so in terms of certain similarities. The cladist agrees, but insists that it would be even more natural to abandon not only Pisces but also Osteichthyes because the latter can be shown to be paraphyletic too. Obviously then, the classification of evolutionary systematists retains elements of arbitrariness: it is up to the systematist, not to consistent method, which groups are formed or retained, and which not. (See also Sober 1993.) Thus, some taxa are formed according to cladistic principles, while others are formed because they correspond to common sense grouping. Instead of attempting to eliminate this prescientific arbitrariness—Simpson's euphemism: the element of *art* in systematics—it is something hailed as a virtue, for certain groups such as reptiles could be recognized at once, even by nonexperts (Mayr and Ashlock 1991, p. 263). By the same token, one could defend Aristotelian physics against Newtonian or quantum physics because the former is more intuitive and more accessible to the layperson.

This is not to deny the legitimacy of different classifications of organisms. After all, classifications are conventions, hence neither true nor false. It is perfectly legitimate (for a home-maker) to classify organisms into, say, "Edibilia" and "Inedibilia". As was explained in Section 7.2.2.1, it is possible to partition a given collection of items according to different equivalence relations. To use a previous example again, in evolutionary systematics the taxon Vertebrata is partitioned into seven classes of the same rank: Agnatha, Chondrichthyes, Osteichthyes, Amphibia, Reptilia, Aves, and Mammalia. But this is not a partition proper because there is no underlying equivalence relation according to which the collection of vertebrates is partitioned. At best, there is an overall similarity relation, perhaps reading "somehow more similar among each other than to any other organism". What happens is that *similar* species are collected into a genus, *similar* genera into a family, *similar* families into an order, and so on. This procedure has been called 'upward classification' (Mayr 1982, 1995) as opposed to 'downward classification' or dichotomous partitioning. Clearly, it satisfies formally the principles of classification laid down in Section 7.2.2.2, but it is not accomplished by means of the partition of a collection with the help of a set of equivalence relations of different power. Rather, the classification is largely intuitive, hence prescientific. Therefore, it is questionable whether condition (4) is satisfied because the precise equivalences in question, if any, are not always made explicit.

We conclude that evolutionary taxonomy occupies a transitional position between Linnean taxonomy and cladistics—*pace* Bunge (1987a) and Leroux (1993). It is not a pretheoretical taxonomy although it still contains remnants of arbitrariness; that is, it does not strive for maximally natural classes. For this reason, it must be regarded as a semiscientific taxonomy—whose days appear to be numbered anyway. It comes as no surprise, then, that the defense of evolutionary taxonomy against the triumphant progress of cladism is evidently much more desperate these days than substantial: see particularly Mayr and Ashlock (1991). Moreover, since evolutionary taxonomy does not consistently acknowledge evolutionary novelties but also contains phenetic residues (e.g., it accepts complements defined by negative attributes, i.e., by nonproperties), it is peculiar that its adherents call it *evolutionary* at all.

Phenetic Taxonomy. Phenetic taxonomy, or *phenetics* for short, is an offshoot of operationalism: it is a strictly empiricist taxonomy. (Phenetics was born under the name *numerical taxonomy*; but this is a misnomer because any systematics can be implemented with numerical methods. The salient point is the phenetic, not the numerical, outlook of this taxonomy. See Mayr 1982; Platnick 1989.) Phenetics originated in the late 1950s along with the rise of the computer (see, e.g., Sokal and Sneath 1963; Sneath and Sokal 1973.) The main epistemological goal of phenetics consisted in achieving an objective and repeatable classification. By observing as many characters as possible, and by assigning to them the same weight, it was hoped to arrive at an objective measure of *overall similarity*.

It is no secret that numerical phenetics has been a failure (Hull 1970; Löther 1972; Ruse 1973; Wiley 1981; Mayr 1982; Bunge 1985b; Rosenberg 1985; Suppe 1989; Mayr and Ashlock 1991; Sudhaus and Rehfeld 1992; Leroux 1993; Sober 1993). However, it still survives among some taxonomists addicted to empiricism and computer systematics. The main problem with phenetics is not so much that its philosophy, namely empiricism plus operationalism, is wrong (see, in particular, Hull 1970.) Indeed, one cannot deny that it is possible to classify organisms according to overall similarity. The question is whether such a classification is a scientifically useful one or not.

How does phenetics fare in this respect? First of all, although pheneticists cherish the scientific values of objectivity and repeatability, all critics agree that they cannot meet this aim, because their classifications depend not only on the features that have to be chosen by the investigator but also on the chosen method for processing the data. (Similar objections can be raised concerning the alleged objectivity of computer cladistics: see Wägele 1994.) What degree of similarity is necessary at all to put two objects into the same taxon? For example, are the (rational) number 1.4 and the (irrational) number $\sqrt{2}$ similar enough to be put together? Second, grouping by mere similarities instead of equivalences corresponding to substantial properties will lead to anything but nested natural classes. Natural kinds may occur only by accident. Thus, not being interested in lawful relationships, a phenetic classification cannot be regarded as scientific: it is an avowedly atheoretical classification. Third, since a phenetic classification is supposed to be atheoretical, although, in fact, not free of extrataxonomic presuppositions, there is no chance of refining it with the help of biological theories, because it would then no longer be a phenetic classification. Its refinement could only consist in turning it into a cladogram by means of cladistic principles. Yet the cladist does not need a phenetic classification to begin with. Fourth, for this reason, phenetics has produced hardly any scientifically interesting or useful results (Rosenberg 1985). The only use for a phenetic classification seems to occur in cases where we find a number of very similar species *prima facie* lacking significant characters suitable for a straightforward cladistic or evolutionary classification. A phenetic sampling may then serve as a starting point for further analysis (Mayr and Ashlock 1991). Yet this already exhausts the scientific use of phenetics. In short: *requiescat in pace*.

7.3 Bionominalism

After having expounded our conceptualist philosophy of taxonomy and shown that it easily fits in with an evolutionary outlook, we proceed with an examination of the nowadays dominant philosophy of taxonomy, namely neonominalism or bionominalism. Since the latter has been regarded as one of the few topics in the philosophy of biology on which "something like a consensus is beginning to emerge" (Sterelny 1995, p. 156), and since it is always hard to persuade believers that the reigning orthodoxy is wrong, it will be worthwhile to examine bionominalism in some detail. In so doing, we shall first study weak, then strong, bionominalism.

7.3.1 Weak Bionominalism

The species problem has haunted biologists since the Darwinian revolution. The main reason for this philosophical problem was apparent from the beginning: if evolution has occurred and is still occurring, and if evolution is the evolution of species, then species cannot be constant and immutable essentialist kinds, but must be changeable entities (Haeckel 1866; Hull 1965; Thompson 1989). On the other hand, the problem remained how to do systematics, that is, how it is possible to classify organisms if they cannot be put into classes. Thus, the requirements of classification and the theory of evolution seemed to be mutually exclusive. Bionominalism seems to provide a solution to this problem.

The first influential step towards bionominalism was taken by the founders of the so-called New Systematics, i.e., Rensch, Huxley, Mayr, and Dobzhansky. The basic tenet was that biological species would be populations of organisms or, more precisely, reproductive communities, also called 'biospecies' (Mayr 1963, 1982). Although the notion of a biospecies as a reproductive community is much older (for a brief history see Grant 1994), this view gained wide acceptance because it fitted in nicely with population genetics and thus the Synthetic Theory. The "biopopulation equals biospecies" doctrine can still be found in current textbooks, although the concept of biospecies has undergone since its inception serious criticisms from biologists and philosophers alike.

The more philosophically oriented and still ongoing debate on the species problem has been initiated by Ghiselin (1974, 1981) and Hull (1976, 1978, 1988, 1989). The claim is that species are not classes but (concrete) individuals, which are sometimes conceived as material systems proper, sometimes as "historical entities". (Having been published in German and in the spirit of dialectical materialism, Löther's earlier book from 1972, which asserted clearly that species are material systems, has been overlooked.) Ever since then, a flood of papers on the "ontology of species" appeared, elaborating on the so-called "species-as-individuals thesis" (henceforth: SAI thesis)—so much so, that this has already been called the 'species plague' (van der Steen and Voorzanger 1986; for an anthology of such essays see Ereshefsky ed. 1992). Only few biologists and (bio)philosophers have resisted the bionominalist turn to begin with or have been or have become—for various and different reasons—critical of bionominalism (e.g., Ruse 1969, 1981, 1987; Bunge 1979a, b, 1981c, 1985b; Kitts and Kitts 1979; Caplan 1981; Heise 1981; Schwartz 1981; Lang 1983; Bernier 1984; Kitcher 1984a, 1987; Bock 1986; Guyot 1986; Løvtrup 1987; Suppe 1989; Leroux 1993; Mahner 1993a, 1994a; Webster 1993; Ax 1995; Gayon 1996).

7.3.1.1 Species as Reproductive Communities

Let us begin with Mayr's famous definition of "biospecies", which reads as follows:

> A species is a reproductive community of populations (reproductively isolated from others) which occupies a specific niche in nature (Mayr 1982, p. 273).

Several objections must be raised against this definition. First, Mayr's is not a definition at all but a so-called operational definition. In fact, it is an *indicator hypothesis*: it does not tell us what a biospecies is but how to recognize it, namely by observing reproduction or else by failing to observe the latter (i.e., by "observing" isolation). Neither reproduction nor isolation are defining properties of a species but, at best, properties of organisms that may be used as symptoms of the latters' membership in a particular species. In other words, two organisms do not belong to the same species because they mate and reproduce, but they only are able to do so because they belong to the same species (Mahner 1994a). That is, having certain properties in common that make reproduction possible to begin with precedes actual mating. Hence, the notion of a species as a class of organisms is logically prior to the notion of a reproductive community as a concrete system composed of organisms. (Incidentally, the same holds for the definition of species as lineages by Wilson 1995, p. 342, who rejects the species-as-kind concept but at the same time logically presupposes it by requiring that the descendant "components" of a lineage be "of the same sort" as their ancestors.) Indeed, the notion of a class or species is an indispensable logico-semantical concept and can therefore not be ontologized or reified without committing a category mistake.

Second, the "definition" in question takes populations instead of organisms to be reproductive units. Unless a mere instance of sloppy language, this is an instance of level-mixing, because populations cannot be said to mate and sexually reproduce: only organisms in a population can do so. Thus, what is at best "defined" is the concept of a species of populations not that of a species of organisms.

Third, depending on the definition of "niche", the occurrence of the term 'niche' in the definiens may render Mayr's "definition" circular. The latter would be the case under any definition of "niche" referring to species (rather than organisms), such as in Futuyma's (1986) definition of a niche as the set of all possible environments in which a species can "survive".

Fourth, the attribution of a niche to species exhibits another instance of level-mixing. We need not comment on this here, for we have explicated in Section 5.4 why only organisms have ecological niches, so that neither species as (alleged) wholes nor species as classes have niches.

However, the failure of Mayr's "definition" of a biospecies does not disqualify the notion of a reproductive community as a material system. In fact, a reproductive community *is* a concrete system whose composition consists of organisms of the same species and whose endostructure is constituted by mating relations. (See also Sober 1993.) In Definition 4.7 we have called such concrete systems *biopopulations*. Consequently, it is mistaken to regard a biopopulation as a class in the way, for instance, Hull (1976) and Caplan and Bock (1988) have done. What is a class is the composition of the biopopulation, but not the population as a cohesive whole.

However, admitting that reproductive communities are real systems does not solve the species problem. As a matter of fact, to equate "biopopulation" and "biospecies" creates more problems than it is expected to resolve. First, as is even admitted by its defenders, Mayr's concept of a biospecies only holds for sexually reproducing organisms. Thus, asexual organisms would not belong to any species, so that there would be speciesless organisms. This not only contradicts biological practice, but it is a clear indicator of the underlying nominalist philosophy, for which properties are anathema. For, if we—as all scientists do when not in a philosophical mood—believe that all things possess properties (representable by predicates), then the expression 'individual b possesses property Q', or 'Qb' for short, is equivalent (though not identical) to 'b belongs to the class of individuals x possessing property Q, or '$b \in \{x \mid Qx\}$'. Second, as is also well known, Mayr's concept of biospecies (as biopopulation) is a "nondimensional" one, which means that neither past nor future organisms are parts of biopopulations. Hence, they do not belong to any species. Consequently, we would not be able to assert that Aristotle is a human being, for he is not part of a biopopulation. Third, if "biopopulation" and "biospecies" were cointensive, geographically separated (i.e., allopatric) populations would have to be regarded as different species. If species were concrete systems, there would have to be as many species as distinct populational systems. In particular, all the organisms that are parts of populations, which, in turn, are parts of different communities and ecosystems, could not be said to belong to the same species. (See also Damuth 1985). Moreover, the splitting of a population into two or more (separate) populations would be identical to speciation. This problem will return when we examine supraspecific taxa, the members of which are, according to strong bionominalism, only individuated by separation and descent, not by any intrinsic properties.

What about the rejoinder that the parts of a whole need not be "physically contiguous" to be parts of the same composite individual? Are Alaska and Hawaii not parts of the concrete individual USA, despite being geographically separated from the mainland (Ghiselin 1974; Mayr 1988, Chap. 20)? Yes, indeed, but this objection misses the point: it is not spatial contiguity that holds systems together, but the existence of bonding relations among the parts (recall Sect. 1.7; see also Guyot 1986.) Thus, whereas Alaska and Hawaii are linked to the rest of the USA by a multitude of political, cultural, and economic relations, which are comparatively distance-independent due to modern means of transportation and communication, two geographically distinct biopopulations are not in any way coupled together unless there is any exchange between them that alters their states.

7.3.1.2 Species as Lineages of Ancestral-Descendant Populations

The problems posed by the "nondimensional" biospecies concept were supposed to be solved by conceiving of species as lineages of ancestral-descendant populations. Usually, these lineages are defined in similar ways and are variously called 'evolutionary species', 'phylogenetic species', or 'cladistic species' (e.g., Simpson 1961; Wiley 1978, 1980; Cracraft 1987; Ridley 1989). According to this view, the lineage *of* a thing is regarded as an entity or individual itself, which is a clear case of reification. Yet let us examine this view in more detail.

The classical concept of an evolutionary species was initially proposed by Simpson (1961). Its refined version by Wiley (1978, p. 18) reads thus:

> A species is a single lineage of ancestral descendant populations of organisms which maintains its identity from other such lineages and which has its own evolutionary tendencies and historical fate.

How is this definition to be understood? According to Definition 7.4, a lineage is the union of the ancestry and progeny of a thing. It is thus a set, hence not a concrete entity that could have a "historical fate". An evolutionary species, then, would be the union of the ancestry and progeny of a biopopulation, i.e., a set of biopopulations, not a concrete system composed of biopopulations. An alternative way of rendering this notion ontologically precise is to conceive of an evolutionary species as the (total) history of a (single) biopopulation. (For the notion of the history of a thing recall Sect. 1.5.) A lineage, then, would be the ordered set of successive states of a biopopulation, where the ordering corresponds to time. However, Wiley's definition speaks of ancestral-descendant populations and thus precludes such a construal: only entities (not states) can descend, and they only descend from other entities, not from themselves. (The relation of descent is irreflexive.) More precisely, the former states of things precede its future states, but they are not ancestors of future states. Still, in either way, we obviously only arrive at sets, not concrete composite wholes.

The bionominalist is likely to object to the conceptualization of lineages as sets that lineages would be "spatiotemporally localized entities", that is, entities with a definite localization in space and with a beginning and ending in time. However, this standard formula rests on a poor ontology lacking definite concepts of thing, state of a thing, change, and history of a thing. Clearly, a biopopulation is a concrete system existing in space and time (where else?). However, the history of the population is not a concrete or real system itself, because the past and future states of the population are not bonded to its present states: the relation of antecedence is not a bonding but merely a temporal relation (see again Sect. 1.7, as well as Guyot 1986.) Therefore, the history of a thing does not exist as a concrete system at any place and time. Similarly, the ancestry and progeny of a biopopulation, i.e., a lineage proper, do not constitute a concrete system together with the present population because a thing and its ancestors and descendants cannot act upon each other unless they all exist contemporaneously in the same region. The ancestry relation is no bonding relation either.

Many authors also say that lineages exist through time, that they change and evolve, and that they can be units of selection as in species and clade selection. We have already rejected these views as being logically impossible (Sect. 7.2.2.4). Histories and lineages of things *are* changes of things in time, so that histories and lineages themselves are unchangeable objects, hence nonentities. Consequently, lineages have neither evolutionary tendencies nor historical fates. Moreover, there is no such thing as a "historical entity" viewed as an entity *sui generis*, somehow "middling between classes and things" (Wiley 1980). Wiley proposed this ontological hybrid because he is very well aware that clades are not cohesive wholes, i.e., concrete systems. Nevertheless, he takes them to be "philosophical individuals", whatever this may be (Wiley 1989). But the fact that organisms are objectively related by descent does not bind them together into some kind of "philosophical individual". Since the notion of a historical entity, then, takes the history of a thing or the lineage of a thing to be real individuals themselves, it is an instance of reification. What we can do, on the other hand, is to collect all objectively related organisms together in a class, which thus is a natural or realistic class. Yet classes, whether natural (or objective or realistic) or arbitrary (or subjective), are conceptual objects, hence not real. There can be no hybrid between a material (or real) object and a conceptual one. The thing/concept distinction is a methodological duality, not an ontological one (see Postulate 1.2). Hence, there is no *tertium quid.*

With regard to the solution of the species problem, the evolutionary species concept fails just like the concept of a species as a reproductive community. First, the concept of an evolutionary species as stated above refers to populations, not to organisms. Yet biologists are interested in species of organisms, not species of populations. Second, suppose that there is a single organism left of a species on the verge of extinction (Mahner 1993a). This lonely creature would no longer be part of any biopopulation, since there is nothing left of which it could be part. Consequently, it would be a speciesless organism. However, the last Moa in New Zealand was still a Moa, and not an individual devoid of any properties. Neither would this last organism of its kind be part of an evolutionary species, because the latter is defined in terms of populations, not organisms, and *ex hypothesi* there is no population left of which it could be part. Third, the evolutionary species concept does not hold for most asexual organisms. Although asexual organisms may come in populations as well, the latter are obviously not reproductive communities. (The internal structure of such a population may, for instance, consist of social or ecological relations instead of mating relations.) However, since only few, if any, asexual organisms live in such biopopulations, the concept of evolutionary species is rarely applicable, if at all. It would only be applicable if such species were construed as a lineage of organisms, not biopopulations.

7.3.1.3 Species-as-Individuals and Classification

How does the SAI thesis fare within the standard or set-theoretic view of classification? (For different accounts of this view see Gregg 1950, 1954, 1968; Beckner 1959; Buck and Hull 1966; Bunge 1967a, 1983a; Løvtrup 1973, 1974; Suppe 1989.) Clearly, if species are individuals, a classification in set-theoretic terms is impossible.

If species were real entities rather than classes of organisms, they would no longer be taxa because taxa are, according to the standard view, classes of organisms. Thus, only the conceptual representations of SAIs could be the units of classification. That is, we would have to classify species, not organisms (Bunge 1981c, 1985b). For instance, if *Homo sapiens* were a real individual rather than a class, it would have to be the element (or unit) of the classification: *Homo sapiens* $\in$ Homo $\subset$ Hominidae $\subset$ Primates $\subset$ Mammalia, and so forth. In this construal, however, it would be impossible to assert that, say, Aristotle is a human being, for this proposition would have to be conceptualized either as "Ha", where a stands for Aristotle and H designates a conjunction of predicates characterizing humanness, or else as "$a \in$ *Homo sapiens*". But now only a statement such as "Aristotle is part of *Homo sapiens*" would make sense; and this statement would have to be formalized as either "Pah" (where P is the binary predicate representing the part-whole relation, which is attributed to the two individuals a and h, where h denotes the individual *Homo sapiens*) or as "$a \sqsubset$ *Homo sapiens*" (where $\sqsubset$ represents the part-whole relation). The classification would thus have to read: Aristotle $\sqsubset$ *Homo sapiens* $\in$ Homo $\subset$ Hominidae, and so on.

Ignoring the obvious differences between the propositions "Ha" and "Pah", we could try harder and assume for the sake of the argument that the proposition "$a \in$ *Homo sapiens*" should, for ontological reasons, be reconceptualized as "$a \sqsubset$ *Homo sapiens*". Yet an immediate problem would emerge: Aristotle is dead, so that obviously he is *not* part of any human population. Neither could we say that Aristotle is a primate or a mammal, because the elements of classification now are SAIs, not organisms, so that $a \notin$ Primates, $a \notin$ Mammalia, and so on. *A fortiori*, no organism whatsoever, whether dead or alive, would be an element of any higher taxon. Instead, we could obtain such quaint statements as "*Homo sapiens* $\in$ Mammalia" or, alternatively, "Mh", which would read "*Homo sapiens* is a mammal"—evidently an absurd expression. We conclude that no classification proper is possible if one adopts the SAI thesis.

7.3.1.4 Species-as-Individuals and Laws

One of the arguments in favor of the SAI thesis consists in the alleged lawlessness of species: it has been claimed that there are no significant laws that have species as their subject (e.g., Hull 1978, 1987, 1989; Rosenberg 1987). According to the SAI thesis, this would not be surprising: since laws would range over classes and since species would be individuals, there could be no species laws anyway.

This argument seems to rest on the failure to distinguish between "law" and "law statement". (See also Leroux 1993.) First, recall that $laws_1$ are properties of things (Sect. 1.3.4). Hence, laws in the ontological sense "hold" for things, not classes. By contrast, law statements are indeed generalizations: they begin with the universal quantifier 'for all' ($\forall$) and thus refer to all the things of a certain kind. But sets and classes can be single-membered, that is, a law statement may refer to the single element of the reference class of a theory. For example, the $laws_2$ of plate tectonics may be true only for Earth, the only known element of the class of all Earth-like planets. Therefore, even the SAI thesis is in principle compatible with species (as individuals) $laws_1$, and the alleged lack of species $laws_2$ in current biology does not corroborate the SAI view. Of course, if the reference class of a law statement is single-membered, we face the methodological problem of distinguishing idiosyncratic from species-specific properties of the thing in question. Still, even the unique thing behaves lawfully and thus must possess $laws_1$. In other (metaphorical) words, there is no thing without a nomological state space.

More important, however, is that the thesis that there are no species laws is plainly false, as can be seen by perusing all of the biological literature, not just that portion dealing with evolutionary biology. When we pick any title from the literature, such as 'Underwater stridulation by corixids. Stridulatory signals and sound-producing mechanisms in *Corixa dentipes* and *Corixa punctata*' (Theiss et al. 1983), we find that the paper is about lawfully related properties of the *organisms* of certain species, as well as about the differences among the *organisms* belonging to different species. Needless to say, the examples could be multiplied *ad libitum*. Suffice it to mention only two. For instance, it is well known that the members of different species often have different allometric growth rates, i.e., growth laws (Thompson 1917; Sudhaus and Rehfeld 1992). Think of the developmental invariants (laws) described in the members of the otherwise extremely variable land snail genus *Cerion* (Gould 1989). Another example is provided by a statistical analysis of a number of mensural characters of the members of a group of species of aquatic bugs, which yields distinct species clusters in the corresponding morphometric space, not a continuum (Sites and Willig 1994a, b; see also Alberch 1982).

Note, however, that most titles referring to particular species (or any other taxa), such as 'Sound-producing mechanism in *Corixa dentipes*' or 'A developmental constraint in *Cerion*', are misleading, if not ill-formed, for the descriptions and law statements contained in the papers do not refer to species (or taxa) as alleged wholes but to their individual members. For example, obviously the possible law statement "Bacteria species *B* metabolizes arsenic" does not hold for the species *B* as an alleged individual but for its members: "All the members of species *B* metabolize arsenic".

In sum, the practice and the writings of ordinary biologists presuppose—at least tacitly—that both empirical generalizations and law statements refer to the *organisms* of a certain species (or any other taxon), which implies that species (and taxa) are classes or kinds. For this reason, the SAI view ignores, and is unable to account for, biological practice. By assuming that species are things and that species $laws_2$, if any, would have to be about such species, the SAI thesis implies that there are no $laws_2$ about organisms. Yet even if species were individuals, there could be $laws_2$ about the composition (i.e., the class of components) of such composite wholes. By further claiming that there are, in fact, no $laws_2$ about SAIs, the bionominalists overlook that SAIs, being concrete things, would also come in kinds or species, even though some (perhaps many) of these kinds might be singletons. Thus, there could be species $laws_2$ after all, although in this case they would be about biopopulations (as concrete wholes) rather than organisms. (For a recent, though partly misguided, attempt to defense species laws see Lange 1995.)

7.3.2 Strong Bionominalism: Taxa-as-Individuals and Classification

Considering the problems with the SAI view and classification, one can reject either the SAI thesis in favor of the set-theoretic view of classification, or the set-theoretic view of classification in favor of the SAI view, and propose an alternative taxonomy. The latter option has been chosen by the strong bionominalists. Indeed, they take not only species to be individuals but they maintain that all taxa are composite individuals in the sense of "historical entities". This contention is motivated by the phylogenetic insight that all organisms, hence vicariously all taxa, are related by common descent from (the members of) a single ancestral species. Thus, all life on this planet would not constitute an objective class or biological kind, but rather a concrete whole in the sense of a "historical entity". This alleged individual, which is sometimes called 'Life', or 'Biota', or even 'Biosphere', would form a hierarchy of nested parts or "taxa" (e.g., Griffiths 1974; Vrba and Eldredge 1984; Eldredge 1985a; Salthe 1985; Hull 1988; de Queiroz 1988, 1992, 1994; Nelson 1989; Wiley 1989). Accordingly, those taxa-as-individuals (henceforth: TAI) would be related by the part-whole relation ($\sqsubset$), not by the set-theoretic relations of membership and inclusion. Such a hierarchy reads, for instance, thus: Aristotle $\sqsubset$ *Homo sapiens* $\sqsubset$ Homo $\sqsubset$ Hominidae $\sqsubset$ Primates $\sqsubset$ Mammalia $\sqsubset$ Amniota $\sqsubset$ Tetrapoda $\sqsubset$ Vertebrata, and so on.

Since these "entities" are no longer taxa proper, i.e., classes (in particular, natural kinds) but concrete individuals, the neonominalists succeeded at last in getting rid of the usual properties characterizing the members of those taxa. They now are able to claim that, no matter what properties any particular organismal part of such a "taxon" may possess, its being part of a TAI would not depend on those properties. For example, a particular whale would be a mammal because it is part of the individual (or clade) Mammalia, not because it possesses certain characteristic properties such as mammary glands or a squamoso-dentary joint. Accordingly, a TAI cannot be defined at all: it can only be described or pointed out. What would be defined are the proper names of taxa, not the taxa themselves (Hull 1965; Buck and Hull 1966; de Queiroz 1988, 1992, 1994; de Queiroz and Gauthier 1990). So much for the main theses of strong bionominalism. Surprisingly, despite the conspicuous differences from the set-theoretic view of classification, one of the fathers of strong bionominalism maintains that "although the change is metaphysically quite drastic, it does not alter any traditional inferences" (Hull 1988, p. 399). Let us see whether this claim withstands closer logical, semantical, and ontological examination.

To begin with, we must reject the idea that descent is a causal relation (Hull 1988, p. 448), which is behind the concept of TAI. Indeed, the causal relation relates events, not things (see Sect. 1.9). Besides, although descendants are produced by their immediate ancestors, this relation does not bind them together in a cohesive whole, for there need not be any interaction between ancestors and descendants. This holds *a fortiori* for mediate ancestors. We therefore insist that, since the relation of descent is not a bonding relation, lineages or clades are not cohesive wholes, and the past and future organisms of a clade are not coupled to its present members. (See also Damuth 1985.) Moreover, neither lineages nor clades have emergent properties, and they cannot be said to be in a certain state at any moment of time. Hence, they can neither change (as alleged wholes), nor can they undergo any processes (as alleged wholes). In particular, they cannot be selected and they cannot evolve. Thus, taxa, whether species or clades, are not material systems, hence not real individuals, and there is no part-whole relation among organisms, species, and higher taxa.

Consequently, taxonomists do not and cannot order real systems in the way some authors (e.g., Griffiths 1974; de Queiroz 1988; de Queiroz and Donoghue 1988) have proposed. These authors think that the ordering of TAIs, being real systems, would be rather a "systematization" than a "classification". Yet only the curator at the museum orders real, though usually dead, things: he or she literally puts and sorts preserved organisms, not TAIs, into different drawers and cabinets. However, this ordering presupposes the existence of a *conceptual* system, i.e., a classification. In other words, the curator must have pigeon-holed organisms in his or her mind before materially sorting them into different drawers and cabinets.

The idea that systematists order real systems, namely TAIs, indicates that strong bionominalists fail to distinguish between a real thing and its conceptual representation. Of course, this is exactly the main flaw of any nominalist philosophy. To bring this point home, let us assume for the sake of the argument that there actually were a hierarchy of TAIs related by the part-whole relation. This hierarchy, however, would not be a classification: although the world is orderly and structured, it is not a classification, because a classification is a purely conceptual operation performed by a classifying subject—an operation in which clearly no ontological part-whole relations occur. After all, classifications are definitions (of classes), hence constructs. To claim, then, that taxa are real individuals is semantically mistaken because it conflates concepts with their referents. Yet the (factual) referents of a (scientific) construct are not part of the construct, but part of the world. If not philosophers, at least some biologists have noticed this distinction and state clearly that a taxon can only be a conceptual representation of a real individual in nature, such as a biospecies-as-individual (e.g., Willmann 1985), or a "clade-as-individual or "closed descent community" (Ax 1985, 1988). In sum, taxa are neither concrete individuals nor "historical entities", but constructs. The problem can only be how these constructs are properly conceived and what they represent: do taxa categorize organisms of the same kind and are they thus class concepts, or do they represent (supraorganismic) individuals, such as clades-as-concrete-individuals (henceforth: CAI), and are they thus to be conceived of as individual concepts? (Recall Sect. 2.1.)

The criticism that most bionominalists fail to distinguish a thing from its conceptual representation affects the claim that TAIs could only be described or characterized, not defined, for one can define signs and constructs, but not things. So if clades were real individuals they could indeed not be defined, but only described. Yet, since taxa are constructs anyway, they could very well be defined, that is, equated with other constructs—at least in principle and if such definitions were useful. For example, before the coelenterate life cycle was known, the polyp and medusoid stages were often classified as different species and even genera. Now that we know better, we must put them in the same taxon, that is, we now say that taxon *A* is the same as taxon *B*, such as in the definition "*Laomedea gelatinosa*" $=_{df}$ "*Obelia commissuralis*". In other cases, we have two different names for the very same taxon, so that we also can define names of taxa, such as in the definition 'Rhynchota' $=_{df}$ 'Hemiptera'.

Let us assume next the strong bionominalist agrees that TAIs can only be descriptions, and thus conceptual representations, of CAIs. Then the problem remains that what is usually described in such a case is not a clade-as-a-whole, but its organismal, and only its organismal, parts. This is particularly obvious when clades are characterized by means of apomorphies. For example, when one describes the (alleged) CAI Amniota as possessing a particular developmental pathway, which is characterized by the development of embryonic membranes and the occurrence of a primitive streak, one does not actually describe properties of the alleged CAI, but properties of its organismal parts. This is because the specific properties of organisms are not resultant at any supraorganismic system level. That is, Amniota-as-a-CAI possesses neither embryonic membranes nor any development. The same holds for all its subclades down to its SAI parts. Therefore, referring to apomorphies in characterizing a taxon does not describe a CAI, but at best its organismal composition: "All organisms developing four embryonic membranes are part of Amniota". Yet since the composition of a system is a collection, not a thing, and since the definition of the class of components of a composite individual is required to characterize it, the definition of a class of amniote organisms would be logically prior to the characterization (description) of Amniota-as-a-CAI. But then a description of the CAI Amniota would have to contain a statement such as "(The composite whole or CAI) Amniota is composed of amniotes". Consequently, the conceptualist's natural class Amniota is coextensive with the organismal composition of the alleged composite individual Amniota. In sum, it is simply impossible to characterize an individual without class concepts. (See also Suppe 1989.)

A further problem with the TAI thesis originates in the properties of the part-whole relation, in particular its transitivity: if $x \sqsubset y$ and $y \sqsubset z$, then $x \sqsubset z$. As mentioned previously with respect to the SAI thesis, in order to obtain the same proposition as, for instance, "$a \in$ Mammalia" in terms of the part-whole relation, i.e., "$a \sqsubset$ Mammalia", we would have to read the latter in the same manner as the relation of set membership. That is, saying "a is part of Mammalia" would have to be equivalent to "a is a mammal". Yet the two relations are neither cointensive nor coextensive. Whereas the part-whole relation is a reflexive, asymmetric, and transitive ontological relation, the set membership relation is an irreflexive, asymmetric, and intransitive logical relation, which is, moreover, *logically prior* to the conceptualization of the part-whole relation. For example, in the proposition "a is part of m", or "Pam" for short, the predicate P (representing the part--whole relation) is attributed to the ordered pair of individuals $\langle a, m\rangle$. Yet this standard construal of a binary relation *presupposes* the set membership relation: if P represents a binary relation, its extension is the set $\mathcal{E}(P) = \{\langle x, y\rangle \mid Pxy\}$, so that a particular pair such as $\langle a, m\rangle$ is an element of $\mathcal{E}(P)$, i.e., $\langle a, m\rangle \in \mathcal{E}(P)$. By contrast, set membership is a primitive concept, not any old relation definable in terms of the set membership relation; that is, set membership is the primitive relation used within the definition of all other relations. Furthermore, the proposition "a is a member of the collection $\{x \mid Mx\}$", or "$a \in \{x \mid Mx\}$" for short, is equivalent to "a is an M", or "Ma" for short, where M is a *unary* predicate. For example, if $\mathcal{M}$ designates the taxon Mammalia and if $\mathcal{M} = \{x \mid Mx\}$, then $a \in \mathcal{M} \Leftrightarrow Ma$. No such equivalence is construable from the predicate "is part of", or from any other binary predicate for that matter.

If one still maintained that the propositions "Aristotle is a human being", or "Ha" for short, and "Aristotle is a mammal", or "Ma", should be reconceptualized as "Aristotle $\sqsubset$ *Homo sapiens*" and "Aristotle $\sqsubset$ Mammalia", respectively (or, in obvious logical notation, "Pah" and "Pam", respectively), we would, due to the transitivity of the part-whole relation, also obtain propositions such as "Mammalia $\sqsubset$ Vertebrata", or "Pmv". For reasons of consistency, these would have to be translated into 'Mammalia *is* a vertebrate'—an obviously meaningless sentence. (Recall that the meaningful proposition "Mammals *are* vertebrates", or $\forall x\, (Mx \Rightarrow Vx)$, is equivalent to "Mammalia $\subseteq$ Vertebrata".) Worse, since according to the transitivity of the part-whole relation parts of organisms would also be parts of higher taxa, we would have to regard Aristotle's brain, for instance, as a human being, as a primate, as a mammal, as a vertebrate, and so on, because it would be part of those TAIs. Clearly, this is not a mere "counterintuitive artifact of reconceptualization that can do no real harm" (Rosenberg 1985, p. 209), but it is a *reductio ad absurdum* of the TAI view.

But perhaps we are not entitled to read the part-whole relation in the same manner as the set membership relation. After all, what makes some CAIs parts of other CAIs is supposedly the so-called "genealogical nexus", i.e., the ancestor-

-descendant relation. However, the statement, say, "Sauropsida and Mammalia are parts of Amniota" just does not have the same meaning as "Sauropsida and Mammalia are descendants of Amniota", because we had to assume an individual Amniota (above its parts) from which its parts were to descend. Yet a whole does not precede its parts. So the genealogical nexus cannot be represented by the part-whole relation. Although this is also true for the set-theoretical construal of classification, for Amniota = Sauropsida $\cup$ Mammalia, the latter allows us at least to make meaningful statements of the form "x is an A", whereas a statement of the form "x is part of a" tells us nothing of interest in the context of classification.

Indeed, statements of the form "x is part of a" are uninformative in themselves, for, as the antiessentialists correctly state (e.g., Hull 1978), it is not necessary that the parts of an individual resemble each other. The same holds for the usual answer to questions of the form 'Why are x and y similar?', namely "Because they descend from a common ancestor z": if common ancestry or descent were all that matters, then the descendants of a common ancestor need not resemble each other. Let us turn these obvious statements against the TAI thesis by inspecting some of their consequences. (See also Leroux 1993.) For example, we could claim that, since some human beings happen to have close relationships to their pets, there is a composite whole consisting of humans, dogs, cats, parrots, and what have you; and we call this composite individual *Homo sapiens*. Since, according to the anti-essentialists, there is neither a human, nor a dog, nor a cat, nor a parrot nature (e.g., Hull 1989), he or she is unable to distinguish this composite whole from one that consists only of human beings.

While the biologist will find this situation unacceptable, the bionominalist philosopher may simply say: "so what?". For example, Hull (1978) seems not be troubled at all by such counterexamples: "If pets and computers function as human beings, then from certain perspectives they may count as human beings even though they are not included in the biological species *Homo sapiens*" (p. 205). But if one is prepared to go so far why not include pets and computers in *Homo sapiens* on top of that? The bionominalist might answer that, although the parts of a TAI need not resemble each other, they have nevertheless something in common, namely common reproduction in the case of populations and SAIs, and common descent in the case of TAIs. However, this move is ineffectual. First, we can clearly dismiss the relation of reproduction to characterize the SAI *Homo sapiens*, because otherwise we would have no reason to regard, for instance, the Pope as a human being. Moreover, this rejoinder would presuppose the possession of certain common properties, namely at least those that allow for mating and reproduction to occur. Second, if we grant that common descent is what matters, the bionominalist cannot satisfyingly answer the following question: why should our children (or, conversely, our parents) be human beings rather than frogs or insects? In order to answer this question, he or she would have to refer to something that is logically (and historically) prior to common ancestry, namely the properties of organisms that make common ancestry, or respectively common descent, possible in the first place.

Indeed, the preceding question can only be answered adequately by saying, for instance, that our children are human beings because they descend from human beings and because the property complex "humanness" is inheritance-dependent (for the notion of inheritance-dependence see Chap. 8). Moreover, besides the properties constituting humanness, we must presuppose a lot more lawfully related properties. We must not only presuppose all the laws holding for *Homo sapiens*, but also the laws of Primates, Mammalia, Amniota, Tetrapoda, Vertebrata, and so on. Thus, we can only hope to produce descendants of our kind if there are certain genetic and developmental laws "governing" reproduction and descent. Yet the existence of such laws, whether higher taxa laws or species-specific ones, is just what the bionominalists deny (see, e.g., Hull 1978, 1989; Rosenberg 1987). Consequently, there should be no such things as developmental and phylogenetic constraints, for they are nothing but the laws_1 to which we refer when we construct the nomological state spaces of the organisms in question. In sum, the relation of descent cannot be the whole story because it cannot explain by itself why our descendants are the way they are. The bionominalists overlook the fact that an explanation of similarity in terms of common ancestry is enthymematical: when we say that x and y are similar (better: equivalent) because they have a common ancestor z, we presuppose that z is also similar to x and y. If these implicit underlying properties and mechanisms of reproduction, inheritance, and development are overlooked, we end up with empty explanations such as "Today is Sunday because yesterday was Saturday". This is exactly what the antiessentialists ask us to accept.

This also re-emphasizes our previous claim that it is impossible to characterize an individual without class concepts. For example, in order to characterize a human population as a composite individual, it is required to tell of what kind of entities it is composed. The same holds for the characterization of *Homo sapiens* as a whole (or SAI): we must say that this (alleged) individual is composed of human beings rather than frogs, insects, or redwoods. This, in turn, presupposes a notion of *Homo sapiens* in terms of a class (see the above example of Amniota). Likewise for the characterization of an individual (organism) as being part of a population: for example, if $\mathcal{H}$ designates the set of all human beings, then the proposition "b was or is or will be a part of some human population" can be stated more precisely as "For some biopopulation p, and for some time t, the composition C of p at t is included in $\mathcal{H}$, and b is part of p at t. That is, $(\exists p)$ $(\exists t)$ (p is a biopopulation & t is a time instant & $C(p, t) \subseteq \mathcal{H}$ & $b \in C(p, t)$, where in turn $C(p, t) =$ $\{x$ is a member of the class of organisms at $t \mid x \sqsubset p$ & $x \in \mathcal{H}\}$ (Bunge 1985b).

Haunted by the purported impossibility of coming up with necessary and sufficient properties for the definition of taxa (see, e.g., Hull 1965), the strong neonominalists also reject the very requirement that taxa (or taxa names in nominalist jargon) are to be defined in terms of organismal characters. At best, the latter would serve as indicators, to recognize the "parts" of TAIs (see also Sober 1993). We have already commented upon this view, but have to add some more criticisms now.

For example, some bionominalists have proposed to define the taxon name '(Recent) Mammalia' as denoting the clade stemming from the most recent common ancestor of Monotremata and Theria (de Queiroz 1988, 1992, 1994). A first problem with this suggestion is that, if taxon names are proper names of concrete individuals, they cannot be defined. A second problem is the following. If you do not know, for instance, to what the name 'Theria' refers, the latter is assigned to the clade stemming from the most recent common ancestor of Marsupialia and Placentalia. And if you now do not know to what the names 'Marsupialia' and 'Placentalia' refer, the whole procedure has to be repeated "upward" or "downward" until one either arrives at Life-as-a-whole at one end, or at (terminal) SAIs on the other. Indeed, in this case of finite regression only the refuge in extensionalism remains, because neither Life nor SAIs are supposed to be characterized by properties either. So what we are offered is not a classification in any comprehensible sense but a set of almost bare individuals, whether SAIs or TAIs or organisms, that are individuated only by the relation of descent from some other individual that descends from some other individual that descends from some other individual, and so on.

This proposal is clearly at variance with any reasonable ontology, because every thing possesses not only relational properties, but also intrinsic ones, which is necessary to qualify as a material object to begin with. Since nominalists abhor properties and classes, they are unable to come up with a detailed description of any concrete individual, for this procedure requires class concepts. For this reason, they sometimes rely on so-called "ostensive definitions" such as "This is a whale". But ostensive "definitions" are not definitions at all, which are identities of either signs or concepts: they are only didactic props (see Copi 1968).

To come back to our previous example, the individual Wolfgang Amadeus Mozart is characterized not only as a son of Leopold and Anna Maria Mozart but also as the composer of the opera *Die Zauberflöte*, as an Austrian, as a violin and pianoforte player, as a person who lived between 1756 and 1791, and so on. Thus, we stipulate that every concrete individual can be characterized by the intersection of a finite number of classes: for every x, if x is a thing, then there is a finite family $F = \{C_i \mid 1 \leq i \leq n\}$ of classes such that $x = (\iota y)\ (y \in C_1 \cap C_2 \cap \ldots \cap C_n)$, or $\{x\} = C_1 \cap C_2 \cap \ldots \cap C_n$. (The symbol ι is called *definite descriptor*, and 'ιy' is to be understood as "exactly this one individual y".) This assumption is consistent with the ontological postulate that every thing is unique.

To summarize, in its attempt to take evolution into account, strong bionominalism misses the essence of evolution: the emergence of qualitative novelty. Qualitative novelty, though not denied, is regarded as merely epiphenomenal. By assuming that all that matters is individuals descending from each other and thereby forming lineages or "historical entities", which are characterized only by spatiotemporality but not by any substantial, let alone essential, properties, what we have is an *empty descentism* but nothing that has anything to do with biological evolution, i.e., the emergence of qualitatively new organisms. In other words, according to strong bionominalism, evolution is descent plus "bare change" *with or without* modification. Furthermore, since neither old nor new species are (necessarily) characterized by qualities, speciation amounts to the splitting of a "lineage", i.e., the formation of a new branch or "part" within the "historical entity" Life. Thus, speciation cannot be distinguished from separation. (Recall: what matters is alleged spatiotemporality, not substantial properties.) We conclude that strong bionominalism does not and cannot provide a metaphysics of evolution but, at best, a metaphysics of empty descentism.

7.3.3 Bionominalism and Some of Its Implications

If the philosopher of biology does not admit the notion of a natural or biological kind as elucidated in Section 7.2, then nothing prevents him from believing that "all biological kinds above the level of the macromolecule will be functional" (Rosenberg 1994, p. 34). The idea that most, if not all, biological kinds are functional kinds, i.e., defined by the equivalence relation "having the same role" (rather than "having the same composition and structure") is not uncommon among philosophers of biology: recall, for instance, the etiologists' conception of functions and roles of biosystems (Sect. 4.6). Nevertheless, it is mistaken because it ignores not only systematics but also the many structurally defined kinds of comparative morphology (see Amundson and Lauder 1994). Example of such kinds (taken from insect morphology) are: the *epimeron*, the *Musculus dorsoventralis secundus mesothoracis*, or the *Nervus frontalis*. Even if the names of certain (types of) organs are descriptive of their roles, as is the case with many muscles, e.g., the *Musculus dilatator cibarii*, the kinds of muscles are defined by the muscles' common *origo* and *insertio*, and perhaps *innervatio*, not by their role ("function"); in short, they are defined structurally, not functionally.

On the basis of the false belief that all biological kinds are functional, as well as from his failure to find $laws_2$ holding for such kinds (e.g., all aquatic animals), Rosenberg (l.c.), concludes that there are no biological laws at all and thus argues for an instrumentalist view of biology. Rosenberg's view implies that biology deals solely with analogies (homoplasies) or, in other words, that all of biology is *Analogienbiologie*. This implication is clearly wrong: much of biology deals with homologous characters and ways of life. This is why biologists are interested in natural (monophyletic) groups of organisms, i.e., in nonfunctional kinds of organisms. Thus, besides investigating the commonalities of the members of a nonnatural kind like "aquatic animals", which is of particular interest to ecologists, they try to come up with natural kinds, such as the aquatic bugs (recall again Sect. 4.6). In this case, we are indeed able to find lawfully related properties. For example, all aquatic bugs share a number of lawfully related apomorphies: among other characters, they all have a pair of tympanal organs in at least the mesothorax; they all are characterized by the fusion of their suboesophageal and pronotal ganglia; they all possess strongly enlarged posterior mesepimeral lobes; and they all have reduced antennae inserting below the eyes (Mahner 1993b). Even if we acknowledge that the kinds thus defined are not natural kinds in the traditional sense, but biological kinds as explicated in Section 7.2.2.2 (Principle 4), it is apparent that these kinds are not functional ones. In sum, if Rosenberg (1985) had not adopted a bionominalist stance, he would have one argument less to make his case for bio-instrumentalism (Rosenberg 1994).

Evidently, our view of biological kinds is tied to our emergentist ontology, in which the notion of qualitative novelty is central. Now, Rosenberg (1994) rejects the concept of emergence as being mysterious and adopts the notion of supervenience instead. However, as we have argued in Sect. 1.7.3, the notion of supervenience cannot account for qualitative novelty: it only relates two sets of properties (or rather predicates) in a static and, moreover, symmetric way. Thus, while bionominalism may be compatible with supervenience, it is incompatible with emergentism, hence with a genuinely evolutionary ontology (see Sect. 9.1).

An area which does not spring to mind immediately when considering the implications of bionominalism is ethics. For example, ethicists concerned with animal welfare, if not animal rights, demand that we ought to pay more attention to the needs, if not interests, of animals. (As will be obvious from Chapter 6, only some animals possessing plastic neural systems of a certain complexity can have interests.) This entails treating them properly according to their species-specific needs. But is this possible if one denies, as Hull (1989) and others do, that species, or any taxa for that matter, have "natures", i.e., nomologically essential properties? Worse, if species lack natures and if, hence, there is no human nature, can there be human rights and duties? May these examples suffice to show that bionominalism has consequences that reach far beyond systematics and even biology. (More on bionominalism and ethics, in particular the role of the concept of human nature in ethics, in Bradie 1994b, Chap. 4.)

7.3.4 Conclusion

Both variants of bionominalism, weak and strong, are logically, semantically, and ontologically mistaken. The popularity of strong bionominalism stems from the well-corroborated hypothesis that all organisms on this planet are related by common descent, and it is thus correlated with the triumph of cladistic taxonomy. The success of weak bionominalism rests mainly on the equivocation of the terms 'biospecies' and 'biopopulation' (Bunge 1981c; Brady 1982; Bock 1986; Mahner 1993a). That is, all the arguments in favor of the SAI view fail to prove that species are things. What they do accomplish is to show that there really are material systems composed of organisms, which we have called 'biopopulations' (Definition 4.7).

Of course, it is possible to call those material systems 'biospecies' rather than 'biopopulations', in order to save the habit of talking of the "evolution of species". But the result of this renaming and reconceptualization is the impossibility of any genuine classification, which presupposes species-as-class or, more generally, taxon-as-class concepts. Worse, it empties the concepts of evolution and speciation: the former is reduced to a form of bare change in which qualitative change is neither necessary nor sufficient, i.e., accidental or epiphenomenal; the latter is equated with separation, in which the emergence of qualitative novelty is again accidental at best.

We conclude that nominalism is not a viable philosophy of taxonomy. (This is not too great a surprise, since nominalism has always been a failure—except as a primitive reaction against Platonism.) And we thus maintain that bionominalism is one of the major misconceptions in the current philosophy of biology.

8 Developmental Biology

8.1 What Is Development?

The development of organisms has always seemed a somewhat mysterious process suggestive of design and purpose. Only the biologico-metaphysical problem of life might have attracted a similar number of mysterymongers. Indeed, development, in particular morphogenesis, remains one of the most fascinating and least understood processes in biology even today. Not surprisingly, it is one of the last strongholds of teleologists. Furthermore, developmental biology is still inspired (or haunted) by the age-old, though somewhat updated, controversy over preformationism versus epigeneticism. And yet, though raising interesting ontological and epistemological problems, developmental biology is not a favorite topic in the standard monographs on the philosophy of biology. With the notable exception of Woodger (1929), there is no chapter on development in Beckner (1959), Ruse (1973), Hull (1974), Rosenberg (1985), or Sober (1993). Only Ruse (1988), in a chapter entitled *Other Topics*, admits that development has so far been neglected in the philosophy of biology. For this reason, and because we shall need certain basic concepts of developmental biology in order to elucidate the notion of evolution (Sect. 9.1), we devote this chapter to examining some of the topics in developmental biology that are of philosophical interest.

8.1.1 Developmental Process and Development

Our first step is to examine what we believe ought to be understood under "development". This is because the concept of development is often emptied by some authors who use "being alive" and "developing", or "life cycle" and "developmental process" interchangeably, just because to live is to change. Moreover, it is unclear what, if any, is the difference between the terms 'development', 'embryogenesis', 'epigenesis', and 'ontogeny'.

We submit that a period (or stage) in the life history of an organism is a process of development only if it is accompanied by the emergence or submergence of at least one generic property (or quality), whether compositional or structural. (Recall

Postulate 1.9, as well as the distinction between generic and individual properties from Sect. 1.3.3. For example, the ability to synthesize a certain protein is a generic property, whereas the precise rate of synthesis of this protein at a given time is a particular or individual property.) This qualitative change, however, does not transform the biosystem in question into a member of a new species.

Moreover, we stipulate that the qualitative change in question must be an internal event or process, that is, one involving some organismic activity or function (see Definition 4.8). In other words, it must not be directly produced by some environmental agent or agents. (Think of a feature such as an open wound inflicted by some predator.) More precisely, it must not be caused by what we called 'strong energy transfer' or '(complete) event generation' by some agent in the organism's environment (recall Sect. 1.9.2), although it may very well be triggered by some external agent. Indeed, many developmental processes are triggered by environmental factors such as temperature or salinity, but they are not directly produced, or caused *in toto,* by them. Think of the temperature-dependent sex determination in most reptiles; also, the formation of a scar is triggered by the injury that, in turn, was directly generated by some environmental agent. Distinguishing, then, the notion of event generation from that of event triggering, we propose:

DEFINITION 8.1. Let $P(b, t)$ represent the set of generic properties of a biosystem b at some time t. Further, call s the state of b at time t and s' its state at a time t', where $t' > t$. Then the event (or process) $\langle s, s' \rangle$ is a *developmental event* (or *process*) of b if, and only if
(i) $\langle s, s' \rangle$ is not (directly) *generated* by some environmental agent(s); and
(ii) $P(b, t') \neq P(b, t)$.

In other words, any period in a biosystem's life history that is not accompanied by qualitative change is not a stage of development but of (mere) living. Accordingly, biosystems that do not change qualitatively (within their species bounds) during their lifetime cannot be said to undergo any development. This seems—*pace* Blackmore (1986)—to be the case with many unicellular organisms which just live and reproduce without undergoing qualitative changes (see also Hall 1992). However, some unicellular organisms do undergo developmental processes. For example, under certain environmental conditions some bacteria may assemble flagella, or form spores or cysts. The members of the Amoeboflagellata may change their form from an amoeban into a flagellate morph (or vice versa), thereby synthesizing a flagellar apparatus. Ciliates absorb and resynthesize their macronucleus during conjugation. After cell division in the ciliate *Stentor*, the daughter cell stemming from the posterior end of the asymmetrical mother cell must regenerate the anterior part containing the so-called oral area of the cell membrane. A final example is provided by the members of the algae genus *Acetabularia* (Chlorophyta, Dasycladaceae), in which the zygote, remaining unicellular and uninucleate until immediately before the formation of the reproductive cells, develops into a thallus characterized by a specific whorl-like cap at the apical end (whence the common name *Mermaid's Cap*).

As embryogenesis is the epitome of development, it is often equated with it. Though this is true in some cases, it is incorrect in general, because (a) an organism's development need not be restricted to embryogenesis, and (b) it need not be a continuous process. For example, holometabolous insects undergo a second conspicuous period of development besides embryogenesis: the pupal stage. To make room for cases like this, we propose:

DEFINITION 8.2. The *development* of a biosystem *b* is the sequence of all developmental events and processes of *b*.

Note the formulation 'developmental processes *of b*'. Not all developmental processes *in* a biosystem, i.e., in some of its subsystems, need result in a developmental change *of* the biosystem itself. For example, in our immune system some cells develop continually from more or less undifferentiated stem cells. Thus, developmental processes at the cellular level take place as long as we live. However, the organism does not undergo a qualitative change as a whole when some of its parts are replaced by parts of the same kind. In other words, the organism remains in the same (subspace of its) nomological state space. When, for instance, some of my bone marrow cells develop into *B*-lymphocytes (or else erythrocytes), I do not develop as a whole: I still have the generic property of possessing *B*-lymphocytes. Only when this property is either newly acquired or lost at some time in my life history does a developmental process at the organism level occur. On the other hand, the acquisition (or else loss) of immunity to infectious agents (e.g., due to the production of antibodies of a new type) is a new quality of the organism as a whole and must thus be regarded as a developmental process.

In sum, "developing" and "being alive" may, but need not, be coextensive. We cannot say, therefore, that all organisms continue to develop as long as they are alive, even though some of their subsystems undergo developmental processes. In other words, the life history of an organism need not be one long developmental process, which is often called *ontogeny*. What, then, about the term 'ontogeny'? Etymologically, the terms 'onto*geny*' or 'onto*genesis*' imply development and are thus often used to designate the concept of "the life history of an organism as one long development". Since we just saw that this concept need not be attributable to all organisms, we shall avoid the term 'ontogeny' and distinguish, whenever necessary, "development" according to Definition 8.2 on the one hand from "life history" (or "life cycle") on the other. Finally, what holds for the term 'ontogeny' also holds for the word *epigenesis*, which we take to be synonymous with 'development'.

8.1.2 Types of Developmental Processes

Usually, developmental processes are partitioned into three types: growth, differentiation, and morphogenesis. This partition distinguishes first of all different aspects of development, not necessarily separate processes, because growth, differentiation and morphogenesis often occur combined with one another. The particular processes (or mechanisms) in these three process classes are, if known at all, so different from one another that the question arises in what respect, if any, they are equivalent. Therefore, we can try to characterize them only in very general, i.e., phenomenological, terms. (For an earlier analysis of these process types see, e.g., Waddington 1970.)

8.1.2.1 Morphogenesis

The very word 'morphogenesis' suggests a process whereby the formless (e.g., an egg) acquires a definite form (or shape or spatial pattern). While this is so at the physical and chemical levels, as in the cases of the generation of waves in a homogeneous fluid or the growth of crystals in solutions, it is not so at the biotic level. Here, morphogenetic processes are those where new forms emerge out of old ones. As we now know, eggs and zygotes are intricately structured biosystems, not homogeneous or formless blobs of cytoplasm. Interestingly, the eggs of more complex animals, such as mammals, are less structured than the eggs of many less complex animals, such as sea-urchins. (For details see any textbook of developmental biology, such as Gilbert 1994.)

Morphogenesis is sometimes defined as a change in the geometrical shape of a biosystem. This definition, however, is too restrictive because a biosystem need not change its overall shape when new "structures", or rather subsystems, are formed. Moreover, forms, shapes, or structures do not exist in themselves: there are only shaped and structured biosystems (e.g., cells, tissues, organs, organisms), so that any change in form involves a change of a given biosystem (or any of its subsystems). We therefore propose:

> DEFINITION 8.3. A developmental process of a biosystem b is a process of *morphogenesis* if, and only if, b acquires a *new* (external) shape or a *new* (internal) structure through the formation of at least one *new* subsystem, that is, one that did not exist before the onset of the process—or through the loss of an existing one.

> DEFINITION 8.4. The *morphogenesis* of a biosystem b is the set of all the morphogenetic processes of b.

According to the preceding definitions, the concept of morphogenesis is a restriction of the ontological concept of self-organization (see Definition 1.10) with reference to biosystems. It might therefore be better termed *biomorphogenesis*. Although it is thus less general than the corresponding ontological concept, it is still phenomenological in that it does not refer to any specific biomorphogenetic mechanisms, such as the influence of morphogens on cell metabolism. Yet the disclosure of the mechanisms of morphogenesis is still one of the most formidable tasks of developmental biology. (More on morphogenetic mechanisms in Odell et al. 1981; Wessells 1982; Goodwin et al. eds. 1983; Gilbert 1994; Gurdon et al. 1994; Hess and Mikhailov 1994; Tabony 1994.)

8.1.2.2 Differentiation

A good example of morphogenesis in cnidarians (polyps and jellyfish) is the development of an interstitial cell into a cnidoblast. Following common parlance, we might be tempted to say that this is a case where a (comparatively) *undifferentiated* cell *differentiates* into a thread cell. However, this habit of speech suggests that morphogenesis is the same as differentiation, which it is not. This is because "differentiation" is a relational concept: it presupposes the existence of a population of systems (or subsystems of a system) whose members (may) become different from each other. In other words, "differentiation" implies "diversification". Thus, the fact that a given cell in state s may be in a very different state s' at a later time is first of all an instance of morphogenesis, not necessarily differentiation. (Incidentally, the degree of diversity within a cell often decreases when it becomes specialized by undergoing a morphogenetic process.) Real differentiation occurs when two systems which are part of a whole (i.e., a system, or an aggregate or population) become *different from each other*, i.e., when they undergo differential morphogeneses. Accordingly, we introduce:

DEFINITION 8.5. A developmental process in a biosystem b is a process of *differentiation* (or *diversification*) if, and only if, the number of kinds of subsystems in b and, thereby, the number of specific functions in b increases. (For the notion of specific function see Definition 4.8.)

Although many developmental processes are irreversible, sometimes the reverse of this process, i.e., a process of *dedifferentiation*, may occur, such as in some cases of regeneration and in the formation of tumors. By analogy to Definition 8.5, a process of dedifferentiation in a biosystem can be defined as the submergence of subsystems and a corresponding decrease in the number of specific functions. As with the case of morphogenesis, we distinguish explicitly between a particular process and a class of processes by formulating:

DEFINITION 8.6. The *differentiation* (or *holodifferentiation*) of a biosystem b is the set of all the differentiation processes in b.

(The term 'holodifferentiation' is borrowed from Grobstein 1962.) Occasionally, the term 'differentiation' is used only to denote the generation of cellular diversity within a multicellular organism (see, e.g., Gilbert 1988). However, differentiation also occurs at higher levels within a multicellular organism, e.g., in particular at the tissue and organ levels. If the distinction of levels of differentiation within a multicellular organism becomes necessary we can, for instance, speak of 'cytodifferentiation', 'histodifferentiation', and 'organodifferentiation'.

8.1.2.3 Growth

Growth in the most general sense is an increase in size of a system. Now, a biosystem may increase its size by different mechanisms. A single cell or the cells of a multicellular organism may, for instance, merely imbibe water. Another growth mechanism is cell division or, rather, multiplication, provided the daughter cells remain connected to each other and are of the same size as the mother cell. That the latter is not always the case can be seen in early embryogenesis when the zygote only divides into smaller and smaller blastomeres. Thus, despite cell multiplication, the early blastula is of the same size as the zygote—whence the term *cleavage*. Whatever specific mechanisms are involved, some incorporation of molecules from the outside and the synthesis of molecules in the inside must take place if the size of the biosystem is to increase. (The minimal conditions for growth seem to be the intake of water and the enlargement of the cell membrane.) This suggests defining *growth* as follows:

DEFINITION 8.7. A biosystem b undergoes a process of *growth* during a time interval τ if, and only if, the rates of intake and synthesis of molecules in b during τ is greater than the rate of breakdown and output of molecules during the same period.

In other words, a biosystem grows as long as anabolism outweighs catabolism (von Bertalanffy 1952). Clearly, the observable result of this process is an increase in the size of the biosystem. Increase in size, however, is a quantitative change, not a qualitative one. Although quantitative change may lead to qualitative change in some cases (e.g., a small cloud may grow large enough to become rainfall, and pieces of certain metals cooled down far enough may suddenly become superconducting), this need not hold in all cases. Hence "growth", if defined thus generally, cannot necessarily be regarded as referring to a developmental process according to Definition 8.1. It is necessary then to distinguish quantitative growth, i.e., mere size increase, from qualitative or developmental growth, i.e., growth resulting in the emergence of new properties. We elucidate the latter concept by:

DEFINITION 8.8. A biosystem b undergoes a process of *developmental growth* during a time interval $[t, t']$, where $t < t'$, if, and only if,

(i) b grows and the incorporation or synthesis of molecules in b during $[t, t']$ involves molecules of a species not present in b before t; i.e., it involves a change in the chemical composition of b; or

(ii) b's growth during $[t, t']$ is combined with a morphogenetic process.

By analogy to Definitions 8.4 and 8.6, we can finally define a biosystem's growth and developmental growth as the set of all its growth and developmental growth processes, respectively.

After this brief foray into the foundations of theoretical developmental biology, let us now turn to some of its philosophical problems.

8.2 Preformationism Versus Epigeneticism

Developmental biology has been both inspired and plagued by two apparently antithetic theoretical frameworks of a partly philosophical and partly scientific nature: preformationism and epigeneticism. The basic idea of *preformationism* is that the formation of new features during development is only apparent: it consists merely in the unfolding or unrolling of characters preformed in the germ (i.e., the sperm, the egg, or the zygote). By contrast, the basic idea of *epigeneticism* is that there is no such pre-existing form: that development consists in the *emergence* of genuinely *new* characters from an *unstructured, formless*, or *homogeneous* germ. As both preformationist and epigeneticist approaches can still be found in modern developmental biology, it will be worthwhile to briefly exhibit the traditional views before we proceed to examine the philosophical presuppositions of modern developmental biology.

8.2.1 Traditional Preformationism

Anaxagoras (499-428 BCE) was apparently the first to formulate a preformationist hypothesis with regard to development. He believed that all the parts of the child were preformed in the paternal semen (Jahn 1990). This idea was revived in the 17th century by the early microscopists, who were able to study embryogenesis in stages of development which had not been accessible to observation before. Thus, Malpighi's studies of chicken embryos and Swammerdam's observations of frog and insect development showed that the embryo was already endowed with some form and hence was not a homogeneous mass. Of course, this does not hold for early stages in which some investigators nevertheless "found" form. For example, even in the 18th century, von Haller believed he had observed embryonic membranes in the chicken egg. Likewise, de Graaf's discovery of ovarian follicles and van Leeuwenhoek's, Ham's, and Hartsoeker's investigations of spermatozoa can hardly have shown much adult structure. Yet two factors seem to have contributed to finding form where there was none: the first was the imperfection of the early microscopes, which easily led to optical artifacts, and the second was imagination. An instance of vivid imagination is Hartsoeker's famous figure of a homunculus preformed in a sperm, which is often used not only to characterize preformationism, but also to caricature it.

However, Gould (1977) has argued that not all preformationists were as naive as this homunculus caricature suggests. It seems appropriate, then, to distinguish two versions of traditional preformationism: strong or naive and weak or critical. Hartsoeker's homunculus exemplifies naive preformationism: the features of the adult organism were assumed to be entirely preformed in either the sperm or the egg. In other words, the latter were believed to contain entire miniature adults. On the other hand, the preformationist view of investigators like Malpighi and Bonnet was much more sophisticated: it is a version of critical preformationism. As a matter of fact, they knew perfectly well that, for instance, the chicken embryo becomes less and less structured in earlier stages until it finally appears as a "transparent homogeneity" (Gould 1977). Further, assuming preformed parts, which later undergo growth and rearrangement, is not the same as assuming an entirely preformed miniature adult. Being realists, not phenomenalists, the critical preformationists conjectured that the young embryo, though tiny and transparent, must have features that just could not be observed with the microscopes of their day. Thus, Bonnet (1762, Vol. 1, p.169) argued:

> Ne jugeons donc pas du tems où les Etres organisés ont commencé à exister, par celui où ils ont commencé à nous devenir visibles, & ne renfermons pas la Nature dans les limites étroites de nos Sens & de nos Instruments. [Do not mark the time when organized beings begin to exist by the time when they begin to become visible; and do not constrain nature by the strict limits of our senses and instruments (Gould 1977, p.20)]

Though partly based on wrong observations, traditional preformationism was far from being a piece of fantasy. In fact, in its time it was a philosophically and scientifically respectable thesis. Philosophically, it was in tune with Leibniz's concept of pre-established harmony, which, in turn, is based on the creationist outlook of his day. Indeed, the anatomist Peyer argued that, since God is the only creator, matter cannot be attributed the ability to create qualitative novelty without any previously implanted form, the *idea realis* (Jahn 1990). Moreover, preformationism was compatible with the prevailing mechanistic world view, which also precluded the emergence of qualitative novelty by assuming that every change can be reduced to mechanical causes and effects and by adopting the principle *causa aequat effectum*. From a scientific point of view, preformationism was able to explain the continuity, specificity, and fidelity of development, such as the fact that acorns give rise to oak trees, not to frogs or elephants. In other words, it helped explain the constancy of species. Furthermore, preformationism precluded spontaneous generation, which was then disconfirmed by the investigations of Redi and, later, Spallanzani.

Of course, preformationism forbids evolution in the modern sense. However, the term 'evolution' was already in use among 18th century preformationists (e.g., Bonnet 1762); but it then denoted development, which was regarded as the unfolding or unrolling (*evolutio*) of preformed parts. Another synonym for 'preformation' is *emboîtement* (*encapsulation*, *Einschachtelung*).

8.2.2 Traditional Epigeneticism

Traditional epigeneticism has been with us at least since Aristotle. In the 17th century it was adopted, for instance, by Harvey. Its revival in the 18th century was a clear reaction to the prevalence of preformationism in the late 17th and early 18th centuries (Jahn 1990). The influence of empiricism and the spread of the experimental method brought about findings that led many embryologists to cast doubts on the truth of preformationism. First, egg and sperm were seen as what they appeared to be: unstructured and homogeneous. Second, experiments on regeneration, notably Trembley's experiments with the fresh-water polyp *Hydra*, in which parts separated from the parent organism were shown to be able to grow again into entire organisms, were inconsistent with strict preformationism. (Bonnet, however, tried to save preformationism by assuming that there existed preformed parts in the entire organism, not only in the gametes. For instance, he held that "une Branche naissante est un Arbre en miniature" [1762, Vol. 1, p. 178].) Third, preformationism could not readily explain the existence of hybrids and some malformations. (Bonnet tried to explain certain malformations by assuming that, for instance, an abnormal pressure on the delicate and still soft and gelatinous preformed parts in the germ may easily result in distortions of form and proportion. He even concluded that, considering this delicacy of the germinal miniatures, it is surprising that monsters are not more common than they actually are [1762, Vol. 2].) Despite careful attempts like Bonnet's to defend preformationism, most embryologists turned epigenetic from the middle of the 18th century on, in particular under the influence of Caspar Friedrich Wolff's scientific work.

However, there was one serious problem with epigeneticism. If the egg is really unstructured or homogeneous, what accounts for the continuity and specificity of development, and where does the increasing complexity in the development of organisms come from? To answer this question, the epigeneticists had to postulate some unobserved force able to direct and guide development. Thus, epigeneticism invites vitalism. For example, Aristotle made use of his concept of *entelechy*, Wolff assumed a *vis essentialis*, Blumenbach postulated a *nisus formativus* (or *Bildungstrieb*), and Buffon stipulated a *force pénétrante*.

Since nowadays vitalism is rightly rejected as an unscientific view devoid of explanatory power, we should point out that 18th century vitalism came in two versions: an animistic and a "materialistic" variety, the latter of which must be regarded as a scientifically respectable hypothesis at its time. The animistic version, which has been called *psychovitalism*, is due to Georg Ernst Stahl, who, in 1708, postulated an immaterial soul-like entity as the steering force of vital processes. Psychovitalism was especially popular in France and it was espoused by the anti-mechanistic medical school of Montpellier (Jahn 1990). By contrast, the vital forces hypothesized by Blumenbach, Buffon, Kielmayer, and Reil were thought to be *physical* forces analogous to Newton's gravitational force. In particular, Reil's view came close to saying that a self-organizing force is an emergent property of living systems (Lenoir 1982). In sum, materialist vitalism was consistent with the naturalistic outlook of the 18th century and hence not a gratuitous assumption. Furthermore, unlike its animist rival, it was testable, and was eventually given up in the mid-19th century, after von Helmholtz had shown that there was no evidence for, and no need to assume, a vital force above and beyond the known physical forces of the time.

8.2.3 Modern or Neopreformationism

Though virtually dead after the triumph of epigeneticism in the late 18th and in the 19th century, preformationism resurfaced in a modern version in mid-20th century. With the advent of molecular biology, in particular since the analysis of the composition and structure of the genetic material in 1953, preformationism is fashionable again. However, in its modern version, encapsulated miniature adults or preformed morphological parts have been replaced by "coded instructions" (Gould 1977; Grant 1978). Neopreformationism, then, is essentially genetic informationism as applied to development.

Moreover, in its strong version, neopreformationism is a form of genetic determinism, hence microreductionism. In this view, an entity called 'genetic information' or 'genetic program' is assumed to contain all the "instructions" for development, determining the timing and details of the formation of every cell, tissue, and organ, and thus the morphogenesis and differentiation of the whole organism. Thus, "the organism is seen as an epiphenomenon of its genes, and embryology is reduced to the study of differential gene expression" (Gilbert 1988, p. 812)—that is, to developmental genetics. Furthermore, in neopreformationism "the organism effectively disappears from biology as a fundamental entity and is replaced by a collection of sufficient causes in the genome: i.e., the phenotype is reducible to the genotype" (Goodwin 1984, p. 224). This reductionist view fits nicely into the population genetic form of the Synthetic Theory, which identifies evolution with the change of gene frequencies in populations, and selection with the differential replication of genotypes. In its extreme version, neopreformationism is a form of outright Platonism: "A gene is not a DNA molecule; it is the transcribable information coded by the molecule" (G.C. Williams 1992a, p. 11). Let us then examine genetic informationism more closely.

8.2.3.1 Genetic Informationism

> *Metaphors [...], with which some branches of biology abound, are often suggestive and maybe harmless enough if they are recognized for what they are. But at best they are makeshifts and substitutes for genuine biological statements, and the fact that recourse is had to them is surely a sign of immaturity.* (Woodger 1952, p. 8)

One of the main problems with genetic informationism lies in the very word 'information', which is so ambiguous that it is being used in the contemporary scientific literature in at least half a dozen different ways (Bunge and Ardila 1987):

- information$_1$ = meaning (semantic information)
- information$_2$ = signal
- information$_3$ = message carried by a pulse-coded signal
- information$_4$ = quantity of order (negentropy) of a system

– $information_5$ = knowledge
– $information_6$ = communication of $information_5$ (knowledge) by social behavior (e.g., speech) involving a signal ($information_2$)

Given all these different senses of the word 'information', it should come as no surprise that it has become an all-purpose term. It sounds very scientific, and seemingly indicates some deep insight, but it is often nothing but a disguise of ignorance, inviting people to proceed according to the rule "If you don't know what it is, call it *information*". So what, if anything, is genetic information?

Let us begin with the notions involved in classical information theory: "$information_2$" and "$information_3$". These concepts do not apply to DNA because they presuppose a genuine information system, which is composed of a coder, a transmitter, a receiver, a decoder, and an information channel in between. No such components are apparent in a chemical system (Apter and Wolpert 1965). To describe chemical processes with the help of linguistic metaphors like "transcription" and "translation" does not alter the chemical nature of these processes. After all, a chemical reaction is not a signal that carries a message. Furthermore, even if there were such a thing as information transmission between molecules, this transmission would be nearly noiseless (i.e., substantially nonrandom), so that the concept of probability, central to the theory of information, does not apply to this kind of alleged information transfer. Indeed, the concept of probability does not occur in considerations on the role of nucleic acids in protein synthesis. In short, many assertions to the contrary notwithstanding, the expression 'genetic information' is unrelated to the concept of information as it is rigorously defined in the statistical theory of information.

Referring to "$information_4$", some authors have attempted not only to estimate the quantity of information locked in the genetic material, but also the information content of entire living systems. In so doing, the method for estimating the amount of information in a biosystem consists essentially in listing all of the instructions necessary to specify how a rational being would assemble a given biosystem from its components. In this way, one may come up with an estimate of about 5×10^{25} bits of information content for an adult human being (Dancoff and Quastler 1953). The problem with such procedure, however, is that it depends critically on what components are chosen as building blocks: elementary particles, atoms, molecules, organelles, cells, tissues, or organs. The information content of the given system will be different accordingly. Thus, taking atoms rather than molecules as components, Dancoff and Quastler arrived at an estimate of 2×10^{28} bits for adult humans. But if such numbers are supposed to measure the degree of order (or negentropy) of a system at a time, which is an intrinsic property of the system, the amount of information cannot vary with the level of analysis and thus can have only a single value. Therefore, such numbers do not measure an objective property of the system in question. What they may be taken to measure is *our knowledge* of the composition and organization of biosystems. Thus, to say that on the average a biomolecule contains n bits of information means that *we* need that much information to specify or characterize or describe the bare essentials of that molecular species: it says nothing about the process of synthesis of such biomolecules. That is, it confuses epistemology with ontology.

In sum, any estimates of the quantity of information of the genetic material or of entire organisms are phoney—so much so that they occur in no biological law statements, and everyone may give his or her own arbitrary estimate without fear of being refuted (Apter and Wolpert 1965; Bunge 1979a, 1985b; Levins and Lewontin 1985; Oyama 1985). However, the believers are still out there and, indeed, they seem to constitute the majority (see, e.g., Brooks et al. 1989).

Although some authors pay lip service to such computational informationism, it is not what most biologists have in mind when speaking of genetic information. To show this, let us assume, for the sake of the argument, that it makes sense to speak of the $information_4$ content of a biomolecule. Then, for example, two different DNA segments (or genes) may have the same information content of, say, 1000 bits. That is, the $information_4$ of the two genes is identical. Yet, since they are two different genes, they are involved in the synthesis of two different proteins. In other words, as having different specific roles, they contain different "informations"—which contradicts the assumption that their information content is the same. Obviously, this contradiction derives from an equivocation: what most biologists have in mind is not $information_4$ but $information_1$, i.e., information in the sense of meaning, or $information_5$, i.e., information in the sense of knowledge or even instruction.

Unfortunately, this worsens the problem instead of solving it. First, "meaning" is a semantical concept not a chemical one (see Sect. 2.3). A biochemical process may produce many things, but it does not produce meaning. In particular, the products of a chemical reaction do not endow any of the reactants with meaning, even if there is a lawful correspondence between molecules acting as templates and others which are the outcome of a template-dependent reaction. Second, the ability to form semantical concepts presupposes the existence of highly evolved brains, which are obviously not components of the genetic material. Hence, the expression 'the meaning of genetic information' can only be a metaphor. The same holds for information in the sense of knowledge: there is no knowledge of any kind outside some brains (recall Chaps. 3 and 6).

The term 'instruction' does not fare better. Instructions are rules or orders. Hence, to speak of instructions "encoded" in molecules makes sense only under the assumption that there is a rational being able to conceive of rules and express them in symbolic form. That is, the notion of an instruction presupposes intentionality or purposiveness. Moreover, being imperatives, instructions may be sources of paradox such as the self-referential instruction "Violate this instruction!". Paradox, however, is a logical item not a chemical one: certain DNA segments either function as template molecules in the cell or they do not. In short, in molecular biology "instruction" too is a metaphor. (See Weiss 1970, 1973; Oyama 1985.)

If the usage of the notion of instruction is illegitimate in molecular biology, so is the notion of a program, because this is nothing but a sequence or perhaps a system of instructions. The analogy with computer programs is of no help here.

For example, a computer diskette does not contain a program in the literal sense. In fact, a diskette is a piece of structured matter capable of interacting lawfully with other material systems, such as a user-computer system. A computer program is such only by virtue of its relation to a programmer or user complete with his intentions. The externalized symbolic representation of instructions on a material object is not the same as the program itself. The latter, like a theorem, exists only in the brain of the programmer. It comes as no surprise, then, that more and more biologists and biophilosophers begin to realize that the genetic program is "an object which is nowhere to be found" and "an entity invested with executive powers, that doesn't exist" (Moss 1992, p. 335).

Another metaphor of informationist preformationism is that of a blueprint. This idea can be quickly disposed of by considering that the notion of a blueprint implies that there is some structural resemblance between the blueprint and the object it maps. Although there is a correspondence between the base sequence of a given gene and the amino acid sequence of a given proteid, there is no resemblance between the structure of DNA and the structure of the vast majority of phenotypic characters of organisms. *Pace* Stent (1981), who criticized the program metaphor on the grounds that a program would be isomorphic with the "enacted play", no such correspondence is involved in the notion of a program, because the latter is a (symbolic) set of instructions, not a map or picture. In short, programs are not blueprints.

Let us finally consider the notion of genetic code. A genuine code is a certain correspondence between two sets of *artificial* signs and, more specifically, *languages* (Birkhoff and Bartee 1970). That is, "code" is a linguistic concept, not a chemical one. Further, the coding and decoding functions of errorless codes are one-to-one, not many-to-one as in the codon (triplet)-amino acid correspondence. What we actually have here is a correspondence between (a part of) a molecule with a given composition and structure, and another molecule with a different composition and structure. Although this correspondence may, of course, be *called* a 'code' for expediency, it is obvious that genes may be said to code, at most, for nRNA or mRNA, but not much else. Particularly, they cannot code for anything that happens after nRNA or mRNA molecules have been synthesized. In other words, any ensuing biochemical processes, such as RNA processing and protein synthesis, and *a fortiori* morphogenesis, are not "in the genes". Thus, the popular expressions of the form 'gene x codes for trait y' are unobjectionable only if they are understood in the sense of 'gene x is a template for the mRNA z, which is necessary for the (developmental) process that ends up in trait y'. But in any case, we emphasize that the use of linguistic and teleological notions to describe the biochemical processes involved in protein synthesis has no explanatory power, however didactically helpful it may be to understand these processes (see also Maynard Smith 1986, p. 100).

It will be obvious from the preceding considerations that 'genetic information' cannot designate any one of the above listed concepts of information. As Lwoff (1962, p. 94) proposed long ago, "for the biologist, 'genetic information' refers to a given actual structure or order of the hereditary material and not to the negative entropy of this structure". Therefore, 'genetic information' can refer only to the specific composition and structure of the genetic material. If used in this sense it is legitimate to continue to speak of genetic information.

However, considering the havoc that the many different senses of the word 'information' have caused within and outside the scientific community, we urge dropping information talk in molecular and developmental biology altogether. (However, a general notion of information in the sense of representation is useful for many purposes, but needs to be elucidated properly: see Kary 1990.) Therefore, expressions like 'information molecule', 'information flow', 'transcription' and 'translation' are purely analogical and should be replaced by strictly physical, chemical, or biochemical expressions in any theory purporting to explain the synthesis and replication of nucleic acids as well as the synthesis of proteins. After all, molecular biologists deal with molecules, not with symbols or letters. The "illuminating" character of the many linguistic metaphors in molecular biology stems from the fact that analogies further our understanding (or rather sense of familiarity), which is, as will be recalled from Section 3.6, a psychological category. Moreover, they undoubtedly provide a handy language of ellipses and metaphors. However, it must be emphasized time and again that they have no explanatory power in the methodological sense. In other words, convenient habits of speech and familiarity by analogy cannot replace scientific explanation.

If there is no information transfer from DNA to proteins, and no instruction in any literal sense, how, then, do we account for morphogenesis and differentiation in organisms? Is it not obvious, after all, that some preformed or pre-existing or encoded "limbness" or "eyeness", for example, must somehow be transferred from the genome to the developing tissues and eventually to the organism? As we shall see below, this intuition is just as wrong (and unnecessary) as the belief that there is an "information" transfer between communicating persons (recall Sect. 3.1.3).

8.2.3.2 DNA: The *Prime Mover* of Development?

Assuming that the molecular biologist admits that linguistic and intentional metaphors ought to be eliminated from biochemical vocabulary need not necessarily affect the basic outlook of reductionism and genetic determinism. The genetic material might still be regarded as the prime cause of development: "ex DNA omnia" (Wolpert 1991, p. 77). We shall show that, given what is known about gene regulation (see, e.g., Gilbert 1988, 1994; Moss 1992; Portin 1993), this view can be upheld only by illegitimately privileging one factor in a network of interacting factors.

First, DNA can hardly be said to be a *prima causa* or a *primum movens* because it is a comparatively inert molecule that does not do anything on its own (Weiss 1973; Lenartowicz 1975; Fox 1984; Lewontin 1991, 1992a; Smith 1992a; Hubbard and Wald 1993). It only "sits" in the nucleus (or elsewhere), moreover safely covered by histone proteins in eukaryotes, and "waits" for some other molecules to act upon it. In particular, the popular attribution of the property of self-replication is wrong and misleading. DNA does not self-replicate but *is replicated* by other molecules involved in cell metabolism, namely certain enzymes. The fact that the copy of a piece of DNA is—mutations aside—compositionally and structurally equivalent ("identical") to its template does not account for the process of replication itself. Though seemingly trivial, we must emphasize this state of affairs considering the frequent portrayal of genes as causal agents and as active components of cells.

Second, according to our construal of causation, DNA cannot be said to be the *cause* of development. There is no "causal power" of genes or alleles—*pace* Sterelny and Kitcher (1988) and Gifford (1990). Indeed, as it will be recalled from Section 1.9, the causal relation is defined for pairs of events (changes of state), not things. For this reason, we cannot admit *causae materiales*, but only *causae efficientes*. Genes always are conditions, never causes, of development. To be sure, they are necessary conditions, and they have codeterminative power, but no causal power whatsoever.

Third, as (in eukaryotes) DNA is covered by histones, it is not ready to undergo chemical reactions without further ado. In any event, genes do not switch themselves on or off, but *are* "activated" by molecules *acting on* DNA. (This also holds for prokaryotes.)

Fourth, the processes following gene "activation" are circumstance-dependent. That is to say, different cells may process differently the same primary gene products (i.e., "transcripts" such as n-RNA or m-RNA). For example, although undoubtedly brain, liver, and kidney cells are cytologically very different, the nRNA of those cells in rats and mice is almost identical. Another example is provided by the processing of a certain mRNA precursor in thyroid and nerve cells. In the former, the processing of the mRNA precursor will yield the hormone calcitonin, whereas in the latter the outcome is a neuropeptide. Further, once a protein is produced, the underlying gene's determining effect, if any, can still be circumvented. For instance, a newly synthesized protein may be inactive without further modification, or it may be selectively inactivated. Other proteins function only at certain places within the cell, i.e., they first need to be transported to those places. Finally, some proteins must join other proteins to form a functional unit at all. (See Gilbert 1988, 1994.)

Fifth, in our immune system, the production of specific antibody molecules involves the creation of *new genes*. Thus, during cell morphogenesis and differentiation, lymphocytes rearrange certain generic immunoglobulin genes, creating any one of up to 10^7 different specific antibody genes and hence proteins, as the need arises (Gilbert 1994).

Sixth, the development of a biosystem depends on environmental items as much as it depends on its composition and structure. For example, the formation of flagellar tubulin in the amoeban morph of Amoeboflagellates is triggered by environmental changes. Another example is embryonic induction, in which the future development of a certain embryonic region depends on the influence of neighboring regions. Finally, many morphological patterns due to mutations can be produced equally well by environmental agents, such as in phenocopies (Horder 1989; Hall 1992). In short, there is no development in a vacuum.

These different modes of gene regulation show that there is no unidirectional regulation or control of development from gene to phene. That is, there are (many) gene-dependent processes but no gene-directed ones. In fact, what we do have is a complex self-regulating system consisting of a network of at least three regulating and interacting components: genetic material, other components of the cell (often collectively referred to as 'the cytoplasm'), and environmental items. This insight deserves to be spelled out in a postulate of its own, to wit:

POSTULATE 8.1. All developmental processes of biosystems are controlled or regulated through the systemic and lawful interaction of (the members of) its genome, its extragenomic composition, and its environment.

This postulate entails:

COROLLARY 8.1. There is no exclusive (or sufficient, or privileged) control system of an organism's development, such as its genic system.

Although all these facts are well known to biologists and biophilosophers (see, e.g., Woodger 1952; Kitcher 1992; Portin 1993; Lander and Schork 1994), most of the latter usually proceed as if only one of these three necessary factors, namely the genetic material, were important, while the others may be relegated to standard conditions. For example, the fact that certain genes fail to produce phenotypic effects in certain individuals is regarded as an instance of incomplete penetrance rather than of ontic irrelevancy—a strategy which leaves the underlying preformationist outlook unscathed. Another example is provided by Sterelny and Kitcher (1988), who attempt to save the causal priority of genes by means of the following definition:

> An allele A at a locus L in a species S is for trait P^* (assumed to be a determinate form of the determinable characteristic P) relative to a local allele B and an environment E just in case (a) L affects the form of P in S, (b) E is a standard environment, and (c) in E organisms that are AB have phenotype P^*. (p. 350)

(A similar strategy has been proposed by Gifford 1990.) To be sure, when describing the development of biosystems, the environment can often be regarded as constant, so that the gene is the variable, that is, the "thing that makes the difference". Woodger (1952, p. 186) called this the 'constant factor principle'. But at the same time he warned: "... if we ... omit reference to the [environment] ... because it is constant and common to all our experiments, we must obviously not slide into the assumption that the [environment] 'plays no part' in the processes involved". Indeed, as Gray (1992) has correctly pointed out, a similar definition could be formulated defining *an environment for* a phenotypic trait given standard genes or genotypes. Moreover, since there is another necessary factor besides genes and environment, namely a certain cytoplasmic constitution, we might as well define a *cytoplasmic constitution* for a certain phenotypic trait given a standard genotype and a standard environment. (See also Smith 1992a; Sterelny et al. 1996.)

To be sure, for the sake of analysis it is legitimate to start by singling out one factor at the expense of others—e.g., in order to simplify the explanation of development by reference to differential gene expression. This procedure is, moreover, justified by the fact that most of an organism's genes may be assumed to be adaptations, whereas many environmental items are not. But even so, it becomes illegitimate as soon as it is no longer seen as a simplifying epistemic strategy but as a claim about the ontic priority of one factor in a network of interacting factors over the other. Since genetic determinism or genetic reductionism, including genic selectionism, confers such ontic priority upon genes, it is a scientifically and metaphysically unsound thesis. Indeed, the limits of explanations of development in terms of gene control are pointed out by many a developmental biologist: "While no aspect of embryogenesis can be entirely independent of the products of the genome, it may nonetheless be the case that control by genes is so indirect that analysis in such terms loses useful explanatory power" (Horder 1983, p. 346).

8.2.3.3 The Genotype-Phenotype Dichotomy

If not only genetic determinism but also the notions of genetic information and genetic program must be given up, then the traditional genotype-phenotype dichotomy, in which the phenotype of an organism is regarded as the "manifestation" or "expression" of its genotype (e.g., Futuyma 1986, p. 43), is called into question. Indeed, that genes as parts of an organism actually belong to its phenotype is not a new piece of knowledge (see, e.g., Woodger 1952; Lenartowicz 1975; Brandon 1990; Lewontin 1992b). Yet let us briefly examine the concepts of gene, genome, and genotype as well as those of phenome and phenotype. (For details see Mahner and Kary 1997, from whom we deviate with respect to some minor conceptual and terminological issues, as they are apposite in the broader context of this book.)

Genome and Genotype. An analysis of the notions of genome and genotype evidently presupposes the concept of a gene. The terms 'gene' and 'genotype' were proposed by Johannsen in 1909 to denote that unobservable entity passed on in the process of reproduction whose sole conjectured property seemed to consist in being some "unit of heredity" (Johannsen 1913). Via the black box of development, this transphenomenal entity was presumed to be *somehow* related to some observable (phenomenal) property of the organism. Thus, differences in phenomenal properties, such as shape and color, were used as indicators of differences in the transphenomenal genes. The investigation of these genes by means of phenotypic indicators, and the search for the laws of their transmission and distribution, is what classical genetics is all about.

Paradoxically, the more we have learned about the genetic material from molecular genetics, the less we seem to know about what exactly a gene is (see, e.g.,

Carlson 1966; Kitcher 1992; Portin 1993; Waters 1994.) As it seems, a particular *gene* is an individual DNA (or RNA) sequence, whether continuous or not, that is somehow involved in a metabolic process, besides its own replication, in the cell of which it is part. Thus, a particular gene is delimited by its base sequence, depending on the role the latter plays in the organism's metabolism. Depending on the level of metabolic involvement, which may range from the production of primary RNA to some final state in some metabolic pathway, different sequences of the same stretch of DNA will have to be regarded as different genes (Waters l.c.). This is because genes may overlap, they may contain introns, they may be processed differently in subsequent metabolic processes, and so on. So, what counts as a gene depends on the circumstances; in other words, "gene" is both a structural and a functional concept.

Whatever definition of "(particular) gene" one accepts, we must now distinguish between an individual gene as a particular DNA sequence and a kind or type of gene. (See also Woodger 1952.) When geneticists speak of a gene "for" this or that character they usually do not have in mind an individual piece of DNA but the equivalence class of such individual genes, i.e., a gene *kind*. Thus, when the papers and journals report the discovery of yet another *gene for* some trait or disease, actually the *kind* of gene involved in the trait or disease is referred to, not any individual DNA sequence.

The obvious distinction between genes and gene kinds helps us clarify the notions of genome and genotype. To begin with, there is the collection of all the concrete individual genes (or alleles), whether cytoplasmic, nuclear, mitochondrial, or plastidial, that are proper parts of an organism: the *total genome* of the organism. (See Bunge 1985b. The qualification 'proper' is supposed to exclude the genes of viruses, bacteria, and eukaryotic parasites and symbionts that may be part of an organism at a certain time.) Thus, the total genome of an organism is a collection, not a thing. What would be a thing would be the material aggregate of all the genes (DNA) of a cell or organism, if extracted and lumped together in a test tube. This concrete aggregate should rather be termed 'genetic complement', for the term 'genome' is also used in a completely different sense, as we shall see in a moment. What would also be a thing or, more precisely, a genic system, are the interacting genes in a cell. However, what is usually called 'gene interaction' does not refer to actually interacting DNA segments but to the interaction of a gene product with another gene. This is because, as mentioned previously, DNA is a rather inert molecule that does hardly anything on its own. That is, most interactions between genes are mediated by cell metabolites, hence they are, at most, indirect interactions.

Now, biologists are not so much interested in the total genome or genetic complement of a multicellular organism. Instead, they are often concerned only with its *cellular* genome or genetic complement, and in particular only with its *standard* or *typical* cellular *genome*, inasmuch as the latter is representative of all the cells in the organism. After all, barring complications such as somatic mutations or the formation of new genes in certain cells of the immune system (see above), an organism's cells are usually genetically identical. For example, whereas the total genetic complement of an adult human consists of an estimated 10^{20} genes, the typical human genome is generally described as comprising only about 50,000–100,000 genes (National Institutes of Health 1990; Mahner and Kary 1997). The concept of a *typical genome*, then, is not a class of concrete genes, but a class of gene *kinds*, where the kinds in question are defined by the equivalence relation "having the same base sequence and the same locus". (Note that reference to the locus is needed to accommodate redundant genes, which, though having the same base sequence, occupy different loci.)

In sum, the term 'genome' designates three different concepts: that of (a) genetic complement—often called the 'physical set' of genes of a biosystem or 'physical genome'; (b) total genome—the mathematical set or, rather, variable collection of genes of a biosystem; and (c) typical genome—the set of gene types, as distinguished by both base sequence and locus.

As with the term 'gene', the terms 'genotype' and 'phenotype' were introduced by Wilhelm Johannsen in 1909. Although he originally defined "genotype" and "phenotype" as statistical concepts (1913), i.e., as character means of populations, he later defined these concepts in terms of properties of individuals (Churchill 1974; Mayr 1982.) Thus he defined the genotype as the "total constitution of the zygote" (1926, p. 166), and the phenotype as the characteristic appearance of an organism at a given time. (Interestingly, he appears to have realized that the entity "responsible for" a phenotypic trait is (disregarding the environment) not the isolated gene but the "total constitution" of the zygote.) Yet, now that the genetic material has been identified, and a distinction is made between genome and genotype, a different notion of genotype is employed by geneticists.

The genotype of a haploid or prokaryotic organism (*genotype*$_H$) may be construed as the collection of all kinds of genes in that organism, where the kinds in question are defined by the equivalence relation "having the same base sequence", disregarding loci and thus, in particular, redundant genes. In other words, the genotype of a haploid organism is the family of equivalence classes that results from partitioning its total genome by the equivalence relation "same base sequence". Adopting the standard distinction between genes and alleles, we could also speak of an *allelotype*. (Note, however, that the term 'allelotype' usually refers to the allele frequency in a population, not to the genotype of organisms: see Rieger et al. 1991.)

In diploid organisms the situation is more complicated, because there are two copies of each chromosome in each cell, so that there may be different alleles at equivalent loci. Nevertheless, corresponding to the concept of genotype$_H$, we could form the class of specific sequence types (or allele types) of a diploid organism: its *allelotype*. But this allelotype would not be what is regarded as the genotype in diploid organisms. Indeed, the genotype of a diploid organism (*genotype*$_D$) is construed as the collection of its allele-pair kinds. For example, if we assume the simplest case of two alleles per locus (type), e.g., the allele kinds *A* and *a* at the *A*-locus, then there are three possible combinations of these alleles, namely *AA*, *Aa*, and *aa*. Any such combination represents an allele-pair kind. Thus, the equivalence relation "same allele pair" partitions the collection of all loci of a diploid organism into a family of equivalence classes of allele pairings, i.e., the genotype$_D$ of the organism in question.

An example will illustrate the differences between all these concepts. Consider, then, a diploid organism in the two-cell stage of development, with two pairs of chromosomes of two kinds, P and Q, in each cell: P_1 and Q_1 as well as their homologues P_1* and Q_1* in cell 1, and P_2 and P_2* as well as Q_2 and Q_2* in cell 2 (see Fig. 8.1). Assume further that the two cells are in the G_1 phase of the cell cycle, so that each chromosome consists only of a single chromatid, and that each chromosome carries only two loci. For notational convenience, we finally stipulate that all the loci are heterozygous, so that the P chromosomes carry alleles of type A and a as well as B and b, and the Q chromosomes alleles of type C and c as well as D and d. Then all together this organism has 16 (nuclear) alleles of 8 kinds, the latter being A, a, B, b, C, c, D, d. Its total (nuclear) genetic complement is the actual 16 alleles. Its total genome (or, rather, allelome) is the composition of this complement, namely the set of those 16 individual alleles, $\{A_1, a_1, A_2, a_2, B_1, b_1, B_2, b_2, C_1, c_1, C_2, c_2, D_1, d_1, D_2, d_2\}$, where the subscripts distinguish what belongs to which cell. The typical genome, on the other hand, is the set $\{A, a, B, b, C, c, D, d\}$, which is at the same time the allelotype of the organism. (Note that, if there were any redundant genes, the typical genome and the allelotype would not be coextensive as in this case.) And the genotype$_D$, finally, is the set $\{Aa, Bb, Cc, Dd\}$.

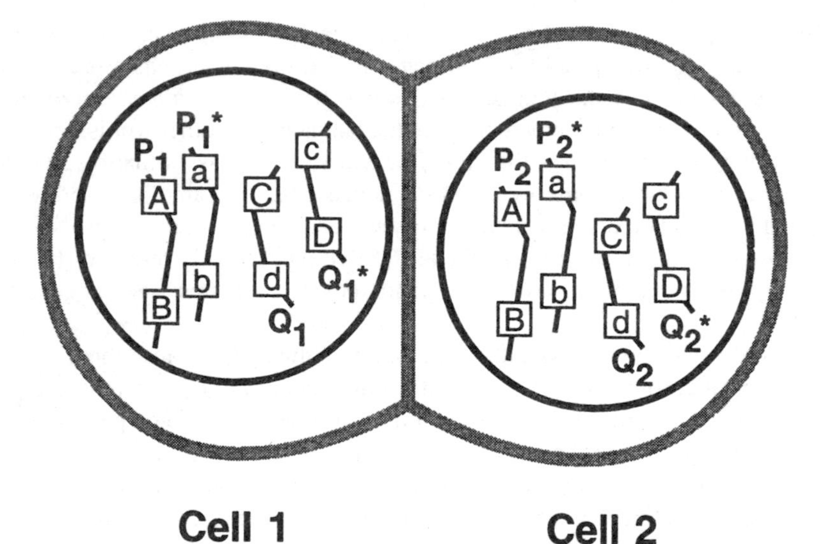

Fig.8.1. A multicellular organism in the two-cell stage of development. The cells are in the G_1-phase and contain two chromosomes (chromatids), P and Q, each. Each chromosome contains two genes. Further explanation in the text

Normally, geneticists are not interested in the total genotype$_D$ of an organism, but only in its genotype at some locus of interest. For example, when saying that *Aa* is a heterozygous genotype they refer only to the genotype at the *A*-locus. Thus, the definition of the genotype$_D$ of an organism can be narrowed down to a definition of its *L-genotype*$_D$, where *L* refers to the locus in question. Aside from practical considerations, focusing on such *partial* genotypes$_D$ (Lewontin 1992b) is necessitated by the enormous genetic variability (diversity) of organisms, which makes it hard, if not impossible, to find two individuals with exactly the same total genotype$_D$. And since scientists are supposed to come up with generalizations, in particular law statements, not descriptions of individuals, they must focus on what organisms have in common, not on what makes them unique.

All this does not exhaust the usage of the term 'genotype' in biology. Particularly in population genetics we find expressions such as 'the differential replication of genotypes', 'the frequency of the genotype *Aa* in population *p* is *x* %', and 'the genotype *Aa* has a higher fitness value than the *aa*-genotype'. The previous concepts of genotype do not apply here, since it is only things, not collections, that replicate and multiply and can have fitness values. Therefore, this makes sense only if the term 'genotype' is a shorthand for 'organism characterized by some genotype$_H$ or genotype$_D$'. This, then, is a third signification of the word 'genotype': *genotype*$_O$, where *O* stands for 'organism'.

We can now collect all the organisms characterized by the same genotype$_H$ or genotype$_D$ in an equivalence class. For example, a population of diploid organisms can be partitioned into a family of equivalence classes of organisms with the same genotype$_D$. (Again, this partitioning is usually restricted to a single locus or to a few loci, because otherwise it might result in a family of singletons.) Think of the Mendelian crossing of two identical dihybrids, *AaBb* × *AaBb*, the offspring of which belong to any of nine possible genotypic classes. We characterize this fourth concept of genotype by the subscript *C*, abbreviating 'class': *genotype*$_C$. (For this concept of genotype see also Lewontin 1992b.)

Incidentally, definitions of genotypes as classes of organisms also occurred in the operationalist period of genetics. For example, Haldane (1929) suggested the following operational definition of "genotype": "A class [of organisms] which can be distinguished from another by breeding tests is called a *genotype*" (p. 485). A similar construal can be found in Woodger's (1952) formalization of genetic concepts. Since, as we have pointed out time and again (Sect. 3.5.7.1, as well as Bunge 1967a, 1983b; Mahner 1994a), operational definitions are not definitions proper but, in fact, indicator hypotheses, we can safely ignore this operationalist notion of genotype.

Phenome and Phenotype. With one exception, the concepts of phenome and phenotype of an organism can be defined in a way similar to those of genome and genotype. The *total phenome* of an organism may be defined as the set of all its (individual) traits, whether at the organic or at the molecular level. (See also Lewontin 1992b.) This set comprises not only the composition of an organism, but also its structure, since also relational properties of organismal subsystems are regarded as characters. (Paralleling the expression 'genetic complement', the concrete aggregate of traits of an organism could be dubbed 'phenetic complement' or 'physical phenome', but these expressions would be synonymous with 'organism', since an organism's parts have no independent existence.)

Note that, quite unlike the typical genome, there is no typical phenome, because the phenome of a cell or any other organismal subsystem is not identical to the phenome of the organism as a whole, unless, of course, the organism is unicellular. For this reason, all the different ways of distinguishing between types of traits are traditionally subsumed under the concept of phenotype. Thus, the *phenotype* of an organism is the collection of all its kinds (or types) of traits, however construed, including also structural properties such as the relative position of its various subsystems. For referring to trait *types*, we distinguish this basic concept of phenotype by the subscript T and label it *phenotype*$_T$. (Note that we can not only conceptualize the phenotypes$_T$ of entire organisms, but also phenotypes$_T$ of their subsystems. For example, we can speak of the phenotype of the brain or even the phenotype of a molecule. We can also form the species-specific phenotype$_T$ of an organism, referring to all the generic traits common to all the members of a species. Yet there is no such thing as a species or species-level phenotype if understood in the sense of the phenotype of a species-as-individual.)

Just like the term 'genotype', the term 'phenotype' is often also used synonymously with 'organism'. Paralleling thus the genotype$_O$, we call this concept *phenotype*$_O$. Whereas a phenotype$_T$ is a collection, hence a conceptual object, any referent of "phenotype$_O$" is a concrete or material object; more precisely, it is an organism characterized by some phenotype$_T$ or phenotypic$_T$ property. When some biologists say, for instance, that "selection acts on the phenotype", they can refer only to the phenotype$_O$, for only material objects can interact. (Likewise, selection can only act on genotypes$_O$; but see Sect. 9.2.) Finally, corresponding to the notion of genotype$_C$, we could define a phenotypic class, the *phenotype*$_C$, as the class of organisms (in a given population of such) defined by some phenotypic$_T$ equivalence relation. Thus while, in the above example of the dihybrid mating, the offspring comes in nine possible genotypes$_C$, it comes only in four possible phenotypes$_C$, provided there is complete dominance.

Incidentally, the concept of phenotype$_T$ allows us to formulate an alternative definition of the concept of developmental event or process. For instance, we could say that any change of state of an organism x during some period $[t, t']$, where $t' > t$, is a *developmental event* or *process* iff the phenotype$_T$ of x at t' is different from the phenotype$_T$ of x at t. Note that this definition requires the precise notion of phenotype$_T$, not the vague, though common, definition of the phenotype in terms of an organism's appearance or observable form.

Conclusion. We must distinguish four significations of the term 'genotype' in biology: (a) genotype$_H$—the set of gene (allele) types of a haploid biosystem; (b) genotype$_D$—the set of allele-pair types of a diploid biosystem; (c) genotype$_O$—an organism characterized by a certain genotype$_H$ or genotype$_D$; and (d) genotype$_C$—a class of organisms sharing the same (usually partial) genotype$_H$ or genotype$_D$. Similarly, we must distinguish three concepts of phenotype: (a) phenotype$_T$—the set of trait types of a biosystem; (b) phenotype$_O$—an organism characterized by some phenotype$_T$; and (c) phenotype$_C$—a class of organisms with a certain phenotype$_T$.

In the light of these definitions it becomes obvious that a concrete string of DNA, that is, an individual gene or an allele, is itself a phenotypic$_T$ character (Lewontin 1992b). It is a (material) subsystem of the organism interacting with other subsystems in the course of development. Thus, the total genome of an organism is a proper subset of its total phenome, and the genotype$_{H/D}$ of an organism is a proper subset of its phenotype$_T$. (As already mentioned above, further subsets of the phenotype$_T$ may be distinguished, such as the karyotype, the enzymotype, the cytotype, the histotype, the organotype, or what have you.) Furthermore, since genes are subsystems of organisms, any qualitative change of such a gene, such as a mutation producing a gene of a new kind, can—according to Definition 8.1—be regarded as a developmental event.

Another consequence of the preceding explications is that it makes no sense to speak of the gene (as well as the genotype) as an abstract informational entity that somehow "expresses" or "manifests" or "instantiates" itself in the form of a phenotypic character. Formulations like these are clear vestiges of Platonism, for which "genetic information" is somehow *ante rem* or, rather, *ante organismum*. This Platonic relic apparently leads some scientists to contend that a classification based on molecules, in particular DNA systematics, is in some sense deeper or more objective or more decisive than a classification based on morphological characters (e.g., Goodman 1989). Clearly, such belief is unjustified because genes *qua* molecules are part of the phenotype$_O$, and gene kinds *qua* DNA sequence types are members of the phenotype$_T$. Therefore, molecular systematics just extends the range of available organismal characters to molecules. It cannot claim any special methodological status. (More on molecular systematics in Schejter and Agassi 1982; Hillis 1987; Patterson ed. 1987; Hillis and Moritz 1990; Moritz and Hillis 1990; Wägele and Wetzel 1994; Ax 1995.)

Besides, definitions of the concepts of genotype and phenotype in terms of blueprints or programs and their manifestations, respectively, are inadequate, because they lead to circularity. Consider the following definitions: (a) 'genotype $=_{df}$ blueprint for the phenotype' and (b) 'phenotype $=_{df}$ manifestation of the genotype'. One can now substitute the term 'genotype' in (b) by the right-hand side of (a), obtaining the circular definition 'phenotype $=_{df}$ manifestation of the blueprint for the phenotype'. Likewise, one could substitute the term 'phenotype' in (a) by the right-hand side of (b), obtaining the circular definition 'genotype $=_{df}$ blueprint for the manifestation of the genotype'.

The most important consequence of the preceding analysis is that developmental biology does not need the genotype-phenotype dichotomy to explain development. What is in fact at issue is how, given a certain environment, the initial state of the zygote is "mapped" into any later state of the organism (Lewontin 1974). Interestingly, this brings us back to Johannsen's conception of the genotype as the "total constitution of the zygote" (rather than of the genetic material), which is what we actually inherit. (See also Simpson 1953.) And it finally leads us back from genetics to developmental biology.

8.2.4 Modern or Neoepigeneticism

If all that remains from neopreformationism is the existence of a highly and specifically structured genetic material in a specifically structured germ cell in a certain structured habitat, then we have to turn to some form of epigeneticism in order to account for development (see, e.g., Løvtrup 1974; Katz 1982). This is not to deny the influence and importance of genes in development: it only de-emphasizes their supposedly predominant role in development. At least two modern epigenetic approaches attempt to provide an explicitly non-gene-centered account for development: the so-called *developmental systems approach* (also called *developmental constructionism*) and the *structuralist approach*, which we proceed to examine.

8.2.4.1 Developmental Structuralism

In Section 8.2.2 we saw that the problem of any epigeneticist approach is to account for the conspicuous specificity and fidelity of development. The traditional epigeneticists had to take refuge in vitalist hypotheses postulating sundry forces that could direct and guide the development of organisms. The aim of the developmental structuralists is to find universal "laws of form" governing development, and to incorporate them into an updated version of idealist morphology, which they prefer to call 'rational morphology'. (See the classical locus Webster and Goodwin 1982; as well as Ho and Saunders eds. 1984, 1993; Goodwin 1982a, b, 1990, 1994; Ho and Fox eds. 1988; Webster 1984, 1993. For reviews and critiques see Smith 1992b; van der Weele 1993; Resnik 1994.)

To this end, developmental and reproductive invariance are understood in terms of invariant developmental or "generative processes". These generative processes are supposedly "governed" by universal "laws of form" or "laws of transformation", as they were implied in the concept of type in idealist morphology. Organisms are regarded as particular forms or "structures" which are "members of a set or system of transformations generated by laws" (Webster and Goodwin 1982). Further, organisms as "structures" are taken to be "self-organizing totalities" or "fields" describable by field equations. Hence, what constrains development and evolution is not the "historical contingency" of common descent, but the postulated universal laws of form. Therefore, a classification according to universal laws of form need not coincide with a classification based on phylogeny. In fact, phylogeny is regarded as "largely irrelevant to an understanding of organisms as transformational structures" (Goodwin 1982a). What is at issue is the "logic of organized transformational relations", the "logical (necessary) origins as governed by law" and, thus, the "rational order" of the living world (Goodwin 1982a, 1990).

All this sounds rather fuzzy and somehow grandiose because it is so. Indeed, developmental structuralism, or *process structuralism* as it is sometimes called (Smith 1992b), is a peculiar mixture of genuine and important insight with philosophical fossils and confusions. It will be worthwhile, then, to examine developmental structuralism more closely.

To begin with, let us quickly dispose of some of the many metaphysically ill-formed expressions and statements to be found in the writings of developmental structuralists. As will be recalled from Section 1.2, we say that a statement is metaphysically ill-formed in the case that either conceptual properties are attributed to material things or substantial properties are attributed to conceptual objects. This is exemplified by expressions and statements like 'logic of development', 'logic of organized transformational relations', 'logical origins', 'organisms as manifestations of rational order', 'rational continuity of order between nature and mind', 'processes generating form during development are rational', and 'theoretical explanations involve formulating a set of laws of generative power' (Goodwin 1982a; Webster and Goodwin 1982; Ho and Saunders 1993). Goodwin (1982a) explicitly muses about a "cognitive view of biological processes", that is, an extension of concepts such as "knowledge" and "rationality" to biological processes. He even admits sympathy with subjectivism (Goodwin 1994). This view clearly presupposes an immaterialist ontology, although Goodwin (1990) denies any Platonic implications. In a scientific ontology, only some very special biological processes, namely processes in plastic neuronal systems, can be rightfully associated with cognition and intentionality (recall Chaps. 3 and 6).

This implicit immaterialist metaphysics is also apparent from the recurrent invocation of "rational morphology" and, with it, the concept of a "universal law of form". It is thus not accidental that already one of the structuralists' models, D'Arcy Thompson (1917), approvingly invoked the spirit of Plato and Pythagoras while musing about the mathematical laws that constrain living as well as nonliving things. Although developmental structuralists regard themselves as realists (Ho and Saunders 1993), their notion of law of form is an idealist concept. This is because these alleged universal formal laws can only be understood as governing development from the outside: they are *ante res*. This view of laws, though not uncommon, seems to rest on the confusion of law and law statement, i.e., substantial laws (objective patterns) and their conceptual representation, particularly in mathematical terms (see Sect. 1.3.4.). Furthermore, it reminds us of the famous discussion of the problem of universals in the Middle Ages, when the Platonists called themselves 'realists'. Laws, however, are properties of things: they are *in rebus*. As soon as it is realized that neither laws nor forms hover above things, there can be no hope of building a "rational" morphology. If there are biological, not just physical, developmental laws, they must depend on the composition and structure of the organisms in question, in particular those of the zygote. However, both the composition and the structure of organisms depend on those of their ancestors. Therefore, developmental laws must be taxon-specific. What makes development a lawful process is that the organisms undergoing developmental processes belong to a certain natural (biological) kind in the broad sense, i.e., a taxon, and thus share certain properties ($laws_1$); figuratively speaking, they share a common nomological state space. The axes of this nomological state space are not constituted by immaterial forms, but by representations of the lawfully related properties of the organisms in question. Therefore, it is true that organisms can undergo only lawful transformations, but this lawfulness is not provided by immaterial laws hovering above things. Moreover, as stated previously, most of these laws will be taxon-dependent. Only strictly physical and chemical laws are independent of considerations of phylogeny. A classification based on such purely physical and chemical laws may actually cut across phylogenetically defined taxa. Yet such procedure does not yield a *biological* classification.

Developmental structuralists might object that properties evolved through phylogeny may not be regarded as laws because they are "historically contingent", not logically necessary. However, no fact is logically necessary; in other words, all facts are contingent. This is because logic is ontologically neutral; that is, neither logical nor mathematical laws have any bearing on matters of fact. Logical (and mathematical) laws are constructs, not objective patterns of being and becoming. Hence, they may occur in our reasoning about and our representations of the world, but they play no role outside our brains. In particular, they cannot "govern" the behavior of things—not even that of our brains. As, in the real world, there are only nomically necessary events, no logically necessary ones, in the factual sciences and in a scientific ontology "lawfulness" always means nomic necessity, not logical necessity. It is mistaken, then, to conclude that, because processes are lawful and perhaps describable in mathematical terms, they are rational or logical. Moral: factual scientists have no use for the concepts of logical necessity and its dual, contingency.

Whereas the authors quoted so far regard themselves as belonging to the realist (i.e., actually cryptoidealist) version of developmental structuralism, there is also a radical or avowedly idealist branch characterized by the adoption of a purely mathematical formalism, which replaces the prevalent reductionist and symbolist formalism (or informationism) of molecular biology. The main representative of this view is René Thom (1972, 1983). Thom claims, for instance, that the mathematical theory of catastrophes would be a universal theory capable of accounting for discontinuities and singularities of all kinds in things of all species, from fluids to cells and organisms to societies. Such generality, however, is gained at the expense of depth and, often, relevance. In fact all of the applications of catastrophe theory are phenomenological (i.e., nonmechanismic) and nonspecific; and most of them are also static. Moreover, none of them involves any biological laws. Consequently, the catastrophe-theoretic models have no explanatory power. In particular, they do not explain how catastrophes come about or what happens afterwards (Truesdell 1982). In other words, the analytical and graphic description (or representation) of a number of possible types of discontinuity does not constitute an explanatory theory of morphogenesis. This is not to say that phenomenological theories are illegitimate. What makes this stance objectionable is the claim that mechanisms are irrelevant, so that a mathematical description is a sufficient theory of morphogenesis. The disregard of stuff (i.e., composition), mechanism, and history, is the main flaw in any (consistent) structuralist approach.

A similar warning must be raised against the attempt to explain morphogenesis in terms of morphogenetic fields—an idea borrowed from field physics. Initially this was only a half-baked speculation, because no field equations were written and no attempt was made to imagine, let alone perform, measurements of the force that such a field would exert on the atoms, molecules, and organelles that are moved around during a morphogenetic process. Goodwin and Trainor (1980) introduced considerable precision by proposing precise equations purporting to represent a "field description of the cleavage process in embryogenesis". They are careful to call theirs a 'field description' rather than a 'field theory', for it contains no field equations—but, having said that, they do write about primary and secondary fields. They would have been even more precise if they had called it 'a field-like description', or 'a physical analog, or even 'a geometrical description', of the cleavage planes.

Goodwin and Trainor admit that their work was suggested by the analogy between the electron distribution (or probability clouds) of a hydrogen atom in various states. However, they actually focus on the nodal lines that appear on the surface of an elastic spherical shell as it vibrates. They take such lines as the analogs of the furrowing process that precedes cleavage. But the nodal lines happen to be more numerous than the furrowing lines. To remove this degeneracy, they assume that a "secondary polar field weaker than the primary field" removes the degeneracy—a sort of epicycle whose only job is to save the initial cycle hypothesis. And they account for the differences in cleavage patterns among species by introducing a third undefined force, namely the "animal-vegetal 'gradient' which acts as a uniform perturbing force on the mitotic spindles". (Note that the animal-vegetal gradient actually exists in most eggs, but the force exerted by the gradient remains undefined in this field-theoretic description.) Consequently, their early approach is still strictly geometric: it does not consider process, let alone any real forces that may drive the process. Moreover, the model does not contain any biological variables other than cell number. (However, Goodwin 1984 lists several candidates for such biological variables.) This model is thus in the same category as Thom's topological account of morphogenesis.

A genuine field-theoretic account of morphogenesis would include, at the very least, (a) field equations relating the field intensities to the densities of the field sources, (b) one or more formulas for the force(s) exerted by the field upon the material being organized, and (c) equations of motion for the particles (molecules, microfibrils, organelles, etc.) involved in morphogenesis. Whereas Goodwin and Trainor do not attempt to do anything of the sort in their 1980 paper, they do propose dynamical equations involving genuine forces in a later paper (1985). So do Brière and Goodwin (1988). Both papers study the formation of the whorl of the acellular alga *Acetabularia* (the mermaid's cap) and propose a mechanism concerning cell wall deformation. This advance was inspired by an earlier dynamical model of morphogenesis by Odell et al. (1981) on the mechanical basis of the folding and invagination of embryonic epithelia. By focusing on the elastic and viscous features of the cytoskeleton, and neglecting the extremely small accelerations, these authors succeeded in proving that the contraction of a single cell propagates to the adjacent cells and generates an invagination in the epithelium.

Such models are philosophically interesting for two reasons. First, because they show that purely mechanical considerations can lead rather far. Second, because they help to dispense with preformationist myths: "[...] once triggered, the morphogenetic process of invagination proceeds on its own, directed solely by the global balance of mechanical forces generated locally by each cell, and with no requirement for individually preprogrammed sequences of patterns of cell shape change" (Odell et al. 1981, p. 450). Moreover, they soften the declared structuralist outlook, as they refer to actual biological (developmental) mechanisms (forces and processes) rather than purely geometrical relations.

However, assuming optimistically that a field-theoretic account of morphogenesis is possible in certain cases and that the postulated field sources and forces will eventually be accessible to measurement, the problem remains that a theory of morphogenetic fields cannot provide a *systemic* explanation of morphogenesis. For example, in the case of tip formation in (the members of) *Acetabularia*, a field-theory of morphogenesis cannot account for the fact that the form of the apical whorl is species-specific. That is, when the nucleus of a specimen of *A. acetabulum* (formerly *A. mediterranea*) is transplanted into the stalk of a decapitated specimen of *A. crenulata* (or vice versa) the regenerated whorl will assume the form of the tip of the donor species (Grant 1978). This example shows that a theory of morphogenesis must not only exhibit formative forces but must also take the composition (not to mention the environment) of the developing organisms into account. Again, matter does matter. To claim that only form or structure matters is therefore a (macro)reductionist approach. Hence, to adopt a structuralist version of macroreductionism in place of neopreformationist microreductionism does not solve the problem of development. On the other hand, molecular models of morphogenesis in terms of morphogens are not systemic explanations of development either, because they disregard the organism as a developing whole.

To conclude, we submit that developmental structuralism would have much more appeal if it reconceptualized some of its ideas in systemic, rather than structuralist, terms. A first step would consist in acknowledging the importance of genes in development. This has recently been done by Goodwin (1990), who says that the role of genes in development would consist in setting the parameter values of the generative processes. A second important step would consist in reformulating the idealistic notion of law in terms of real biological laws. To this end, it is necessary to understand that development is lawful not because it is guided by immaterial laws of form but because the members of a biological taxon (at any level of inclusion) are nomologically equivalent. These (substantial) $laws_1$ constrain the development and evolution of the corresponding organisms. So, although we are very sympathetic to the developmental structuralists' attempt at finding developmental laws and mechanisms, we question their underlying philosophical outlook, for it involves structuralism, holism, (crypto)idealism, and subjectivism. Whatever heuristic and inspirational value this philosophical outlook may have had to the structuralists, we suggest dropping it soon and replacing it with a materialist and systemist one.

8.2.4.2 Developmental Constructionism

Both developmental structuralism and developmental constructionism are reactions to genetic reductionism in developmental and evolutionary biology. Thus, their basic common goal is to account for development and evolution from a non-gene--centered, i.e., a consistently epigenetic, perspective. Whereas developmental structuralism is a product of developmental biologists and mathematicians, the developmental systems approach is mainly a brain child of developmental psychologists. It is inspired by the nature-nurture controversy in psychology and the behavioral sciences. (See, e.g., Lehrman 1970; Lewontin 1983a, b; Oyama 1985, 1988; Johnston and Gottlieb 1990; Gottlieb 1991; Gray 1992; Griffiths and Gray 1994. For a critical review see van der Weele 1993; Sterelny 1995; Sterelny et al. 1996.)

The central topic of the developmental systems approach is a systemic view of development. In this view, no causal or determinative priority in development can be assigned either to the genes, i.e., to internal factors, or to the environment, i.e., to external factors. It is emphasized that "phenotypes" are not *transmitted* from one generation to the next, e.g., in coded form in the genetic material, but that they are *constructed* anew in each generation through organism-environment interactions during development. (Hence Gray 1992 proposed the name *developmental constructionism*.) Thus, the control of development does not reside in the genome alone. Rather, in a self-regulating multilevel system such as an organism, control is exerted by all the components of the gene-in-a-cell-in-an-organism-in-an-environment system.

As a consequence, the traditional inherited/genetic and acquired/environmental dichotomy of characters is rejected. Instead, it is emphasized that *all* the features of a developing organism are both genetically and environmentally determined, because both are necessary but none is sufficient for a developmental process to occur. The fact that in some cases one of the two necessary parameters is invariant, while in some cases the other is, does not imply that the variant factor, though making the difference in the given circumstance, is sufficient. For example, the differences in the development of a duck's egg and a hen's egg in the same incubator, i.e., in a standardized environment, will be attributed to internal factors rather than environmental differences. Still, the developing features are not independent of these environmental conditions. By contrast, incubating one of two eggs of the same crocodile below a certain temperature threshold and the other above it will result in a male and a female animal (or vice versa, depending on the species). Yet, such cases of environmental sex determination do not imply that the genetic constitution of the embryos is irrelevant. It comes as no surprise, therefore, that the inherited-acquired dichotomy has been doubted by biologists for quite a while: "...it is fair to say that [most characters] are never strictly inherited nor strictly acquired, but are both or neither, depending on the point of view" (Simpson 1953, p. 61).

The constructionist view of development also calls for an expanded notion of inheritance (Oyama 1985). If phenotypic traits are not in any way transmitted but constructed anew during development, then the question arises what, if anything, do organisms inherit. Of course, organisms inherit genes, but they also inherit cytoplasmic factors (actually the entire initial organization of the cell) and, after all, a certain environment (Gray 1992). To regard a certain habitat as inherited is at first sight shocking. Yet biologists know very well that, for instance, many animals carefully deposit their eggs only in specific sites. The latter may be certain host plants or animals, or places characterized by a certain moisture and temperature. In other words, inheriting the wrong environment may turn out to be as fatal as inheriting a lethal mutation. (Incidentally, Woodger 1929, p. 385f., was even more radical, suggesting dropping the notion of heredity altogether; see also his 1952, p. 181.)

These considerations are summarized in Oyama's definition of "developmental system":

> The developmental system [...] encompasses not just genomes with cellular structures and processes, but intra- and interorganismic relations, including relations with members of other species and interactions with the inanimate surround as well. (Oyama 1985, p. 123)

Thus, it is claimed that what develops is not the organism, but the whole developmental system (Gray 1992). Accordingly, the usual organism-environment distinction, in which the organism either adapts to a given environment or perishes, is called into question. Rather, it is emphasized that organisms and environment "construct" each other (Lewontin 1983a, b; Griffiths and Gray 1994). A similar view, by the way, had already been expressed by Woodger (1929), who held that "...the characters of the organisms are really characters of the organism *and* its environment" (p. 346; italics in the original).

Although there is much truth in developmental constructionism, several critical issues must be raised. The first is a fundamental problem with the very notion of a developmental system. If what develops were not the organism but the developmental system, i.e., the organism-environment system, we would have to give up systemism in favor of holism. To clarify this concern we have to recall our view of a system as being analyzable into its composition, environment, and structure, or CES for short (see Sect. 1.7.2). A CES analysis of a system comprises a characterization of the system's environment, i.e., the collection of items other than the system itself which may act upon the system and upon which the system may act. That is, this concept of environment is a relational one, so that there are as many environments as there are systems—disregarding the universe as a whole, which has no environment. Finally, the external structure of the system is the collection of relations between the system and all items in its environment as defined precedingly. In sum, according to this definition, every system is characterized by its own specific environment, which is delimited by the external structure of the system.

If we characterize an organism in terms of the CES triple, the organism's environment as well as its internal and external structure are taken into account—hence all the aspects mentioned in Oyama's above-quoted definition of a developmental system. Therefore, we do not need to delimit a developmental system in addition to the organism. If we had to characterize a developmental system in terms of the CES triple, we would have to determine the environment of the developmental system, i.e., the environment of the organism-environment system. After all, not only the organism, but also the developmental system as a whole, does not exist in a vacuum. So why not say that this developmental supersystem is *the* developing entity? But then this developmental supersystem could hardly be said to develop on its own. In a CES analysis of this supersystem we had to take the environment of this developmental supersystem into account, that is, we had to look for the developmental super-supersystem of the developmental supersystem. Again, this developmental super-supersystem would not exist in a vacuum either. Expanding further and further nested developmental systems would lead us directly to holism, that is, to the assumption that the entire universe is *the* developmental system. (Incidentally, already Oyama 1985, realizing the vagueness of the notion of a developmental system, was worried by this kind of objection, but did not dispel it.) This holistic view, however, would leave developmental biology without explanatory power, because we would have to know the whole universe in order to account for the development of a single organism. (See Sect. 1.7.1, as well as Sterelny et al. 1996.)

In view of these difficulties, we suggest dropping the notion of a developmental system altogether and retaining the organism as the developing entity. After all, the characterization of a developing organism in terms of a CES triple comprises all the items relevant to account for development. Furthermore, we need not exile the organism again from developmental biology as has already been accomplished by genetic reductionism. However, while genetic reductionism is a form of microreductionism, the replacement of organisms by developmental systems, as advocated, for instance, by Gray (1992), who regards organisms as just one developmental resource among others, is an instance of macroreductionism.

Another advantage of dropping the notion of a developmental system is that it spares us any musings about the ontology of developmental systems. An attempt to clarify the ontology of a developmental system has recently been carried out by Griffiths and Gray (1994). They claim that the central entity in developmental constructionism is the developmental process or life cycle:

> The developmental process is a series of interactions with developmental resources which exhibits a suitably stable recurrence in the lineage. (p. 292)

> The developmental process or life cycle is a series of developmental events which forms a unit of repetition in a lineage. Each life cycle is initiated by a period in which the functional structures characteristic of the lineage must be reconstructed from relatively simple resources. (p. 304)

(Incidentally, a somewhat similar position was advocated in a forgotten book by Lenartowicz 1975.) Clearly, all this is a reification of processes, histories, and lineages. The first of the quoted sentences does not even designate a proposition, because no objects of the alleged interactions are specified: they are interactions in themselves. (In other words, "interacts" is an at least binary predicate: x interacts with y.) Second, interaction is conflated with development. Thus, the concept of development is reduced to a merely ecological notion. The main meaning of "development", namely the emergence of qualitatively novel features of organisms, is missing (see Definition 8.1). In short, Griffiths' and Gray's treatment of development is untenable.

Another problem with the notion of a developmental system lies in the difficulty of distinguishing evolved traits from individual (or idiosyncratic) ones. Griffiths and Gray (1994) give the following example: whereas the thumbs on their hands are evolved traits, the scar that one of them has on his left hand is an individual trait. Evolved traits would be part of the developmental system, while individual (idiosyncratic) traits would not be. Since they believe that

> the fact that a developmental outcome has an evolutionary history is not an intrinsic property that can be determined by inspection of the outcome, or of the process that constructs it (p. 287),

they have to contrive a different account of the distinction. They suggest doing so by distinguishing traits that have evolutionary explanations from those which do not. In other words, they suggest regarding a certain feature as evolved because "it fits into a particular pattern of explanation" (p. 287).

Unfortunately, this proposal rests on a category mistake. More precisely, it is a confusion of ontological and epistemological categories, so much so that its (inadvertent) consequence is that evolution is a product of evolutionary biology—an idea that makes sense only in ontological constructivism, itself a version of subjective idealism. However, it is a matter of ontology whether a feature is the result of evolution. Therefore, such a feature must be characterized in ontological, not epistemological, terms. Whether we can come up with an evolutionary explanation of this feature (or not), can have no bearing on the ontological status of that character. Thus, we must be able to *define* the notion of an evolved character versus an idiosyncratic one without recourse to adaptive-historical just-so stories, i.e., what often figures as "evolutionary explanation". To have such an explanation is undoubtedly interesting and important but it will certainly not change the ancestry of the organisms, and thereby the characters, in question.

We submit that it is possible to draw a distinction between evolved and *some*—though not all—idiosyncratic traits when we examine the origin of the characters in question and give up the notion of a developmental system, i.e., if we presuppose the traditional distinction between the developing organism and its environment. By so doing we can distinguish between features which are the results of developmental construction and those, such as mutilations, which are *directly produced* by some environmental item(s). (Recall Sect. 8.1, as well as our conception of causation as energy transfer from Sect. 1.9.2.) In the latter case, an environmental item *acts on* an organism: it directly produces a certain feature, such as a wound. This is not a developmental process. Consequently, there *are acquired* characters, even though they appear to consist merely in mutilations rather than in valuable new features. By contrast, in all developmental processes environmental items provide either conditions (e.g., materials, temperature, and humidity) for development, or they trigger the onset of alternative developmental pathways (e.g., temperature or certain chemicals). (See also Lenartowicz 1975.) Note that only the open wound is an acquired character. The healing of the wound and the formation of a scar must be regarded as developmental processes, for they involve organismic functions, which were triggered by the injury. Still, *pace* Griffiths and Gray (1994), in this case, the inspection of the processes that construct the feature allows us to determine that the trait is a nonevolved one.

We emphasize that our developed-acquired dichotomy is not the same as the innate-acquired and the genetic environmental dichotomies: it only introduces a distinction between traits which are the outcome of developmental processes in an organism, and those which are not, i.e., which are directly produced by some item(s) in the organism's environment. Still, there is no development in a vacuum: we can go on maintaining that evolved features are not transmitted but constructed anew during the organism's epigenesis in a certain environment. Thus, except for the concept of a developmental system, several basic tenets of developmental constructionism remain unscathed.

However, our developed-acquired dichotomy is still insufficient to distinguish evolved traits from developed-individual qualitative novelties, such as extra fingers or any other deviant features, whether functional or dysfunctional. The same holds for nondeviant (evolved) traits, such as that of having *a* fingerprint pattern versus the individual feature of having a *particular* pattern. After all, all these features are constructed anew during an organism's development. How developed-evolved traits may be *defined* will be shown in Section 8.2.4.3. Yet we can already say how they may be *recognized*, namely by studying their comparative morphology and embryology. In other words, we need to determine whether a given feature is taxon-specific, which allows us to sort out (statistical) abnormalities. Although a plausible hypothesis about the feature's function and role as well as its possible adaptive history may help us understand it, our distinction between evolved and nonevolved traits will be based predominantly on the result of our comparative studies.

A final problem with the notion of a developmental system as portrayed by Griffiths and Gray (1994) consists in its implications for evolutionary biology, which would be disastrous if they were true. Calling into question the organism--environment distinction leads them to abandon the traditional notion of adaptedness or fitness. Therefore, they have to conceive of selection as mere differential replication. Worse, they then equate selection with evolution and claim that the unit of evolution is the unit of self-replication. Evolution is thus reduced to the differential replication of developmental processes. The latter notion is finally alleged to provide maximum explanatory power.

The main flaw in this view is that conceiving of selection as differential replication amounts to returning to the operationalist concept of selection, which is responsible for the charge that the concept of natural selection is tautological. Indeed, differential replication has no explanatory power, but is a fact in need of explanation (see Sect. 9.2). The observable fact of the differential replication of organisms in a population can only be explained by means of the transempirical notion of differential adaptedness. Yet this concept is rejected by Griffiths and Gray. To say that fitness values only measure the self-replicating power of the developmental system or process has the same explanatory capacity as the famous *virtus dormitiva* of opium.

To conclude, the attempt of Griffiths and Gray (1994) to render the notion of a developmental system more precise and to examine its ontology must be judged as an utter failure. In fact, their attempt is a disservice to developmental constructionism, as associating it with a severely flawed ontology that can be traced from Whitehead (1929) through Woodger (1929) and Løvtrup (1974) to Hull (1989). Fortunately, this ontology is not necessary for adopting the core tenets of developmental constructionism.

So far the developmental systems approach is no more than that: an approach providing a new general outlook for developmental biology. This outlook is an alternative to the gene-centered reductionism in contemporary developmental biology. It reminds us that, since development does not occur in a vacuum, the characters of an organism cannot be divided into those with a genetic base and into those which are environmentally (including culturally) acquired: all organismic features have both genetic and environmental roots. Compared to developmental structuralism, however, developmental constructionism does not suggest a mechanismic explanation of development (see also van der Weele 1993). Whereas developmental structuralism attempts to explain development with the help of (putative) laws of form and morphogenetic fields, developmental constructionists can only account for the specificity and continuity of development by saying that similar initial conditions will produce similar developmental results. Of course, this is only possible if the underlying developmental processes are lawful. Yet the notion of law is conspicuously absent from developmental constructionism. We shall attempt to remedy this situation in the following.

8.2.4.3 Epigenetic Synthesis

In our view, a sound developmental biology will adopt the systemic outlook of developmental constructionism and will attempt to explain the development of *organisms* by studying all relevant levels, from the molecular to the environmental. (See, e.g., Hall 1992.) And it should adopt the developmental structuralists' conviction that there are laws of development, although these laws cannot be abstract laws of form. Assuming that the laws of development may be partitioned into physical, (bio)chemical, and biological, it is clear that, for example, some physical, in particular, mechanical laws hold for organisms *qua* physical, not biological, objects (see, e.g., Thompson 1917; Odell et al. 1981). However, since organisms happen to be biological entities, any list of physical laws of development, such as laws of morphogenetic fields in the sense of developmental structuralism, will be insufficient to explain development.

Since the hypothesis that biosystems develop lawfully is central, we shall formulate it explicitly. However, we need not formulate a new axiom here because the statement is just a special case of Theorem 1.1. Indeed, the latter entails, by supposing that every biosystem is a thing:

COROLLARY 8.2. Every organism can undergo only lawful transformations.

This corollary should come as no surprise to developmental biologists. After all, what is discussed under the label 'developmental constraints' in evolutionary biology are nothing but laws in the context of development (see Bonner ed. 1982; Levins and Lewontin 1985; Kauffman 1985; Maynard Smith et al. 1985; Gould 1989; Hall 1992; Amundson 1994). Furthermore, that developmental laws restrict the logically possible forms of organisms is evident in the notion of morphospace used by some developmental biologists (e.g., Alberch 1982; Lauder 1982; Gould 1989). The conceivable morphospace of an organism is that subspace of its total state space concerned with its morphological and anatomical form. As with the total state space in general, the actual morphospace of an organism is always a subset of its conceivable morphospace. In tune with our notions of conceivable and nomological state space (Sects. 1.4.2-3), we recommend the use of the expression 'nomological morphospace' for the set of really possible morphologies of organisms. This may reassure those biologists who still regard the concept of a law with suspicion.

Let us now define a restricted concept of inheritance according to a developmental constructionist outlook. In so doing, we shall view the notion of inheritance from the perspective of the descendant rather than that of the ancestor. For, what matters from the perspective of the developing organism is the set of available resources as well as its own composition and structure, i.e., the situation it finds itself in, rather than the items the parent actually bequeaths to its offspring. Accordingly, we propose:

DEFINITION 8.12. Let x represent an organism with composition C, environment E, and structure S, and let t_0 denote the time of its origin. Then all and only the properties (or characters) of x at t_0, i.e., those of $C(x, t_0)$, $E(x, t_0)$, and $S(x, t_0)$, are said to be *inherited*. All properties (or characters) of x that are not present at t_0, or from t_0 on, but are so only at any later time $t_i > t_0$, are said to be *noninherited*.

To determine what counts as t_0 is the task of biologists. We suggest, though, that among sexually reproducing organisms the formation of the zygote may count as the origin of a new organism. More precisely, the moment after the fusion of the maternal and paternal pronuclei may be regarded as the initial state of a new organism. Among parthenogenetically reproducing organisms the final product of oogenesis—the ovum capable of development—may be the candidate. Finally, among asexually reproducing organisms, the moment of detachment of a bud from the mother organism or that of a daughter cell from some mother cell may count as the origin of a new organism.

From the Definitions 8.1 and 8.12 we infer that the noninherited characters of organisms can be grouped into two classes: those which are the result of some developmental process(es) of the organism and those which are generated (i.e., directly produced) by some environmental agent, as explicated in the previous section. We spell this out in:

COROLLARY 8.3. All non-inherited characters of organisms are either the outcome of some developmental event(s) or process(es), or generated (i.e., directly produced) by some environmental agent(s).

The corresponding property classes deserve names of their own. Thus, we suggest:

DEFINITION 8.13. Let p designate a noninherited property (character) of an organism x, and let $s(x, t_0)$ represent the initial (or inherited) state of x at time t_0. Then p is called *inheritance-dependent* if, and only if, p is the outcome of some developmental event(s) or process(es) with initial condition $s(x, t_0)$. Otherwise, p is called an *acquired* property.

Note that inheritance-dependence is not the same as gene-dependence. Since we refer to the initial state of an organism, whether as a zygote or as a newly separated bud from some mother organism, we make room for heritable changes in the extragenomic composition and structure of the cell, as is the case with phenocopies and dauermodifications. Moreover, there are nongenic properties of the zygote, such as its cytoplasmic organization, that determine organizational properties of the adult, such as its dorso-ventral or anterior-posterior axes (Løvtrup 1974; Wolpert 1991).

Naturally, inheritance-dependent characters are of particular interest to geneticists and developmental biologists. Since those features depend on the inherited state of the biosystem in question, they are usually regarded as inherited or transmitted, too. Yet they can be said to be inherited only in the wider sense of *mediately* inherited. To avoid any ambiguity, we opt for the term 'inheritance-dependent'.

Our notions of both inherited and inheritance-dependent traits seem to be similar to Wimsatt's (1986) concept of *generatively entrenched* traits, that is, those features that are followed by some later developing traits depending on them. This conception has been criticized by Burian (1986) on the ground that it does not allow one to distinguish between genetic and nongenetic aspects of the traits in question. Indeed, the same holds for our construal of inherited and inheritance--dependent characters. But as should be obvious from the preceding, the genetic-

-nongenetic distinction is artificial. It is only a matter of analytical focus on, not of ontological priority of, one necessary component of the developing biosystem.

Since the later states of a biosystem depend more or less on its inherited state, any later state can be represented by a mapping of that initial state into some later state in question. This is what geneticists have in mind when they speak of the mapping of the genotype into the phenotype (e.g., Lewontin 1974). With regard to the notions of genotype_H and genotype_D, such mapping may satisfy the geneticist. However, for developmental biologists, such mapping is only useful if it involves Johannsen's broad conception of the genotype as the total constitution (state) of the zygote. Since not all organisms start out as a zygote, and since the genotype thus conceived must be regarded as a subset of the phenotype, modelers in developmental biology can only be interested in the mapping of the initial state of a biosystem into some later state. The history of the organism in question will then be represented by a trajectory in the organismic state space.

As mentioned above, when talking of acquired characters we could only think of injuries or mutilations. Thus, acquired characters are rather "negative" traits and, moreover, likely to be disvaluable or at least neutral traits. However, what depends on the initial state of the organism is, for instance, the regeneration of lost parts or the healing of wounds. That is, acquired features may trigger developmental processes. Still, the resulting feature is an individual trait, not an evolved one.

We are now ready to state the central hypothesis of any epigeneticist developmental biology. This hypothesis is implicit in the previous corollaries and definitions:

COROLLARY 8.4. All inheritance-dependent characters of an organism are *lawfully constructed de novo* in the course of its development.

Note that *de novo* should not be mistaken for *ex nihilo*. Since the noninherited and nonacquired features of organisms depend on its initial state as well as on its environment, they are not built out of nothing. Yet they are constructed *de novo* because they are neither preformed nor somehow pre-existing in the zygote. (See also Horder 1989.) Only the information-Platonist can say that features pre-exist in the form of information or instructions, before they become "embodied", "realized" or "instantiated" during development.

A sound developmental biology combining the constructionist approach with the search for laws would be able to eventually get rid of any teleological residues. To show that this is indeed the case, we need a nonteleological definition of the conspicuous equivalence and equifinality of most developmental processes in conspecific organisms. To begin with, we say that two processes are *equivalent* iff they lead from the same initial to the same final states, where "initial" and "final" are taken to be relative, not absolute, concepts; that is, they are applicable to any earlier and later states of an organism. Note that "final state" is not to be equated with "goal". As for the equivalence of developmental processes, we propose:

DEFINITION 8.14. Let $s(x)$ and $s(y)$ designate some initial states of two organisms (or any subsystems of such) x and y respectively , and let $s'(x)$ and $s'(y)$ designate some later (or final) states of x and y respectively. Further, let $P(x, s)$ and $P(y, s)$ represent the set of *generic* properties (or some relevant subsets of them) of x and y respectively when in (initial or prior) states s, and let $P(x, s')$ and $P(y, s')$ represent the set of generic properties of x and y respectively when in (later) states s'. Finally, let $\langle s(x), s'(x)\rangle$ and $\langle s(y), s'(y)\rangle$ represent developmental events or processes of x and y respectively. Then the developmental events or processes $\langle s(x), s'(x)\rangle$ and $\langle s(y), s'(y)\rangle$ are said to be *equivalent* iff $[P(x, s) = P(y, s)]$ & $[P(x, s') = P(y, s')]$.

Note the following points. First, only generic properties, hence generic states, are involved. This is because two entities cannot be in exactly the same state, i.e., have exactly the same individual properties (they would otherwise be identical, i.e., the same thing), and because we must make room for random disturbances in development. Such random processes that account for some of the variability among conspecific organisms are called *developmental noise* (see Lewontin 1983a, b). Second, equivalence is not the same as identity. That is, the initial and final states of two equivalent processes may be reached through different intermediate states. Third, no explicit reference was made to the times at which x and y are in the states s or s', because neither the initial nor the final states of x and y need be simultaneous. Indeed, x and y need not even be contemporaneous individuals. Hence, all that was needed was the relative notions of antecedent and succedent state.

Fourth, the concept of process equivalence is stronger than the concept of process equifinality. The latter is elucidated by:

DEFINITION 8.15. Two processes are said to be *equifinal* iff the same end state is reached from either the same or from different initial states.

Thus, all equivalent processes are equifinal, but not conversely. As with the concept of process equivalence, when applying this definition to developmental processes we must restrict the notion of state to generic properties.

From Corollary 8.2, Definition 7.2, which defines a class of nomologically equivalent entities as a species, Definition 8.14, and Definition 8.15 as applied to developmental processes we obtain (enthymematically):

COROLLARY 8.5. The development of organisms belonging to the same species is equivalent and, hence, equifinal.

We finally turn to the definition of the concept of an evolved trait. We may say that an inheritance-dependent feature of an organism is an *evolved* trait if that organism descends from ancestors in which equivalent developmental processes produced inheritance-dependent features of the same kind. Whereas in Definition 8.14 process equivalence was defined only in relation to any earlier and later states, the notion of inheritance-dependence now connects the equivalent developmental processes in question to the *absolute* initial states s_0 of the organisms in question, e.g., to the initial states of the zygotes. This provides the connection between the members of the ancestor-descendant lineage. More precisely, we suggest:

> DEFINITION 8.16. An inheritance-dependent feature of kind K of an organism x of taxon T which is the outcome of a lawful developmental process d is said to be an evolved *trait* of x if, and only if, x descends from ancestors in which inheritance-dependent features of kind K were produced by developmental processes equivalent to d ever since the emergence of the first feature of kind K in the first member of taxon T.

It goes without saying that evolved traits are likely to be adaptations. However, "evolved trait" and "adaptation" (according to Definition 4.15) are neither cointensive nor coextensive, because an evolved trait may cease to be an aptation with regard to a changed habitat. In this case, though malaptive, it still is an evolved trait, even if it is selected against.

To conclude, no directive agency is required to explain why a biosystem, such as a zygote, with a given initial composition, environment, and structure—which is what remains from preformationism—undergoes lawful developmental (epigenetic) transformations. Nomic necessity, not immaterial and preformed instructions or programs, accounts for the developmental transformations of organisms.

Here the evolutionary biologist is likely to remind us that, while all this may indeed be so, we must not overlook the fact that an understanding of the developmental transformations of organisms requires not only knowledge of the underlying laws but also of their evolution. After all, the species-specific development of an organism is itself a result of evolution. Although this is true, we shall see in the next chapter that the concept of a developmental process is nevertheless logically prior to that of an evolutionary process in that the former helps define the latter.

9 Evolutionary Theory

If we are to understand evolution, we must remember that it is a process which occurs in populations, not individuals. Individual animals may dig, swim, climb or gallop, and they also develop, but they do not evolve. To attempt an explanation of evolution in terms of the development of individuals is to commit precisely that error of misplaced reductionism of which geneticists are sometimes accused. (Maynard Smith 1983, p. 45)

9.1 Evolution and Speciation

9.1.1 The Ontological Concept of Evolution as Speciation

The concept of evolution is not restricted to biology. In fact, it is an ontological concept, for it applies to various natural processes: we speak, for instance, of cosmic, stellar, chemical, biotic, and cultural evolution. (See also Vollmer 1989.) What is common to all these specific notions of evolution is, first of all, the ontological concept of change. Change, however, can be quantitative or qualitative, whereby every qualitative change is accompanied by some quantitative change, but not conversely. Admitting quantitative change renders one's ontology dynamicist as opposed to static, but it does not make it evolutionary. For an ontology to be *evolutionary*, we must posit in addition that there is also *qualitative change*. But the latter, though necessary, is still not sufficient for evolutionary change proper. For example, a developing organism undergoes qualitative changes, but we do not regard them as evolutionary. For a qualitative change to be considered *evolutionary*, we must finally assume that it consists in the *emergence of things of a new kind* or (ontological) *species*. In other words, a proper concept of evolution involves the concept of speciation in its ontological sense of the coming into being of a thing of a new kind. Thus, the ontological concept of evolution applies to all qualitative changes that result in speciation (recall Definitions 7.2 and 7.5).

However, in many sciences (e.g., from quantum mechanics through thermodynamics to the social sciences) the term 'evolution' is used as a synonym for 'change' *simpliciter*. Here, talk of the evolution of a system *may* involve qualitative change, but it does not *explicitly* require it. The same holds for the general state space models of the evolution of a system, in which the latter is represented by any old trajectory. It comes as no surprise, therefore, that this general notion of evolution which does not explicitly require qualitative change is also used in biology, where it underlies, for instance, the widely accepted definition of the concept of bioevolution in terms of changes of gene or genotype frequencies in populations. Indeed, several elucidations of the concept of evolutionary change of populations in terms of the state space approach, such as those of Lewontin (1974), Bunge (1977a), and Burns (1992), allow for, yet do not explicitly require, qualitative changes to occur. Before we suggest amending this situation, we had better recall briefly the notion of state space from Section 1.4.

The state of any system is representable by a point in an n-dimensional mathematical space the axes of which are the attributes (or state variables) representing the properties of interest of the given system (Fig. 1.1). Changes of state, i.e., events and processes, are representable by a trajectory connecting two or more points in the given state space; and the (total) history of a system is representable by the complete trajectory connecting the initial state of the system to its final state (Fig. 1.2).

In a genetic description of a biopopulation, then, we can regard the occurrence of certain genotypes$_O$ (i.e., organisms with some genotype$_D$ at one or more loci of interest) in a biopopulation as state variables, the values of which are the frequencies of the genotypes$_O$ in the population. (For the different concepts of genotype recall Sect. 8.2.3.3.) Thus the genetic state of a biopopulation at a certain time is representable by a point in the corresponding state space. Considering the deeply entrenched conceptualization of evolution as change in gene or genotype frequencies in populations, it is tempting to think that, in such state space representation, the movement of the point through the state space represents the evolution of the biopopulation (Lewontin 1974; Bunge 1977a; Burns 1992). Yet this, though partially true, is insufficient: not any old history of a thing ought to be regarded as an evolutionary process.

To conceptualize a genuine evolutionary change by means of a state space model, we need to recall that the state variables of the system in question do not take values in its entire conceivable state space, because the laws of the system only allow for certain (lawful) states and thus certain (lawful) changes (trajectories). In other words, the laws of the system "determine" the latter's nomological state space. (The latter can be further restricted by external constraints.) Thus, every change is actually represented by a trajectory within the nomological state space of the system. For an evolutionary change to occur we must assume that the system in question undergoes a change in kind. In a state space model, such change in kind of a system amounts to the acquisition of new state variables and therefore of a new nomological state space. Figuratively speaking, we could say that the evolutionarily changing thing jumps into a different nomological state space. (See Fig. 1.3. Note that from a mathematical point of view we need not speak of a *new* state space, since all the axes of the state space may be conceived of as being present right from the beginning, so that qualitative change is represented by the movement of the trajectory in a different region of the total state space. After all, mathematics is Platonic and static.) As all things sharing the same nomological state space are said to belong in the same (ontological) species, the acquisition of a new nomological state space represents (ontological) speciation.

To conclude, evolutionary theory is and should indeed be about the *origin of species* or, more precisely, about the origin of biotic entities belonging to a new species. Yet which are those evolutionarily changing, i.e., speciating, entities? In other words, which are the units of evolution: organisms, biopopulations, communities, ecosystems, or perhaps the entire biosphere or ecosphere, as some authors have suggested (e.g., Dunbar 1972; Walker 1985)? (Obviously, in our ontology it would be absurd to regard species as speciating entities.) The answer(s) to this question will reveal whether the term 'evolution' refers to a single individual process, or whether it refers to a class of equivalent processes occurring in separate entities of the same kind, or whether it refers perhaps just to a collection of completely different individual processes occurring in separate entities of different kinds. These, then, are the main ontological topics in evolutionary biology that will be tackled next.

9.1.2 Speciation in Biology

Although one should naturally expect that a theory of biotic evolution is about living systems, the standard view, i.e., the view of the Synthetic Theory, has it that it is not organisms but biopopulations or species (as alleged individuals) that evolve. Sometimes it is also said that, whereas the organism is the unit of selection, the biopopulation is the unit of evolution. According to this standard view, it is actually nonliving entities that are the supposed qualitatively changing objects in biotic evolution.

On the other hand, the evolutionary novelties or innovations to which we refer in evolutionary biology, paleontology, and systematics are *properties of organisms*, not of populations, species, or higher taxa—a fact that is very important to bear in mind. For example, the possession of a notochord, a neurenteric canal, and branchial arches is neither a property of any vertebrate biopopulation nor a property of the taxon Vertebrata. To agree with the latter point, it does not even matter whether one regards taxa as either individuals or classes, because neither a taxon as an (alleged) "historical entity" nor a taxon as a class of organisms has branchial arches, notochords, or any other specific organism-level properties. (Note that by 'organism-level properties' or 'organismal properties' we mean the specific properties of entities belonging to the organism biolevel: see Sect. 5.3. We do not refer to the generic properties that are shared by systems on different levels, such as composition and structure, in particular coordination or functional integration.)

Our emphasis on the proper distinction between specific organism-level properties and properties of higher-level entities is not just philosophical pedantry.

Rather, it is a necessity considering the orthodox view on evolution, as expressed, for instance, in the above quotation by Maynard Smith. It is also necessary with regard to the fact that some philosophers have explicitly attributed organism-level properties to populations in order to defend the importance of population thinking in biology. For example, Mary Williams (1981) has claimed that the uterus would be a characteristic of a population, since it is useless for the organism itself but valuable to the population. Yet, even if the uterus were valuable to the population rather than the organism, it would be an organismal character: there is no such thing as a populational uterus that breeds, and gives birth to, entire populations as superorganismic wholes. Our emphasis is further necessitated by the fact that we also encounter plenty of (apparently) inadvertent level-mixing of properties in biology. An example of such confusion is the following characterization of the concept of evolution:

> Organic evolution is a series of partial or complete and irreversible transformations of the genetic composition of populations, based principally upon altered interactions with their environment. It consists chiefly of adaptive radiations into new environments, adjustments to environmental changes that take place in a particular habitat, and the origin of new ways for exploiting existing habitats. These adaptive changes occasionally give rise to greater complexity of developmental pattern, of physiological reactions, and of interactions between populations and their environment. (Dobzhansky et al. 1977, p. 8).

The only populational property referred to in this characterization is genetic composition. All the other intrinsic and relational properties referred to are organismal, not populational, properties. This only goes to show how easily habits of speech may obfuscate conceptual clarity and proper theorizing.

According to this standard view, then, we must assume that, if what evolve are biopopulations ("species"), the population is the entity that undergoes a qualitative change as a whole and that consequently possesses qualitative novelties. But what (emergent) qualitative novelties can populations have? One answer might be that such qualitative novelty of a biopopulation is its being reproductively (i.e., genetically) isolated from other populations, in particular its mother- or sister population. This answer will of course not do, because the reproduction referred to is sexual reproduction, which is either an organismal (parthenogenesis) or mating-pair property (gonochorism), not a biopopulational one. Although it might be argued that a population can in some sense reproduce as well, e.g., by splitting off propagule populations, this is certainly not what is referred to here by the notion of reproductive isolation. Moreover, genetic and reproductive isolation would be negative relational properties. Yet since there are no negative properties—*pace* Bock (1986, p. 33) and others there is no such thing as the "property of no gene flow"—, reproductive isolation is a nonrelation, hence no property at all. (For biological criticisms of the concept of reproductive isolation see, e.g., Paterson 1985; Cracraft 1989.)

The actual properties involved here are certain genetic, physiological, morphological, and—among animals—ethological *properties of organisms* that allow them to mate and reproduce successfully only with organisms sharing equivalent properties. (One of the few to point this out explicitly is Bock 1986, although his view on species seems inconsistent.) These properties, in turn, enable those genetically and reproductively equivalent organisms, i.e., the organisms of the same species, to take part in a reproductive community or biopopulation. So the reproductive relations of organisms constitute the cohesion of the biopopulation. That a population is thus cohesive is *indicated* by observing whether or not the organisms of different populations interbreed. In other words, reproductive isolation is a *symptom* of populational cohesion.

From this, we must conclude that organisms do not belong to the same species because they are parts of some biopopulation, but that they can only be parts of a biopopulation because they belong to the same species. The concept of a species as a biological kind, e.g., defined by a certain complex of organismal properties allowing for common reproduction, logically precedes the concept of a biopopulation as a reproductive community (Bunge 1979a; Mahner 1994a). But is it not just the other way round? Do the organisms in question not possess certain common properties only because they are part of the same biopopulation and, particularly, because they descend from organisms that had been part of the same population before? Although this seems to be a problem of the hen-egg type, it can easily be resolved evolutionarily: sexual reproduction is itself an evolutionary innovation, which cannot have evolved within a reproductive community, i.e., by the evolution of a biopopulation. Thus, the property of sexual reproduction or, for that matter, the disposition thereto not only logically but also historically precedes the formation of biopopulations. (Recall also Sect. 4.5 and Sect. 7.3.)

If, at a closer look, even the "property" of reproductive isolation boils down to a complex of properties of organisms and mating-pairs, then there are hardly any significant properties left which could be regarded as evolutionary novelties of populations as reproductive communities. (One such property could be social structure; but even so it presupposes the existence of social behavior, i.e., a relational property of individual animals, to begin with.) Indeed, *evolutionary innovations are properties of organisms*. So every qualitative novelty must first occur in a single organism. Thus, Cracraft (1990) defines *prospective innovations* as "singular phenotypic changes that arise in individual organisms within a population as a result of a modification in one or more ontogenetic pathways" (p. 27). If we read 'phenotype' in the sense of "phenotype$_T$" as elucidated in Section 8.2.3.3, i.e., in the sense of the set of types of traits of an organism, and if we recall that an organism's genotype$_{H,D}$ is a proper subset of its phenotype$_T$, then any qualitative change in an organism, whether at the molecular or at the organ level, is a phenotypic change, hence a developmental event or process. (See also Johnston and Gottlieb 1990.) So Cracraft's definition can be seen as being consistent with our Postulate 1.9, according to which all processes of development and evolution are accompanied by the emergence or else submergence of generic properties.

However, Cracraft's definition is insufficient to distinguish mere developmental from evolutionary change. During its development an organism undergoes qualitative changes, but these changes occur "within its nomological state space", i.e., the developing organism does not change in kind or species. (In principle, one could, of course, conceptualize developmental changes as changes in kind, but it would be theoretically and practically cumbersome to regard every developmental stage as a thing of a new kind.) To make room for changes in kind, we must say that the phenotypic change in question results in a character of a kind that has not yet been present in the organisms of this species before. In other words, some developmental process transforms the organism in question into an organism of a new kind or, more precisely, of a new (ontological) species. This developmental event or process, then, is an event or process of speciation in the ontological sense. (Caution: biologists would speak only of variants or varieties here, for they are not so much interested in ontological as in taxonomic species.) These insights allow us to define the following notion of organismal speciation:

DEFINITION 9.1. Let s represent the state of an organism b of species B at some time t, and s' the state of b at a time t', where $t' > t$. Finally, let B' designate a species different from B. Then the event $\langle s, s' \rangle$ is an *organismal evolutionary event* (or *process*), or an event (or process) of *organismal speciation* if, and only if,
(i) $\langle s, s' \rangle$ is a developmental event (or process), and
(ii) $b \in B$ at time t & $b \in B'$ at time t'.

Note that, if we symbolize a speciation event by '$A \rightarrow B$', this notation does not imply that species A (as a kind) changes into species B. Neither is it implied that species A has become extinct after giving rise to species B. Rather, what this notation signifies is the fact that some thing of species A either turns into or else produces a thing of species B. A speciation event can also be of the form $A + B \rightarrow C$, which means that a thing of kind A and a thing of kind B assemble together, forming a thing of (a new) kind C. Examples of such speciations by assembly are hybridization and endosymbiosis, which is assumed to have occurred in the evolution of the eukaryotic cell. Finally, speciation might be of the form $A \rightarrow B + C$, which signifies that a thing of species A gives rise (either simultaneously or subsequently) to things of species B and C. Again, this notation implies neither that the ancestral species has become extinct, nor that the formation of the (members of the) two new species occurred simultaneously (see Fig. 7.6).

Note further that so far we have been concerned only with ontological speciation and thus with a very general notion of qualitative novelty. Accordingly, we have been concerned with neither whether the qualitative novelties in question are functional or adaptive, nor whether the new organism in question is viable for more than a short period of time—in principle, it could be a totally hopeless monster. Moreover, our ontological notion of qualitative novelty does not coincide with the biological notion of "key innovations", such as "the" notochord, "the" feather, and so on, for these complex features can hardly have been evolved in a single step, but will depend on several coordinated changes (Mayr 1960; Thomson 1988). What is important, however, is the idea that new qualities come into existence by developmental processes, which are qualitative changes in organisms, not in populations—a view which is being increasingly emphasized by biologists (e.g., Ho and Saunders 1979; Gould 1980; Thomson 1988; Cracraft 1989; Horder 1989; Johnston and Gottlieb 1990; Müller 1990; Raff et al. 1990; Müller and Wagner 1991; Atkinson 1992; Hall 1992; Gilbert 1994).

From an ontological point of view it is also important to note that what constitutes speciation in this case is an organismal change. Hence, *organisms are the speciating entities*. Biologists will readily acknowledge such organismal speciation in certain cases, such as allopolyploid hybrids among plants. But they might feel uneasy about Definition 9.1, because, in most cases, a new variant does not constitute a new species in the biological and taxonomic sense unless the variant spreads and becomes fixed in the population, and unless it is accompanied by, or consists in, reproductive "isolation". Thus, Cracraft (1990) distinguishes prospective innovations from *evolutionary innovations*, which he defines as "singular phenotypic changes that, subsequent to arising in individual organisms, spread through a population and become fixed, thus characterizing that population as a new differentiated evolutionary taxon" (p. 27).

Indeed, the new qualities and the species in the ontological sense with which we have been so far concerned need not be significant from the point of view of evolutionary biology. Yet what matters for the ontologist is the qualitative change involved in speciation. The fact that biologists acknowledge only a subset of all qualitative novelties as evolutionary novelties and only a subset of ontological species as biological species is beyond the ontologist's concern. But it is his or her task to point out that, even if the new variant becomes fixed in the population, it is still not the population as a cohesive whole that has become the new evolutionary taxon. This is because, although the biopopulation as a whole derivatively undergoes a qualitative change as well when its components do, the novelties in question are organismal, not populational, properties. Moreover, as we took pains to explain in Chapter 7, biological taxa are classes (of organisms), not cohesive wholes. Therefore, the concept of species employed is a concept of species of organisms, not a concept of species of populations. Furthermore, as we stated before, the general notion of a species as a class is logically prior to the notion of a biopopulation. So the concept of populational evolution usually employed by biologists presupposes the concept of organismal speciation. This becomes apparent in the following definition, which is an attempt to come closer to the concerns of biologists:

DEFINITION 9.2. Let B and B' designate two different species of organisms. Further, let s represent the state of a biopopulation p with organismal composition $C_O(p) \subseteq B$ at a time t, s' the state of p at t', and s'' the state of p at t'', where $t'' > t' > t$. Then the process $\langle s, s', s'' \rangle$ may be said to be a process of *populational evolution* if, and only if,

(i) there is at least one organism $x \in C_O(p, t) \subseteq B$ undergoing a process of organismal speciation during the interval $[t, t']$, such that $C_O(p, t') \subseteq B \cup B'$, and

(ii) there is a process of selection during the period τ, where $\tau = [t', t'']$, among all $x \in C_O(p, \tau)$, such that $C_O(p, t'') \subseteq B'$.

Since the composition of the biopopulation p has changed during the period $[t, t'']$, we are entitled to regard the population as a qualitatively new system too, that is, one having undergone a change in kind. The same holds for the community and the ecosystem of which the population is part. So one is justified in speaking of the evolution and coevolution of biopopulations and communities, even though this evolution only derives from organismal evolution. Yet we must bear in mind that it is organisms that are the smallest units of biological speciation and thus the units of biological classification, not biopopulations or any other supraorganismal systems. Therefore, it would be wrong to believe that the population is *the* speciating entity and thus *the* new evolutionary taxon. Although we indeed have a new kind of population, no one—especially not the systematist—is interested in kinds of populations. What matters are kinds of organisms, whether or not the latter compose a population. This is the reason that the existence of biopopulations is of only subordinate interest to systematics, however important they may be for evolution, in particular for the processes of selection and adaptation. Of course, the ecologist is also interested in kinds of supraorganismic entities, such as tundra ecosystems, tropical rainforest ecosystems, and so on. But none of these occur in biological systematics, whose goal is a classification of organisms, not populations, communities, or ecosystems.

It seems that, if biologists think this matter through, they are bound to arrive at quite similar conclusions. Thus, again, Cracraft (1989, p. 47):

> ...if we consider populations to be the individuals of a theory of differentiation (speciation), and examine the content of this theory from the standpoint of the part/whole relationships of populations, it is apparent that not even populations, as discrete cohesive entities, themselves speciate. Instead change is generated within individuals, at the levels of the genome and developmental pathways. Sometimes these changes become fixed in populations as diagnostic character variation through processes having individual organisms as their entities. If this view of evolutionary change is correct, then populations and species do not function as entities in speciation theory, rather they are the effects of lower-level processes.

Still, not all biologists might be persuaded yet. After all, Definition 9.2 assumes that a biopopulation is at some time composed of organisms belonging to two different species. Moreover, it seems to presuppose that speciation consists of saltational events rather than small cumulative changes that eventually add up to a more conspicuous qualitative difference (Mayr 1960). A few further remarks may therefore be in order.

The two preceding definitions are compatible with both "gradual" and saltational evolution. As for gradual change, we put the term 'gradual' in quotation marks, because only quantitative change, such as motion, can actually be gradual or continuous. Every qualitative change, i.e., the acquisition or the loss of a property, is discontinuous or saltational (recall Fig. 1.3). Thus, when we speak of gradual evolution we can only mean an overall change through a sequence of more or less small qualitative jumps. For example, in the standard view, the evolution of a biopopulation by a series of allelic substitutions is regarded as gradual, although every allelic substitution is a qualitative jump. The same holds, by the way, for the (re)combination of alleles and their emergent effects, if any, and, *a fortiori*, it holds for macromutations, such as homeotic mutations, as well as for any changes in developmental pathways, which may indeed be rather drastic yet surprisingly functional due to the plasticity of developmental mechanisms (Müller 1990).

Even in a gradual model of evolution there must be a first individual in the population in which a further mutation or perhaps a combination of alleles inherited from its parents has a more or less conspicuous emergent effect (Mayr 1994). For instance, it may be the last step necessary to bring about reproductive "isolation" with regard to individuals outside its own population. Note, however, that this qualitative change is the speciation event, not—*pace* Mayr and his followers—the test of the "isolating" mechanism in sympatry.

Of course, the biologist will not speak of a new (biological) species yet, but will wait until the new variant has spread through the population. After all, he or she is not interested in the occasional monster but only in novelties that last for a significant period on the evolutionary time scale. To this end, the novel organisms in question—traditionally relegated to the status of "raw material for evolution"—must undergo natural selection and thus most likely a process of adaptation. Finally, if the organisms in question form lineages, biopopulations, and lineages of biopopulations, so that they "persist" for an (evolutionary) period without undergoing further speciations, biologists will be ready to acknowledge a new biological (or taxonomic) species.

Let us, for the sake of clarity, restate what happens in terms of the state space approach (see Fig. 7.3). To begin with, we have a number of organisms which share the same nomological state space, that is, which belong to the same species. These organisms may be organized in a single biopopulation, in several populations (which, in turn, may be part of different communities), or in none. Recall that being or not being part of a biopopulation is, in principle, of no relevance to membership in a species or higher taxon. Imagine now an organism in one population undergoing a qualitative change that "throws" it into a different nomological state space. Though different, this new nomological state space will have a significant overlap with the nomological state space of the other organisms in the population. Otherwise, no sexual reproduction with any other organism in the population would be possible. (Recall from Sects. 7.2.1.7.8 that biological kinds come in degrees.)

Adding further and further qualitative changes to the organisms amounts to further and further changes of their nomological state space. Eventually, after the new "variant" has spread through the entire population, all the organisms in the population will share—at least for some time being—a final and stable nomological state space. (We may disregard balanced polymorphism here.) Thus, from an ontological point of view, we will have an ancestral species, a number of transient species with few members, and a "final" species to which all the organisms in the population eventually belong. To visualize this evolutionary process, imagine a transformation of the two-dimensional nomological state space depicted in Fig. 7.3a into a higher-dimensional one by adding new axes. (See also Fig. 1.3.)

In sum, biologists just disregard the initial and intermediate (ontological) species, as being evolutionarily inconspicuous, and focus on the "good" and "stable" ancestral and descendant species, to which are assigned the status of a taxonomic species. And since all this occurs in biopopulations, at least among sexually reproducing organisms, it is not surprising that they confuse changes of organisms, i.e., speciation proper, with the resulting changes of populations. Thus, processes in populations, which are actually subordinate to speciation proper, are seen as the main speciation events. To be sure, population thinking is necessary to understand the concept of natural selection. But just as developmental biology cannot fully explain bioevolution (see Maynard Smith 1983), neither can population genetics nor the theory of selection. However, population thinking or, if preferred, the *variational model* of evolution (Lewontin 1983a) turned out to be rather misleading with regard to speciation, as well as related issues such as the concept of species and biological taxonomy in general. In particular, we submit that the population-theoretic approach is responsible for one of the most misleading ideas in evolutionary biology, namely the belief that species must be mutable for evolution to occur. Yet, as we have attempted to show, the very concept of evolution presupposes the concept of species as (natural or, at least, biological) kinds (Ch. 7.2), and such kinds are constructs, hence neither mutable nor immutable (recall Definition 1.6).

9.1.3 Speciation Proper and Some of Its Consequences

As we sometimes speak, for instance, of *the* evolutionary process or of *the* evolution of mammals, we should briefly explore the consequences of the preceding analysis for such expressions. If organisms are the primary units of evolution (in the sense of speciation), then the term 'evolution' does not denote a single (individual) process. Rather, there are as many evolutionary processes as there are speciating organisms. The same holds for biopopulations and communities as (derivative) units of evolution. Only if the biosphere as a whole were a unit of evolution would we have a single evolutionary process or, in other words, one evolutionarily changing entity. Thus, in most cases "evolution" is a class concept: it refers to the *collection* of all evolutionary processes, whether in organisms, biopopulations, or communities. And of course, there is no such thing as the evolution of a lineage, because lineages are collections, not changeable entities (recall Definition 7.6).

Since processes in populations depend on speciation proper (i.e., organismal speciation), we did not need to distinguish what has been called *phyletic speciation* from *cladogenetic speciation*. The latter refers to the splitting of biopopulations into (at least two) daughter populations which, upon such geographical separation, undergo different evolutionary processes until they finally become genetically divergent enough to be reproductively isolated against their sister population(s). (See Fig. 7.5.) It is also believed that this process of allopatric "speciation" would account for the increase in the number of species during evolution. By "phyletic speciation", on the other hand, biologists mean the gradual transformation of a biopopulation through time without splitting and hence divergence. So, in this case, species would not multiply, but only succeed each other.

The concept of phyletic or anagenetic speciation was commonly accepted for a long time. However, due to the predominance of the allopatric model of speciation, which is moreover regarded as *the* mode of speciation that justifies cladistic taxonomy, many biologists now contend that, whatever qualitative changes a population may undergo through time, it always remains the same species unless it splits into two or more daughter populations, which subsequently evolve differently. They further argue that it makes no sense to distinguish different species in a continuum, because we could do so only arbitrarily.

Obviously, both the concepts of phyletic and cladogenetic speciation rest on the common confusion of "population" and "species". What actually happens in both cases is that organisms belonging to one species give rise to organisms belonging to another, new species. Whether the population (or the "gene pool") has been split or at least disturbed (e.g., "bottle-necked") before speciation proper, is of secondary interest. Of course, the splitting of populations, in particular the branching off of a small population, seems to provide favorable conditions for the fixation of new variants within the daughter population. But the fact remains that, from an ontological point of view, the essential speciation event is the coming into being of new variants, not their subsequent distribution. Thus, allopatric speciation is nothing but (facilitated) phyletic speciation in two daughter populations after some separation event.

As for the denial of the existence of phyletic speciation, it overlooks the fact that the qualitative changes of organisms are objective, so that the organisms that constitute some long-"lived" population must indeed be assigned different nomological state spaces. Since we are talking about nomological state spaces of organisms, not of populations, the ontologist will speak of a new (ontological) species as soon as he or she finds a single organism of a new kind. The biologist, on the other hand, will speak of a new (biological or taxonomic) species only if either the new variants occur as a stable morph in the population (polymorphism) or the entire population consists of organisms of the new kind. There is no harm in this procedure, which is, moreover, reasonable from the biological point of view, as long as populations are not equated with species.

Some biologists contend that, even if only one of two daughter populations were to evolve "isolating mechanisms" or any other new features, the sister population, though not having changed qualitatively, would also have to be regarded as a new species because of the new relation of isolation between the two sister populations (see, e.g., Willmann 1985). If this were true, a nonevent (no qualitative change) together with a nonrelation (reproductive isolation) would combine to bring about speciation. This quaint idea has been proposed to back up ontologically the methodological requirement of cladistic taxonomy that, whenever possible, collections of organisms should be dichotomously resolved into sister taxa. Although this is a sound methodological principle, we can hardly expect that nature will always comply with it in that ontological speciation is predominantly dichotomous. (See also Nelson and Platnick 1981; Mayr 1988; Splitter 1988; Kitcher 1989a; Mahner 1994a.)

Our definition of the concept of evolution in terms of speciation amounts to what is known under the labels 'trans-specific evolution' or 'macroevolution'. This, by the way, holds for both the ontological and the taxonomic (or biological) concept of speciation: whereas microevolution in the biological sense, that is, the gradual accumulation of small qualitative changes in a biopopulation (or, more precisely, in the organismal components of the population) may very well produce a sequence of ontological species, the biologist takes only "significant" and somehow "persistent" changes as grounds for speaking of a new biological species, and hence of trans-specific evolution. Still, whether we are concerned with either ontological or taxonomic species, there can actually be no microevolution proper, since microevolutionary processes are by definition intraspecific changes. In fact, as far as the processes of selection and adaptation among organisms in populations are concerned, macroevolution precedes microevolution (Mahner 1994b). That is, the coming into being of organisms of a new kind clearly must precede their sorting in the population, which may eventually lead to their fixation in the population. This is our (free) reading of Gould's statement that "speciation, by forming new entities [...], provides raw material for selection" (Gould 1980, p. 124). And to selection we turn now.

9.2 The Theory of Natural Selection

Before we can tackle the theory of selection proper, we need to recall the notion of adaptedness, which we shall distinguish from the concept of fitness. (See also Byerly 1986.)

9.2.1 Adaptedness and Fitness

In Section 4.8 we have elucidated the notions of adaptation and adaptedness. In so doing, our definition of "adaptedness" (Definition 4.18) refers to the physiological and ecological performance of a biosystem. Thus, we have not identified the adaptedness of an organism with its fitness, which we take to be its reproductive capacity (see also Endler 1986; Lennox 1991, 1992b). Adaptedness should be distinguished from fitness because adaptedness is a relational property, whereas fitness is an intrinsic dispositional property (Mills and Beatty 1979; Sober 1984; Beatty and

Finsen 1989). Whether or not this disposition is a chance propensity, so that the concept of fitness can be elucidated in terms of probabilities, is a matter of debate (Richardson and Burian 1992; Rosenberg 1994). We believe that fitness is not a chance propensity, although its actualization is subject to external accidents. This, however, does not turn the disposition itself into a probabilistic property. Another reason for distinguishing adaptedness from fitness is that not all adapted organisms are at the same time reproductively fit (see Sect. 4.2). That is, even well-adapted organisms may have zero (replicative) fitness, as is the case of sterile organisms like mules and certain castes among eusocial insects. By contrast, a biosystem cannot have zero adaptedness because this would amount to being dead. Therefore, the adaptedness value of an organism may be represented by a positive real number ranging between $0 < \alpha \leq 1$, whereas the range of fitness values ϕ is $0 \leq \phi \leq 1$. (See M.B. Williams 1970. For distinctions and analyses of different fitness concepts see Dawkins 1982; Byerly 1986; Endler 1986; Byerly and Michod 1991; de Jong 1994; van der Steen 1994.)

Since we distinguish adaptedness from fitness, we must *postulate* a relationship between the degree of adaptedness of an organism capable of reproduction and its degree of fitness. This is:

> POSTULATE 9.1. An organism's (degree of) adaptedness, (co)determines its (degree of) reproductive capacity or *fitness*. More precisely, the greater an organism's (degree of) adaptedness, the greater its (degree of) reproductive capacity or *fitness*.

This dependence of fitness on adaptedness also holds in the case of sexual reproduction. For example, the ability to discriminate between more or less fit mating partners is relevant to individual adaptedness. However, the selectively relevant fitness value is the compound fitness value of the mating pair; similarly with animal societies that consist of mostly sterile morphs and only a few reproducing individuals. Here the fitness value of the few reproducing individuals depends not only on their own adaptedness but also on that of the sterile morphs that are part of the social system. However, we should bear in mind that the members of the sterile castes belong to the environment of the reproducing organism(s). This shows that an organism's adaptedness and fitness can be increased by manipulating or even producing valuable environmental items (see Lewontin 1983a, b).

According to Definition 4.18, the degree of adaptedness of an organism in a given habitat, i.e., its performance, is determined by the biovalues of all its aptations, adaptations, nullaptations, and malaptations; and so is its fitness value, according to Postulate 9.1. Thus, the adaptedness value of an organism at a given time is not just the sum of the biovalues of its subsystems (including their functions and roles) but a systemic value due to the functional integration of all subsystems in the organism. So much so, that the members of two very different taxa A and B may have the same adaptedness (and also fitness) value due to different features and roles in their (common or different) habitats. For this reason, it has become fashionable to say that an organism's adaptedness, or fitness, respectively, "supervenes" on its "physical" properties (Rosenberg 1978, 1985, 1994; Sober 1993). If here 'physical' is the adjective to 'physics', then the ecological property of adaptedness would supervene not only on the organism's physical properties, but also on its chemical *and* biological ones. If, on the other hand, 'physical' just means 'material', then the notion of supervenience makes no sense to us, because all properties of concrete things, whether intrinsic or relational, are material. Anyway, as have argued in Section 1.7, we have no use for the notion of supervenience.

Note, though, that the so-called supervenience of both adaptedness and fitness is neither an obstacle to speaking of adaptedness and fitness in general nor to the existence of a generic theory of selection containing a general concept of adaptedness or fitness. Although it is likely that there are different (specific) laws of adaptedness and selection for each family, genus, or even species of organisms—e.g., the specific adaptedness of an insect is different from that of a bird—, we still need a generic theory of selection, if only of the gray or black box type (recall Sects. 3.4.3 and 3.5). There are many analogous cases in other fields. For example, there are about ten million chemical reactions, each of them with its own mechanism, but all of them satisfy the basic law of chemical kinetics (concerning the rates at which reactions proceed). Thus, the rate law is the universal feature common to all specific reactions, and it makes perfect sense to have a general theory of chemical kinetics. However, what would be gained by saying that this law "supervenes" on the various specific laws? Or, to ask pungently, what insight is gained by saying that universal "sockness" supervenes on the particular properties that characterize your socks and ours?

Since, in a generic theory of selection, the notions of adaptedness or else fitness are phenomenological (or generic) concepts, i.e., unspecified in terms of particular traits and environments constituting the particular adaptedness or fitness of the organisms concerned, Rosenberg (1985) has claimed that the concept of fitness (which he equates with "adaptedness" as defined here) would have to figure as a primitive (i.e., undefined) notion in any such theory. Although it is true that "fitness" is an unspecific concept in a general theory of selection, it is mistaken to say that therefore it must be a primitive concept. Indeed, if we wish to refer to the particular traits of organisms of a particular species, we have to specify or particularize the general theory of selection; that is, we have to enrich the general theory with subsidiary assumptions specifying the particularities of the specific referents under consideration. (In other words, we have to generate a theoretical model from the general theory: recall Sect. 3.5.3.2). But this has nothing to do with definition proper: in so doing, one does not "define" the concept of fitness in terms of particular traits. Whether or not a concept figures as a primitive or as a defined notion in a theory is a matter of theory organization or structure, not a matter of its degree of generality. For example, if a general theory of selection were to contain our Definition 4.18, "adaptedness" would be a defined concept in this theory. However, it would still be a generic concept in a general theory, from which we must generate a (bound) theoretical model in order to refer to anything in particular.

Let us return to the adaptedness-fitness distinction. This distinction is also in order with regard to the concept of *relative adaptedness*. Yet before we can define this concept, we need the notion of a common environment of two or more organisms. For obvious reasons it only makes sense to compare the performances of two or more organisms with regard to those environmental items that are common to both of them (Brandon 1990, 1992; Bell 1992). Moreover, we will need the notion of common environment to define the concept of selection.

As will be recalled from Section 1.7.2, the (proximate or immediate rather than total) environment of a system is the collection of all environmental items that may act upon the given system, or upon which the system in question may act. In our analysis of the concept of selection at least three notions of environment will be relevant. First, there is the environment of the individual organism: if b denotes an organism, then $E(b)$ designates b 's (immediate) environment. Note that the fellow organisms of b in the given population, whether conspecific or not, belong to $E(b)$. Second, there is the common environment of the organisms in a population p, which is construed as the intersection of the individual environments (see Definition 9.3). This environment common to all the organisms in p may, but need not, coincide with the environment $E(p)$ of the population as a whole, where p is an aggregate or a system. Obviously, the organisms composing the population p do not belong to $E(p)$. The following example will help illustrate these distinctions.

Suppose a system s is composed of two components 1 and 2, and there are four items a, b, c, and d in the total environment of s, i.e., $E_T(s) = \{a, b, c, d\}$. Suppose further that component 1 interacts with component 2—they constitute a system after all—as well as with the items a, b, and c. That is, the immediate environment E of 1 is the collection $\{a, b, c, 2\}$. Assume, on the other hand, that component 2 interacts with 1 as well as with the items b, c, and d. That is, the immediate environment E of 2 is the collection $\{b, c, d, 1\}$. Then the immediate environment that 1 and 2 have in common is $E(1) \cap E(2) = \{b, c\}$, and the immediate environment of s equals its total environment rather than the common environment of its individual components. If, by contrast, $E(1) = \{a, b, c, d, 2\}$ and $E(2) = \{a, b, c, d, 1\}$, then $E(1) \cap E(2) = E_T(s) = \{a, b, c, d\}$. Thus, the common (immediate) environment of all the components of a system must be distinguished from the (immediate) environment of the system as a whole.

The notion of a common (immediate) environment of the organismic components of a population is elucidated by:

> DEFINITION 9.3. Let p denote a population of organisms i with (immediate) environment E_i, where the organismal composition C_O of p is a finite collection of organisms containing at least two members. Then the common environment of all organismal components i of p, i.e., the environment shared by all organisms in p, is
>
> $$E_C = \bigcap_{i=1}^{n} E_i.$$

As mentioned above, this common environment E_C of all the components of p may but need not equal $E(p)$. We are now ready to define the notion of relative adaptedness:

DEFINITION 9.4. Let b and c name organisms of the same or of different species but with a common environment $E_C = E(b) \cap E(c)$, and let α denote an organism's degree of adaptedness or adaptedness value. Then b with regard to the items in E_C is *better adapted* than c relative to E_C iff $\alpha(b) > \alpha(c)$.

As we are hardly able to compare adaptedness values directly, we have to rely on indicators. An indicator of the relative adaptedness of two organisms b and c with common environment E_C may be the observation that b leaves more (viable and fertile) offspring than c. Since the relative reproductive success of two or more organisms is an *indicator* of their relative adaptedness, we cannot *define* the concept of adaptedness in terms of reproductive success, for this would be an operationist mistake (Rosenberg 1985; Lennox 1992b; Mahner 1994a). From Postulate 9.1 and Definition 9.4 we infer that, if b with E_C is better adapted than c with E_C, the fitness value ϕ of b will be higher than that of c. This inference is made explicit in:

COROLLARY 9.1. The greater the relative adaptedness of two organisms b and c with common environment E_C, the greater their relative fitness. That is, $\phi(b) > \phi(c)$ if, and only if, $\alpha(b) > \alpha(c)$.

Although Definition 9.4 and Corollary 9.1 may seem rather trivial, they are not, because they help clarify the relations of the concepts of adaptedness (fitness_1), fitness (fitness_2) and actual reproductive success or Darwinian fitness (fitness_3). To measure the latter and to distinguish it from the results of mere random drift we must hypothesize that reproductive capacity or fitness proper determines actual reproductive success. That is, we make:

POSTULATE 9.2. An organism's degree of (total) reproductive success depends on its (degree of) fitness. More precisely, the greater an organism's fitness, the greater its (total) actual reproductive success.

We can now obtain:

COROLLARY 9.2. The greater an organism's (degree of) adaptedness, the greater its (total) actual reproductive success.

COROLLARY 9.3. The greater the relative adaptedness of two organisms b and c with common environment E_C, the greater their relative (actual) reproductive success δ. That is $\delta(b) > \delta(c)$ if, and only if, $\alpha(b) > \alpha(c)$.

Note that the (deterministic) correlation between fitness_2 and Darwinian fitness is still weaker than that between adaptedness and fitness_2. This is because accidental sorting factors may exert an influence on the number of offspring of two equally fit individuals. The proverbial example is the pair of identical twins, that is, two equally adapted and thus fit individuals, one of which is killed by a flash of lightning before it can actually reproduce. Therefore, actual reproductive success is a highly ambiguous indicator of fitness and adaptedness. The ambiguity is further increased by the fact that we cannot directly observe the dependencies of adaptedness, fitness, and Darwinian fitness in individuals. Rather, we have to operate with average adaptedness and fitness values of the organisms in a population, and we infer these averages from the statistical correlation of the average reproductive success of organisms (of a certain kind) in a population with the possession and distribution of certain features of a certain kind—aptations and adaptations—of those organisms. (More on the relation of character, performance, and reproductive success, as well as their measurement, in Arnold 1983.)

For this reason, the above hypotheses are usually formulated in statistical terms, that is, in terms of organisms in populations of such and in terms of averages. However, these statistical notions may well be mere heuristic devices, and do not entail that the theory itself is irreducibly statistical or probabilistic (Rosenberg 1994). Thus, in order to emphasize that it is really organisms that are adapted, we stick to our idealized formulations in terms of individuals. Also, for the same reason, it appears that the notion of fitness as reproductive capacity plays no role in selection theory (Byerly 1986). Still, if perhaps not for practical reasons, we need this concept of fitness for theoretical reasons because it relates the notion of adaptedness to that of actual reproductive success.

Whereas morphologists, physiologists, and ecologists are interested in adaptedness *per se*, evolutionary biologists are mostly interested in the concept of adaptedness inasmuch as it relates to fitness and Darwinian fitness, that is, inasmuch as it is the basis for the concept of natural selection.

9.2.2 Concepts of Selection

Since the word 'selection' occurs in ordinary language, the technical concept of selection in biology is notoriously as tricky as those of function and adaptation. In particular, the question of what kinds of entities are capable of undergoing a selection process has been the object of debate over many years. This problem is known as the units of selection controversy. (See, e.g., Lewontin 1970; Brandon and Burian eds. 1984; Sober 1984, 1993; Lloyd 1988, 1992; Sober and Wilson 1994.) However, it is not only controversial which the units of selection are, but also which kinds of processes can be said to constitute selection (see, e.g., Endler 1986).

As the unit of selection problem is one of the most intensely discussed topics in contemporary biophilosophy, we shall not attempt to give a detailed analysis of the various positions in this debate. We shall only expound our view of the matter and state some of the implications for the various competing positions in this controversy. Let us begin by distinguishing several concepts of selection, the first of which is so general that it must be regarded as an ontological concept.

9.2.2.1 An Ontological Concept of Selection

Every individual process of selection must be a sequence of changes in some thing or other or, if preferred, a sequence of events (recall Sect. 1.5). Since there are no events or processes in themselves, but only things that may undergo successive changes, we must assume a changing thing or system to begin with. Since every system (except the universe) is characterized by some environment *E*, we can say that, given a number of environmental items in *E*, a complex thing (or system) *b* —be it a molecule, a crystal, an organism, or what have you—will be able to exist or not relative to *E(b)*. (Note that since we have defined the environment *E* of a thing as a collection, we cannot properly say that a thing exists *in* its environment *E*. If we do so, we must be aware that this is elliptical.) The changing thing, then, is the system composed of *b* and the items in *E(b)*. In other words, the interaction of the individual *b* with environmental items in *E(b)* is, in the most general sense, the concrete selection *process*, which results either in the continued existence or in the elimination of *b* with regard to *E(b)*. (See also Bock and von Wahlert 1965; Tuomi 1992.)

Again, since the environment of a system has been defined as a collection, not a thing, we cannot properly speak of a thing-environment interaction. We must therefore emphasize that the expression 'thing-environment interaction' is shorthand for the interaction of a thing with any or all of the items collected in its environment *E*. We can now (elliptically) say that selection in the ontological sense is any *b-E* interaction that affects the continued existence of *b* in *E*. The continued existence of a thing in its environment *E* is usually called 'survival'. However, this is very often a misnomer because only living beings can truly be said to survive. (Interestingly, neither biologists nor philosophers seem to care about this ontologically significant distinction, but are trapped in ordinary language, which allows for nearly anything to "survive".)

We submit that all individual selection processes involve a thing-environment interaction, which can be said to be the *basic process of selection*. In other words, there are as many selection processes as there are thing-environment systems. Consequently, the basic concept of selection as a thing-environment interaction holds for every thing in the universe and is thus an ontological concept. Since the result of every individual selection process is either the continued existence or the elimination of a thing in the given environment, we call it (individual) *all-or-nothing selection*, or *1/0-selection* for short. More precisely, we propose:

DEFINITION 9.5. Let *b* name a thing with environment *E*. Then all interactions between *b* and things in *E(b)* that affect *b*'s subsistence with regard to *E* are called *individual all-or-nothing selection*.

Thus far, 1/0-selection concerns the interaction of a single thing or individual with items in its environment. However, this notion still lacks an important connotation of the concept of selection in biology, namely the aspect of *sorting*, which presupposes the existence of more than one thing in a given habitat. (See

Vrba and Gould 1986.) The next step in our analysis, then, is to expand our basic concept of individual 1/0-selection to aggregates (not necessarily systems) of things. That is, we introduce the concept of a population as an aggregate of individuals, whether uni- or multispecific. (Note that an aggregate has been defined as a thing, not a collection: see Sect. 1.7. What is a collection or a class is the composition of the aggregate.) Now, selection consists in the shrinking of the composition C of a population p of things x with a common environment E_C, to some subset C_A of p during some time interval, namely the collection of minimally apted parts of p (see also Bunge 1979a).

We are now ready to formulate:

> DEFINITION 9.6. Call $C(p, t)$ the composition of a population p of systems (of the same or of different kinds) with a common environment E_C at some time t. Further, call s_E: $C(p) \rightarrow C_A(p)$ the inclusion function from $C(p)$ into $C_A(p)$, where $C_A(p) \subseteq C(p)$. Then *populational 1/0-selection* is that set of interactions among all the members of $C(p)$ and items in E_C during the interval $[t, t']$, where $t' > t$, that produce the sorting s_E: $C(p, t) \rightarrow C_A(p, t')$.

In other words, the members of the common environment E_C exert a selective action on all the members of $C(p)$ during a time interval so that, at a later time, p consists only of the members of $C_A(p)$. Though now a populational concept, this is still a concept of all-or-nothing selection. That is, the members of $C(p)$ are either able to subsist in E_C due to their possessing certain properties (aptations), or are eliminated. If the members of such a population are organisms, we say, in the case of survival, that their adaptedness value is positive, i.e., that they are minimally apted, or, in the case of death, that their adaptedness value is nil. (Note that we do not yet need the concept of fitness.) Now, the result of this selection process on things in a population of such is *differential existence*. Again, only if living beings are involved, should we speak of either differential survival or differential mortality.

All-or-nothing selection, resulting in differential survival, is the concept underlying the classical notions of the "struggle for existence" and the "survival of the fittest". Darwin defined "natural selection" thus: "This preservation of favourable variations and the rejection of injurious variations, I call Natural Selection" (1859, p. 81; see also Simpson 1953; Goudge 1961). However, the modern theory of natural selection contains a different concept of selection: it focuses on selection processes that result in differential reproduction rather than differential survival. Although differential survival may also result in differential reproduction, the former is not necessary for differential reproduction to occur. This wider concept of selection, which is basically independent of the concept of differential survival, is introduced in the following.

9.2.2.2 Natural Selection

Let us assume next that the components of a population p *differ* in certain properties. That is, suppose that the composition of p is characterized by what biologists call either 'variability' or 'variation'. (Note that, while in other research fields "variability" usually means changeability over time, the biological concept of variability is a populational notion in the sense of "diversity": it is an attribute of the collection of components of a population, neither a [substantial] property of its individual components, nor a property of the population as a concrete whole or individual. Hence, strictly speaking, there is no such thing as "individual variability". What is a property of the population, though, is its having a variable composition.) Variability comprises both the possession of different generic properties and different individual properties (i.e., different values of generic properties). In other words, variability may be qualitative as well as quantitative.

If we now subject the components of such a population p to the process or, rather, the processes of 1/0-selection, only the members of $C_A(p)$ will remain. That is, those individuals subsist that are minimally apted with regard to their common environment E_C. If p consists of organisms we say, as above, that their adaptedness value is positive. In principle, the result of this selection process can be predicted if we know the relevant properties of the individuals and of the items in their environment. Needless to say, its prediction in practice is a different matter.

Let us further assume that the members of the collection $C_A(p)$ are organisms varying in their degree of adaptedness. Thus, as noted in Sections 4.8.3 and 9.2.1, their adaptedness value α may range between 0 and 1, more precisely, $0 < \alpha \leq 1$. Accordingly, there will be differences in the continuing organisms-environment interactions. However, these differences do not affect the continued existence, i.e., the survival, of the members of $C_A(p)$ but the *performance* of the organisms in E_C. In other words, differential adaptedness leads to *differential performance*. (This is an instance of the state-process distinction.) Note that 'differential performance' does not refer to an individual process in a population, because the population need not be a system: what actually occurs are different individual processes in an aggregate of organisms, which are compared to each other. Thus, it is not the population as a whole that undergoes a selection process but *individuals* in a population of such. If some population p as a whole were to undergo a process of selection, p would have to be part of a population of populations—a metapopulation—and it would have to interact as a whole with some item(s) in its environment $E(p)$. (See also Brandon 1990.)

Contrary to 1/0-selection, the differential performances of the biosystems in a population are necessary but not sufficient for natural selection to occur, because differential performance does not lead to a *sorting* of the members of $C_A(p)$. If there is any sorting involved it is the sorting of the offspring of the biosystems in question. To arrive at this sorting, we must consider the ability of biosystems to reproduce. Their differential performance may then affect their reproductive capacity or fitness. In other words, their fitness may depend on their performance, which, in turn, depends on their adaptedness; and their fitness may eventually result in actual reproductive success or Darwinian fitness (see Postulates 9.1 and 9.2, as well as Arnold 1983). We can now call those differential performances or interactions that affect fitness and eventually reproductive success *performance-selection*, or *p-selection* for short. Performance selection, then, results in differential reproduction, and selection has become a trans-generation sorting mechanism. More precisely, we lay down:

DEFINITION 9.7. Let C_A designate the composition of a population p of differentially adapted organisms of the same kind with a common environment E_C, and let D designate the collection of immediate descendants of the members of C_A. Then *p-selection* is that set of organism-environment interactions of the members of C_A that result in the sorting s_E: $C_A \rightarrow 2^D$ of the offspring of the members of C_A.

(2^D is the power set of D, i.e., the family of subsets of D, including D and $\emptyset$. Hence, every value of the sorting function s_E is a subset of D.) Evidently, in the case of sexual reproduction the composition C of the population must be partitioned into males and females, i.e., C_M and C_F, so that the sorting function reads s_E: $C_M \times C_F \rightarrow 2^D$.)

To summarize, the term 'selection' designates several different concepts. These are the general concept of 1/0-selection in its individual as well as populational (or statistical) form, and the more restricted, though still general, concept of p-selection, which is a statistical concept. The concept of 1/0-selection is used when the survival or the death of a biosystem is at issue. Furthermore, this concept seems to be presupposed when some biologists say that "an organism is always subject to selection". Indeed, every organism is subject to 1/0-selection at any state of its history, particularly during its development. And every subsystem of an organism is subject to 1/0-selection with regard to its own environment, i.e., the rest of the organism. This is sometimes called 'internal selection'. However, 'internal selection' is a misnomer because it is not the organism as a whole that directly undergoes a process of selection, but some of its subsystems. Of course, since organisms are highly integrated systems, the failure of any of its subsystems is very likely to affect the organism as a whole, so that it will eventually be subject to 1/0-selection itself.

On the other hand, entities *in a population of such* which (a) vary in their degree of adaptedness, i.e., which are differentially adapted, (b) possess the property of replicability (either on their own or as mating-pairs), and (c) vary in their reproductive capacity or fitness, i.e., which are differentially fit, may undergo a process of p-selection (see also Lewontin 1970; Endler 1986). To repeat, variability, differential adaptedness, and differential fitness are neither features of isolated individuals nor properties of the population as a concrete whole: they are attributes of the population's composition, i.e., of the collection of its organismal components. What is a property of the biopopulation as a whole, though, is its being composed of differentially adapted or differentially fit organisms, respectively. Furthermore, it is not necessary for selection to occur that the population be a system: it suffices that it be an aggregate of individuals. Yet special cases of selection, such as density-dependent and frequency-dependent selection, presuppose bonding relations among the parts of a population, so that populations need be systems for such cases to occur.

In sum, there are only individual *p*-selection processes: the interaction of some entity with some environmental item, the reproduction of this entity, and the ensuing change in the composition of the population of entities in question. The general term 'selection' designates the collection of such individual selection processes. Thus, selection is not a force—*pace* Sober and Lewontin (1982) and Sober (1984); and neither is it an agent: not being a changing thing, it cannot "act" on anything.

9.2.2.3 Natural Selection as a Mechanism of Populational Evolution

Neither 1/0-selection nor *p*-selection have so far anything to do with evolution proper. That is, a theory of selection is, in principle, independent of any theory of evolution. However, the aforementioned trans-generation sorting mechanism may become a mechanism of evolution or, more precisely, of populational evolution (see Definition 9.2), if the performance of the offspring is "somehow" dependent on the performance of their ancestors. The "somehow" of this dependence is usually assumed to consist in the inheritance of the aptations and adaptations in question. More precisely, it is said that the degrees of adaptedness and fitness of an organism must be heritable (see, e.g., Lewontin 1970; Endler 1986; Brandon 1990). Yet since we saw in Section 8.2.4 that only the initial state of the zygote can be said to be genuinely inherited, not any later states (in particular relations to environmental items), we can only say that adaptedness and fitness must be inheritance-dependent. If the latter is the case, we have to do with what is usually called 'the concept of evolution by natural selection'.

It should be noted, however, that natural selection (as a mechanism of populational evolution) is not basically different from selection that does not lead to evolution. In both cases, the basic process is the same: it consists in a thing-(in a population)-environment interaction resulting in the differential replication of the things in question. The difference lies in *heredity*, so that repeated trans-generation sortings may bring about continuous and directional changes in the composition of successive generations. Therefore, the theory of natural selection explains only the continuous change of genotype$_O$ and phenotype$_O$ frequencies in populations, and thereby also the change of the frequencies of all the organismal subsystems, such as genes.

However, the theory of natural selection does not and cannot account for speciation proper, i.e., the emergence of qualitatively novel organisms. For this reason, the claim that natural selection is a "two-step process" (Mayr 1988), the first of which consists in the production of variation, and only the second in selection proper, is incorrect. The same holds for the popular claim that natural selection is creative. The fact that natural selection favors or eliminates certain phenotypes$_O$ in populations of such, has nothing to do with creativity proper—a concept which implies the production of something new. After all, we do not attribute creativity to the public that either favors or ignores an artist's works: only the artist herself can be said to be genuinely creative. Indeed, as has been said aptly, "natural selection is only the editor and not the author of evolutionary change" (Simon 1971, p. 177; see also Wassermann 1981).

Any genuinely creative process in evolution can be only a developmental process (see Definition 9.1). Yet since developmental biology has never been part of the Synthetic Theory, it is not surprising that it has been attempted to smuggle in the alien notion of creativity into the theory of natural selection by claiming selection to be a "two-step process". So we insist that the fact that the theory of selection presupposes the existence of variation does not make selection itself creative. Indeed, as we saw in Section 9.1, speciation in the ontological sense must precede selection. However, the theory of selection accounts for the prevalence or the frequency of characters (or, more precisely, of organisms possessing the characters in question) in a population (Sober 1984; Endler 1986).

Focusing on selection as an evolutionary mechanism, evolutionary biologists are often more interested in the results of selection than in the ecological process of *p*-selection itself. Thus, many of them (see, e.g., Ayala 1970; Futuyma 1986) do not use at all the notion of ecological process with regard to selection, but equate selection with its result, namely differential reproduction. (For a review of such approaches see Bradie and Gromko 1981; a few such examples are also provided by Rosenberg 1985, p. 127.) This, however, is mistaken.

First, although the result of a (populational) selection process always consists in a sorting of the organismal parts of a population, not every sorting need be the result of selection (Vrba and Gould 1986). Random drift, for instance, also leads to the sorting of the parts of a population. Second, to define "selection" merely in terms of sorting is an operationist mistake, because we need the concept of selection as referring to an ecological process (together with the concept of differential adaptedness and differential fitness) if it is to *explain* the observed sorting pattern, such as differential reproduction (Brady 1979; Bradie and Gromko 1981; Vrba 1984; Damuth 1985; Rosenberg 1985; Vrba and Gould 1986; Lloyd 1988; Darden and Cain 1989; Brandon 1990; Lennox 1992b). Differential reproduction is an *indicator* of selection, not selection itself (Mahner 1994a). This also holds for those cases in which differential reproduction appears to be all there is. For example, the differential growth of two strains of bacteria in an excess nutrient medium seems to depend only on the different intrinsic division times of the two strains, not on any ecological process. Yet the division rates of the two strains depend on the nutrient, temperature, pH, and so on. Change any or all of these and the other strain may grow faster, because its members are better adapted to this new environment than its competitors, which were better adapted under the previous conditions.

9.2.3 The Units of Selection

9.2.3.1 What Is a Unit of Selection?

As 1/0-selection is an ontological concept, it holds for all things. Not so *p*-selection: here we had to specify several properties enabling certain entities to undergo a process of *p*-selection, namely (a) existence in a population of like entities, (b) (differential) adaptedness, (c) ability to reproduce, and (d) (differential) fitness. These properties are evidently possessed by most organisms, which is why we have formulated our definition of *p*-selection with reference to living beings. However, we can now ask whether there are other entities besides organisms that also possess these properties. If so, all entities possessing the necessary properties can be said to be *units of selection* or, more precisely, units of *p*-selection. As the expression 'unit of selection' is at the center of an ongoing controversy, it will be convenient to start with a clarification of the notion of a unit.

We distinguish two kinds of unit: *unit of a process* and *unit of analysis or description*. A unit of a process is, of course, a material unit: it is just that thing undergoing the process in question. As we mentioned in Section 1.5, abstracting processes from the things undergoing the processes and thereby forming process classes may lead one to lose sight of the changing things in question. So they must later be recovered as the "units" of the processes in the given process class. By contrast, a unit of analysis or description is either a referential or a conceptual unit. If referential, it is any system belonging to that system level that is at the center of our research interest. If conceptual, it is a concept that cannot be defined in terms of simpler concepts in the given context. For example, in logic the concept of identity is taken as basic (or primitive), and in systematics the concept of an organism is taken to be undefined, and is used to define that of species.

We submit that what is at issue in evolutionary biology is, first of all, the material unit(s) of selection, i.e., that thing or those things that are capable of undergoing a process of *p*-selection. Any such entity capable of undergoing a process of *p*-selection has been called an *interactor*. An interactor is "an entity that directly interacts as a cohesive whole with its environment in such a way that this interaction causes replication to be differential" (Hull 1988, p. 408; see also Hull 1980, p. 318). This definition correctly recognizes selection as a thing-environment interaction. However, there are two problems with this definition. The first is that a single entity cannot be said to replicate *differentially*. Only individuals in a population of such can do so. The second is that, on our view of causation, the thing--environment interaction referred to cannot be said to *cause* replication: it can only be said to *condition* the reproductive success of the individuals in question (recall Sect. 1.9.) We can say that some event *causes* the replication of some organism only if there is an environmental stimulus triggering a reproductive process in the organism in question; but this is at best a very special case of Hull's definition of an interactor. These (minor) flaws aside, it is obvious that interactors are the material units of selection processes. (See also Lloyd 1988.) Therefore, the units of selection problem is about the *selection of* (material) objects, not about the properties underlying the selective success of the individual in question, i.e., *selection for* in Sober's terms (1984). More on this anon.

9.2.3.2 Genes, Gametes, Cells, and Organisms

What things can undergo a *p*-selection process? First, pieces of "junk" or "selfish" DNA seem to be able to interact with cell metabolites, that is, some of their environmental items, in such a way that they are reproduced differentially within the cell. The (composition of the) population in question is the set of genes in the cell, i.e., the cellular genome. Note that, in this case, self-replication is not an intrinsic property of the genes in question: they are passively reproduced by the cellular enzyme machinery. Second, and similarly, some genes and some parts of chromosomes are differentially replicated in what is called *meiotic drive* (Lewontin 1970; Futuyma 1986; Pomiankowski and Hurst 1993). That is, rather than being equally distributed during meiosis, certain genes are able to interact with cellular metabolites in such a way that more copies of themselves are enclosed in the gametes. Third, in principle, certain molecules, organelles, and cells as wholes are able to replicate differentially with respect to other molecules, organelles, or cells in a multicellular organism. A dramatic example are cancer cells. Further, antibody formation in the immune system may involve selection processes (Darden and Cain 1989). Fourth, organisms are undoubtedly units of selection—a fact that is conceded even by such staunch defenders of genic selectionism as Dawkins (1982). It should be noted, however, that only cells and organisms, i.e., biosystems, possess the intrinsic property of self-reproduction. That is, the environment provides only the conditions for reproduction to occur, while in all other cases, in particular in the case of genes, it is environmental items that do the reproducing. The reproduced entity thus serves only as a template for replication. Still, the final result is the same.

9.2.3.3 Groups or Populations

Whereas the previous examples of genic and organismal selection are now seen as rather unproblematic, the notion of group selection is still controversial because the expression 'group selection' refers to (at least) two different cases of selection (Arnold and Fristrup 1982; Maynard Smith 1984; Sober 1984; Mayo and Gilinsky 1987; Damuth and Heisler 1988; Lloyd 1988; Brandon 1990). Indeed, this difference is reflected in the existence of group selection models of two kinds. In the models of the first kind the unit of selection is the individual organism, but the adaptedness and fitness values of the individuals depend on their being part of a system (of individuals), i.e., a "group". Think of frequency- and density-dependent selection. That is, the adaptedness and fitness values of the organisms in question are *structurally* emergent properties of theirs (see Sect. 1.7). Recall that, for *p*-selection to occur, we already had to assume that organisms live in populations, but it was sufficient that these populations be just aggregates, not systems. So the groups (populations) referred to in this first kind of group selection models are systems. Still, the material unit of selection is the individual organism.

One of the reasons that this first type of group selection models refers to groups at all, is the difficulty of properly distinguishing individual from group properties, and properties from attributes, in particular statistical artifacts (recall Sect. 1.7). An example of the former is the confusion of group-wide or species-wide—so--called *monomorphic*—characters of organisms with group or species properties (see, e.g., Vrba 1984, Stidd and Wade 1995). Yet even if all the organisms in a certain population possess the same property, e.g., color, it is ontologically mistaken to attribute this property to the population as a whole. The population (or "species") neither has a color, nor does it develop, hatch from an egg, eat, digest, excrete, or mate. In short, even if a property is species-specific or, more generally, taxon-specific, it is never a property of the species or of the taxon.

On the other hand, the confusion of collective properties with genuine group properties may be blamed on the mathematical models of group selection operating with character means (e.g., average height, average fitness values, and so forth). These averages are said to be collective or aggregate or group properties and are sometimes even called 'group phenotypes', so that these models are supposedly models of group selection. However, such collective attributes represent neither properties of the individuals of the population, nor resultant or emergent properties of the group as a whole (*pace* Ereshefsky 1988 and others). For instance, a particular organism has an individual height, shape, length, or what have you, but no average characters, even if the latter were to coincide with the average. (See also Horan 1994.) And groups (as wholes) are neither long nor tall, and *a fortiori* not of average length or height. Rather, such collective attributes are mere statistical artifacts that have no ontological status at all. In other words, these statistical attributes (or predicates) do not represent substantial properties. Moral: not every analytical or conceptual tool—however important and indispensable it may be—need have a counterpart in the real world. (However, *pace* Rosenberg 1994, this does not entail an instrumentalist view of biological theories.)

The models of the second type refer to genuine group selection in that the unit of selection is indeed the group as a whole. That is, the groups in question, whether called 'groups', 'demes', 'populations', or what have you, occur in a population of such (sometimes called the 'metapopulation'), and the adaptedness and fitness values are (genuine) properties of the group. In this case, the group (as a whole) must interact with some environmental item(s), so that the reproduction of the group as a whole, e.g., by producing propagule groups, is differential compared to the other groups in the metapopulation. In other words, there must be a differential group performance, which does not (ontologically) reduce to the performances of the individual components.

Consequently, we agree with those authors who maintain that there must be some emergent (global) group property, i.e., some group aptation, for genuine group selection to occur (see, e.g., Vrba 1984). In other words, we must presuppose group adaptedness. Yet, considering that statistical artifacts do not count, what significant group properties are there besides the rather trivial property of population density? For example, if the groups in question are social systems, a significant emergent property is social structure, such as the division of labor, as is the case with the eusocial insects. Thus, social groups might perhaps outperform nonsocial groups.

Be that as it may, the evidence for genuine group selection appears to be rather inconclusive (see, e.g., Mayr 1988; Cracraft 1989; Brandon 1990). Moreover, the analysis of possible cases of group selection is plagued and hampered by the fact that type I and type II models of group selection are often not properly distinguished from one another. For instance, is it really an instance of group selection when groups containing altruistically behaving individuals (usually) outperform groups containing only selfish individuals (Wilson 1989)? After all, behavior is an individual, though relational, property but not a group property, let alone a group aptation. In conclusion, the problem of group selection will be with us for a while, although the distinction of type I models applying to organisms with structurally emergent properties from type II models applying to groups as wholes with globally emergent properties should help clarify the issues involved (recall Definition 1.9).

9.2.3.4 Species and Clades

Obviously, the notions of species and clade selection presuppose that species and clades are concrete individuals. Since, in Section 7.3, we have shown this thesis to be seriously flawed, and since we take species and higher taxa to be classes, not things, it goes almost without saying that they cannot be involved in any selection process. (See also Bock 1986.) Indeed, in our view, the idea that species and clades can be units of selection "would not be false but sheer nonsense" (Hull 1988, p. 400).

However, even if we assumed for the sake of the argument that species and clades are material objects, the application of the concept of selection would be questionable. The fact that, for example, one clade contains a hundred species and the other only five, has been regarded as an instance of the "differential success" of clades and, hence, selection. But this only shows how misleading the operationist definition of selection in terms of differential reproduction is. The differential success of clades, if any, has to be explained in terms of the ecological process of *p*--selection among a population of species or clades. This, in turn, presupposes species or clade aptations (Vrba 1984.) Since there are hardly any significant aptations of groups (as wholes) within the same species, we submit that the success of clades, if any, can only be due to the properties of the individual members of the clade. (See also Damuth 1985; Mayr 1988.) As long as organisms are well adapted and thus perform well, the species or higher taxon they belong to can be said to be "well adapted", too. Yet all this is metaphorical talk devoid of any explanatory power. The fact that the different numbers of subtaxa within a higher taxon can be formulated in selection-speak, or even modeled by group selection analogs, does not guarantee that such exercise actually refers to some real process in nature. Witness all the metaphorical theories of "conceptual evolution" or the "evolution of science".

Although some authors have realized that species and clades are not interactors (e.g., Damuth 1985; Eldredge 1985a; Wiley 1989), they believe that at least species can be such in the case where a species consists of a single population. Yet even this is impossible, as resting on the confusion of a singleton (i.e., a set with a single member) with its sole member; but $x \neq \{x\}$. Besides, the members or elements of species, hence of higher taxa, are organisms, not populations.

Others have attempted to defend species selection by reference to species-wide properties (Stidd and Wade 1995). If a character is monomorphic, there is no intraspecific variation—so the argument goes—and hence no selection within the species (population). Therefore, the monomorphic organisms of a species can, at most, compete with organisms of a different species (perhaps with a different monomorphic character). Thus, selection is shifted to the species level. Clearly, all this rests on the confusion of organismal properties with species properties. A species-specific property of an organism is still an organismal property, not a species one. Thus, at most there is *selection for* species-specific properties of organisms. However, we are not (yet) concerned with "selection for" properties but with the *p*-selection *of* individuals. But then it is not required that there be intraspecific variation. What is necessary is intrapopulational variation, where the population in question may as well be a multispecific population, in which the varying organisms perform and reproduce differentially with regard to a common environment. (Note that it is not necessary that the populations in question be reproductive communities.) Indeed, for selection to occur, it is irrelevant whether the variants in question belong to the same or to different species, or even to different higher taxa; and it is also irrelevant whether the variants form two monomorphic subpopulations. We still have either organismal or group selection but no species selection.

In sum, what is believed to be species and clade selection may be accounted for by models of organismal, group (population), and perhaps even community selection, but there can be no such thing as species and clade selection. However, there can be "selection for" species-specific or clade-specific properties of organisms, but this does not concern the material units of selection.

9.2.3.5 Units of Description

The mathematical models of evolutionary processes in population genetics are usually stated in terms of genes or genotypes. The fact that all selection processes may indeed be modeled in terms of selection coefficients of genes or genotypes is one of the reasons behind Williams's (1966, 1992a) and Dawkins's (1976, 1982) claim that genes are the (ultimate) units of selection. (See also Sterelny and Kitcher 1988.) In addition, Dawkins claims that the unit of selection problem is not about interactors but about those entities that "benefit" from selection in the long run, namely genes. In other words, the problem of the units of selection would not be about interactors but about so-called "replicators". What would be at issue, then, is the long-term (differential) "survival" and "replication" of genes (read: gene kinds), genotypes, or gene lineages. So the phenotype$_O$ would be nothing but a transient vehicle of the (potentially) immortal genes.

We reject this view for the following reasons. First, the fact that some selection models refer to genes does not guarantee that this reference captures the actual objects in question (Sober and Lewontin 1982; Sober 1984; Lloyd 1988; Godfrey-Smith and Lewontin 1993). For instance, the model may be an oversimplification biased by a microreductionist outlook. Furthermore, any subsystem or feature of an organism may, in principle, be the referent of a selection model. After all, there are plenty of statements in the biological literature such as "In the savanna a standing foot is of greater selective value than a prehensile foot". Yet neither genes nor feet are autonomously reproducing entities. Therefore, it makes no ontological sense to attribute fitness values to them. (If we do so, this is again a case of simplification.) Second, to speak of survival in the context of replication or reproduction is wrong, because replacing an individual with a daughter individual in a lineage has nothing to do with survival. Even if "survival" is construed as mere "continued existence" or "perpetuation", the problem is that what are said to be perpetuated are not complex things, but structures (see also Hull 1980; Rosenberg 1994). However, structures are not physically separable from concrete systems: they may only be distinguished analytically. Third, the assumption that it is structures that are replicated reveals the kind of metaphysics that lurks behind genic selectionism or replicator selection. This ontology becomes apparent when we look at a recent statement of G.C. Williams (1992a, p. 11): "A gene is not a DNA molecule; it is the transcribable information coded by the molecule." This is good old Platonism in modern informationist garb. (For more of this see, e.g., Gliddon and Gouyon 1989.) Thus, what supposedly "benefits" from replication in that replication secures its "survival" is information, although, as Williams and Dawkins admit, through its material "carriers" or "vehicles" only. However, how nonentities can be said to survive, to be the beneficiary of something, or even to have interests, as many sociobiologists are fond of saying, remains the mystery of the genic selectionists. In sum, Williams's attempt to dissolve the units of selection controversy by distinguishing "two mutually exclusive domains of selection", one dealing with material entities, the other with information, fails.

If we reject informationist Platonism because it not only assumes that structures can be detached from things ("vehicles"), but that they are somehow *prior* to or more important than things, then the *kinds* of gene and the geno*types* that population genetic models deal with have to be conceived of as conceptual units, not material ones. To avoid misunderstandings: referential and conceptual units are not informational units in a realm of their own, that is, they have no ontological status as in Dawkins's or Williams's versions of information Platonism. They are only units of analysis or description. That is, population genetic models *abstract* from the living things actually undergoing the process of selection and *describe* selection in terms of the replication and fitness of genes or genotypes, as if the latter were actually to undergo the selection process. But this does not render them more adequate or realistic. (A physical analog is the geometric study of the trajectories of bodies. But here nobody claims that it is trajectories, rather than bodies, that actually move.)

Such instances of confusion between ontological and methodological abstraction show that genic selectionist models are merely idealized models using a unit of analysis, whereas ecological models refer to a material unit of a process, such as the organism. By conferring an ontological status upon their abstractions, genic selectionists commit the fallacy of reification. Only material units (at different system levels) can undergo a selection process. As for genes, then, only a concrete piece of DNA, which is an interactor, can be a unit of selection—if we count passive replication as replication. In sum, we see no use for the notion of a replicator as different from that of an interactor. (For a recent defense of the replicator/interactor distinction and the concept of an extended replicator see Sterelny et al. 1996.)

So far, we have dealt with the selection *of* entities as interactors, and rejected genic selectionism on the ground that it does not refer to material units of selection but just models selection in terms of some unit of analysis. (The latter strategy is also known as the *representability* argument: see, e.g., Sober and Lewontin 1982; Lloyd 1988). However, Sober (1984) has claimed that there might be an alternative view of the units of selection, namely in the case that a gene (or any other component of an organism) in question is the *cause* of the selection *of* the higher-level entity of which it is part. In other words, Sober has attempted to solve the units of selection problem by distinguishing the selection *of* entities from the selection *for* the possession of some character, which would be the "cause" of the selection process. After all, an organism is subject to selection because it possesses certain properties, e.g., aptations, adaptations, malaptations, and so on. Thus, if a gene x "causes" some phenotypic property y, then there will be selection *for* (or against) x, even though the whole organism possessing y is the unit of selection *of*. In this sense, the gene x is said to be a unit of selection *for*.

This suggestion is open to the following objections. First, according to our account of causation, which is completely different from Sober's, and on what we saw in Section 8.2.3.2, neither genes nor any other things can be said to cause anything: only events qualify as causes or effects. Second, even if they were to qualify, the causal chain from gene to phenotypic character would consist in many intermediate steps, all of which could be regarded as causes of the final stage. (The causal relation is transitive.) Moreover, all intermediate developmental steps could be said to "benefit" from the selection process, that is, there would be selection *for* all the intermediates as well. And, just as the gene in question "replicates itself" by being "transmitted" to the next generation, so also are the intermediate developmental steps: they, too, are "replicated" during the development of the descendant organism. (Beware, all this is elliptical: recall Sect. 8.2.4.3.) So the entity or the step of the developmental process on which we focus is just a matter of interest.

Therefore, the units of *selection for* are still units of analysis or description. This does not mean that it is irrelevant to know which genes or organs are involved in the organism-environment interaction we call 'selection': on the contrary. However, the fact that we can distinguish selectively relevant parts of organisms from irrelevant ones does not imply that the organismal subsystems can be detached from the living whole. After all, to paraphrase Lewontin (1983a), only living systems are the "subject and object of evolution".

To conclude, the distinction between *selection of* (entities) and *selection for* (properties) is useful if interpreted as a distinction between material units of selection and units of analysis or description. The latter may or may not describe the "causes" of selection. But we maintain that the units of selection problem concerns only the material units of selection. Our conclusion is thus congenial to Walton's (1991), who suggested distinguishing what we call the (material) *units* of selection from what he called the *bases* of selection, i.e., the features underlying the process of selection.

9.2.3.6 "Screening Off" and the Units of Selection

The genic selectionists argue that, since the genotype "causes" the phenotype which in turn "causes" an organism's reproductive success, and since the causal relation is transitive, the genotype is, after all, just as directly subject to selection as the phenotype. Moreover, since genes are the "ultimate causes" involved, an explanation of selection in terms of genes would be deeper than an explanation in phenotypic terms. However, these arguments notwithstanding, many organismic biologists maintain that "selection acts on phenotypes, not genotypes" (Mayr 1963). This is to say that, in a process of individual selection, environmental items usually interact with the organism as a whole but not directly with its DNA, for the latter is safely contained within its interior. (Irradiation may be an exception to this generalization. Even in this case, however, the organism's fitness would not be immediately affected by a change in the DNA molecule itself, but only by the latter's metabolic effect, if any, on the rest of the organism.)

To render precise the view that the phenotype is the (most relevant) unit of selection as well as to defend it against genic selectionism, Brandon (1990) has suggested employing the probabilistic notion of *screening off*, which supposedly allows one to determine why the proximate cause—in this case the phenotype—is a better explainer than the remote one—the genotype. In other words, it would allow one to determine the actual level of selection. Thus, he claims that the following relation would hold among an organism's phenotype (p), genotype (g), and reproductive success (n): $P(n \mid p \,\&\, g) = P(n \mid p) \neq P(n \mid g)$ (Brandon et al. 1994, p. 479). If these conditional probabilities hold, then an organism's phenotype would screen off its genotype with respect to reproductive success, so that the unit of selection is the phenotype, not the genes or the genotype. We shall not examine the debate that ensued from this proposal (see Sober 1992; Brandon et al. 1994;

Sober and Wilson 1994; McClamrock 1995; van der Steen 1996), but only state our view of the matter.

To begin with, we must ask which probability interpretation underlies the notion of screening off (recall Sect. 1.10). Brandon (1990) and Brandon et al. (1994) emphasize that the interpretation involved is the propensity interpretation, and they criticize Sober (1992) for analyzing the notion of screening off in terms of subjective probabilities. This implies either that the capacity to produce a certain number of offspring is not a causal disposition, but a chance propensity, or that actual reproduction is a random process, or both. Now we can certainly rule out that fitness is a chance propensity (for the latter recall Section 1.3.1). The process of reproduction, on the other hand, involves some elements of chance, although it is not a fully random process. For example, whether a pollen grain transported by the wind (anemophily) ends up on some pistil or on the ground is partly deterministic, partly random. And although many animals actively search for mates, there are also chance encounters, which add to their total reproductive success. Nevertheless, reproduction is not a pure chance process.

Even if reproduction were a pure chance process, and we had a stochastic model for it, the problem would remain that the allegedly objective probabilities in the formula '$P(n \mid p \,\&\, g) = P(n \mid p) \neq P(n \mid g)$' are unknown. Still, if they are indeed objective, then the formula is a factual hypothesis. Hence, the corresponding probabilities should be obtainable by empirical means such as measurement, not by fiat. (Note that, given the complexity of the process of reproduction, we cannot propose a reasonable estimate *a priori*, as we can, for instance, in the simple event of a coin flipping. But even in this case, the estimated probability of 0.5 must be confirmed empirically.) However, as long as the probabilities involved have not been measured, the claim that either the formula in general or any of its particular cases actually holds is unsubstantiated. Worse, the probabilities appear to be unobtainable, because an organism always has both a phenotype and a genotype, so that, at best, the value of $P(n \mid p \,\&\, g)$ might be obtained, but not those of $P(n \mid p)$ and $P(n \mid g)$.

In any case, the notion of screening off seems superfluous. For one, the factual situation is quite clear: genes are contained within biosystems and thus usually do not directly interact with the items in the biosystems' environments. Thus, the system level or, if preferred, the interactor level involved in a process of individual selection is to be determined by a system analysis, not by any probabilistic considerations. Moreover, a phrase such as 'g is irrelevant to n', whether true or false, is clear enough to state what is at issue. So we see no need for a technical notion like that of screening off.

For another, the standard belief that the genotype would "cause" the phenotype is not only highly elliptical, as van der Steen (1996) has rightly observed, but also false; in particular, it rests on a rather simplistic view of development. Recalling Section 8.2.3.3, the only notions of genotype at issue here can be the concepts of genotype$_H$ or genotype$_D$. But these are sets, not concrete events, and hence not causally efficacious. Likewise, the alleged effect—the organism's phenotype$_T$—is a set, too, namely the set of its trait types. Even if one attempted to circumvent this objection by referring to particular genes and individual organismal traits, one would not arrive at a causal relation proper, as we pointed out in the preceding section. (The same holds for any of the other concepts of genotype and phenotype analyzed in Sect. 8.2.3.3.) For this reason, there is no need in the first place to distinguish remote from proximate causes of selection by means of the notion of screening off. (See also van der Steen 1996.) Finally, explanations of both development and selection exclusively in terms of genes are not deeper, but simply reductionistic: recall that we require a satisfactory explanation to be a *systemic* explanation, not a purely microreductive one.

9.2.4 Conclusion

Selection may occur at different levels of systems. The differentially performing and thus reproducing entities belonging to these levels are the (material) units of *p*-selection. Therefore, it is no longer controversial that *there are several levels of selection*. However, given the multitude of possible levels of selection, what does remain a subject of theoretical and practical controversy is (a) the problem of how to find and model the real units and processes at the adequate level, i.e., how to distinguish material units from conceptual ones, and (b) the determination of the frequency of those different possible units of selection and thus their relevance to evolution. The available evidence suggests that the most important unit of selection is the organism (Brandon 1990).

The fact that there are different (material) units of selection at different system levels is also reflected in the structure of the theory of selection. In our terminology (for which see Sect. 3.5.3.2) the theory of selection is a *hypergeneral* theory referring to rather unspecific "interactors". (Darden and Cain 1989 regard it as an abstract theory, but it is not: what is at issue here is the degree of generality of a factual theory, not its degree of abstraction.) Interpreting this hypergeneral theory in terms of molecules, cells, organisms, or groups will result in a family of *general* theories of selection. Enriching any of these general theories of selection, say, a theory of organismal selection, with specific assumptions concerning the organisms in question, results in a specific theory or model$_3$ of a general theory of natural selection. For instance, specifying a general theory of selection to *Biston betularia*, the famous peppered moth, we obtain one model of selection. If we now analyze the referents of the model we will find at least three referents (or classes of referents), namely (aggregates of) moths, birds, and birches. The relevant properties of these referents, such as the color of the moths' wings, which is the relevant aptation in this case, and the color of the birches' barks, which is the relatum of the moths' adaptedness, are not themselves referents of the model, because they are not things. Hence, they do not qualify as units of selection. What qualifies as the (material) units of selection in this case are the referents of the concepts of (differential) adaptedness, (differential) fitness, and (differential) reproduction occurring in the model: that is, the moths. What "benefits" in this case are not only the moths but necessarily all their subsystems, whether wings or genes, as well as certain supersystems such as the biopopulation and perhaps even the ecosystem. To single out any of the subsystems involved in the production of the aptation(s) in question is, of course, of analytical relevance, but it does not change their status as units of description rather than as units of the process of selection.

9.3 The Structure of Evolutionary Theory

Having dealt mainly with some ontological problems of evolutionary theory (henceforth ET) in the preceding, we turn now to some of its semantical and epistemological problems.

9.3.1 What, If Anything, Is "Evolutionary Theory"?

To begin with, we must ask whether there is such a thing as *the* theory of evolution at all. More precisely, we have to ask whether there is a general or perhaps hypergeneral unified theory of evolution. Some authors (e.g., M.B. Williams 1970, 1973a; Rosenberg 1985, 1994) believe that the theory of natural selection is at the center of ET, hence *the* theory of evolution. Others (e.g., Ruse 1973) are more inclined to view population genetic theory as the kernel of ET. As will be obvious from the preceding, neither of the two theories is a complete ET. Population genetic theory holds only for eukaryotic organisms. That is, it cannot account for the evolution (i.e., the origin) of eukaryotic organisms, but only for further evolution within the Eukaryota. This is the reason that it cannot be at the center of a general theory of the evolution of all organisms. (See also Rosenberg 1985, 1994.)

Both population genetic theory and selection theory may account for the distribution and fixation of variants in a population, i.e., speciation in the population genetic sense. But they do not contain any notion of speciation proper, i.e., any notion of emergence of qualitative novelty. Moreover, population genetic theory is evidently simplistic because it focuses on the frequencies of genes and genotypes as if they were self-existing entities. It is even reductionist if seen not just as a simplification of the actual processes, but as providing a sufficient explanation of evolution. (More on the explanatory power, or lack thereof, of population genetic theory in Horan 1994.) Even adding our knowledge of molecular genetics and cytogenetics, we can only explain mutations at the gene and chromosome levels. Unless we also make use of the whole of developmental biology, we remain either simplistic or reductionist (or both), that is, we are not even dealing with living entities: we are not doing biology proper.

The theory of natural selection is also insufficient, as focusing on the differential survival and reproduction of either genes, organisms, or populations. It certainly explains the differential trans-generational distribution and the prevalence of these entities in a given habitat; but it takes the concept of (ad)aptedness for granted—in other words, it treats (ad)aptedness as a black box—and thus leaves the very essence of selection unaccounted for, namely the organism-environment interaction. The latter can only be explained with the help of functional morphology and ecology, both organismal and populational.

To conclude, only all the previously mentioned theories together are necessary and (hopefully) sufficient for a fully-fledged explanation of biological evolution. Therefore, the so-called Synthetic Theory is far from being a synthesis of all we need to account for biotic evolution. Moreover, neither the Synthetic Theory nor a combination of all the above-mentioned disciplines and theories seems to constitute a single unified general theory. Worse, some of these disciplines, such as functional morphology, ecology, and developmental biology, seem to be more descriptive than theoretical, in particular mechanismic. In other words, at present they may not even have a general theory proper, that is, one capable of genuine explanations. Thus, it appears that there are several partial theories of evolution, each dealing with a different aspect of the evolutionary process(es). Yet there seems to be no (hyper)general theory of the evolution of all organisms. Consequently, the theory of evolution appears to be rather a collection of theories or, at best, a system of theories (see, e.g., Beckner 1959; Caplan 1978; Lewis 1980; Lloyd 1988; Thompson 1989).

9.3.2 What the Structure of Evolutionary Theory Is Not

The lack of a unified hypergeneral theory of evolution, together with the seeming lack of law statements in biology, has encouraged some philosophers, critical of the so-called received view of theories, to claim that this is evidence for the inadequacy of the view that scientific theories are hypothetico-deductive systems containing law statements. Thus, Beatty (1981, 1987), Thompson (1983, 1987, 1989), and Lloyd (1988) have argued that an alternative view of theories should be tried in biology, particularly since they believe that this alternative view has succeeded in physics. This alternative is the so-called *semantic view* of scientific theories. As only few philosophers of biology have examined this view critically (e.g., Sloep and van der Steen 1987; Ereshefsky 1991), we shall take a closer look at this approach. As a matter of fact, the "semantic" view consists of at least two subviews, namely the structuralist conception of theories and the "semantic" conception proper. Before we tackle their applications to biology, let us examine these views with regard to their alleged general virtues as well as with regard to their alleged success in physics.

9.3.2.1 The Structuralist (Suppes's) Conception of Scientific Theories

The structuralist view goes back to the 1953 paper on particle mechanics by the logician J.C.C. McKinsey and his then students P. Suppes and A.C. Sugar. Although this paper was only published with a disclaimer by its communicator—the maximal living authority in classical physics, Clifford Truesdell, who in 1984 demolished this kind of foundations of physics once and for all—Suppes's later version (1957) is quoted up until today as a successful exercise in structuralism (e.g., by Thompson 1989, p. 83). Though thus actually flogging a (*de jure*) dead horse, we shall first analyze the technical flaws of Suppes's axiomatization and then the general problems of such construals of scientific theories.

Let us consider the often quoted definition of a "system of particle mechanics" as a mathematical structure (system) satisfying certain purely mathematical axioms. Suppes's (1957, p. 294) formulation begins thus: "DEFINITION 1. A system $\mathcal{P}= \langle P, T, s, m, f, g\rangle$ is a system of particle mechanics if and only if the following seven axioms are satisfied"—here an incomplete and incorrect formulation of Newton's axioms follows. The first three axioms come under the heading 'Kinematical Axioms', the remaining four are termed 'Dynamical Axioms'. Some of the objections against this axiomatic definition are the following.

First, the partition of the axioms into kinematical and dynamical ones is, to put it mildly, unusual. Indeed, Axiom 1, "The set P is finite and non-empty", is a purely mathematical assumption: it does not state anything about motion. The same holds for the other two "kinematical" axioms. Ditto Axiom 4, the first of the "dynamical" axioms: "For p in P, $m(p)$ is a positive real number". Obviously, this statement says nothing about dynamics. This is not quibbling: the differences between mathematical, kinematical, and dynamical assumptions, obvious to any physicist, are of philosophical interest. A mathematical formula is a pure construct devoid of any relation to the world as long as neither of the concepts occurring in it is assigned a factual interpretation. A kinematical formula *describes* change of some kind (e.g., motion), and a dynamical one *explains* it in terms of forces, stresses, fields, or what have you. All of this is conspicuously absent from Suppes's axiomatization.

Another defect of this axiomatization is that it omits the key concepts of reference frame and unit. Without them, the basic variables, the position coordinate and the force, are ill-defined; consequently, so are the derived variables, such as the velocity and the acceleration. To be sure, the axioms should not be tied to any particular reference frames or units: they must make room for all of them in order to allow for specification at the theorem level. (The way to do so in the case of a classical particle coordinate X is to stipulate that X is a function from the Cartesian product $P \times F \times T \times U_L$ of the collections P of particles, F of frames, T of time instants, and U_L of length units, to the set of triples of real numbers. In other words, $X(p, f, t, u) = \langle x, y, z\rangle$, where $p \in P, f \in F, t \in T, u \in U_L$, and $x, y, z \in \mathbb{R}$. See Bunge 1976 for additional criticisms, and 1967c for an alternative axiomatization of elementary mechanics.)

Of course, the technical flaws of a particular example do not invalidate the whole enterprise. But they suffice to refute the claim that Suppes's axiomatization is a *successful* application of structuralism to physics. Yet there are not only technical flaws: the conception advocated by Suppes and his followers has, from a realist perspective, also the following serious methodological and philosophical defects. The first is the ambiguous expression 'system of particle mechanics' occurring in the definition, for it may mean either "system of particles"—a concrete thing, such as a gas or a rigid body—or "particle mechanics"—a theory or model dealing with individual particles as well as with systems of particles.

As for the reading "system of particles", the realist must object that the method of axiomatic definition, adopted by Suppes and his followers, is taken from pure mathematics. For instance, a lattice may be defined thus: a lattice is a set S of elements equipped with two binary operations in S, designated by $\cap$ and $\cup$, such that the following axioms hold for all elements of S—and here the six relevant axioms are listed. This method of "creative definition" works well in pure mathematics, but not in factual science. The reason is that, unlike concepts, concrete things—the referents of factual science—are undefinable: they can only be described or characterized (recall Sect. 3.5.7.1). In fact, every explicit definition is an identity, and the two sides of a definition can only be constructs, as in the case of "2 = 1 + 1". It is impossible to equate two concrete things: if two such objects were identical "they" would be only one. Nor can one equate a concrete thing with a concept, for things have only physical (or biological, or other) properties, whereas constructs have only (conceptual) attributes. For these reasons, one must not say, for example, that a body *is* a set, or a field *is* a differentiable manifold, but that a body is *representable* as a set, and a field is *representable* as a differentiable manifold. Thus, an amendment to Suppes's example would consist in replacing the statement "a system $\mathcal{P} = \langle P, T, s, m, f, g\rangle$ *is* a system of particle mechanics" by "a system $\mathcal{P} = \langle P, T, s, m, f, g\rangle$ *represents* a system of particle mechanics" (in the sense of "system of particles"). Yet this option is not available to the structuralists for they insist on *defining abstract* or *ideal* systems without regard to factual correspondence.

However, even if this amendment were possible, it would not solve the major defect, which also affects the second reading of 'system of particle mechanics', namely "(model or theory of) particle mechanics". This defect is the complete lack among the axioms of any interpretation in factual terms: the latter occurs, if at all, only in the extrasystematic remarks (and even so it is incomplete in Suppes's case). Now, any mathematical formula can be interpreted in alternative ways—or in none. What makes a particular set of mathematical formulas a part of mechanics, or of biology, is the interpretation of the concepts in factual terms—and such interpretation is absent from structuralist axiomatizations. Therefore, those structuralists who want to make contact with reality after all claim that their abstract models have an "intended application", that is, a possible interpretation in either empirical or factual terms. However, this "intended application" is clearly extra-theoretical, so that structuralist theories proper have no factual content (or reference) whatsoever (see also Sloep and van der Steen 1987). Without such content, neither the reading "system of particles" (or perhaps "ideal system of particles") nor "(model of) particle mechanics" makes any sense. (Incidentally, Giere's 1984 version of structuralism seems to be exempt from many of our objections, because what he calls 'theoretical models' are not at all abstract systems but factually interpreted ones. Still, he believes that theoretical models are definitions.)

In sum, the structuralist view of scientific theories is of no use to scientific realists. So much so that Suppes's and his followers' "reconstructions" of physical theories are ignored by physicists (see Truesdell 1984; for a witty deconstruction of Stegmüllerian structuralism see Weingartner 1990).

9.3.2.2 The "Semantic" View

The nucleus of the "semantic" view of scientific theories (see, e.g., van Fraassen 1972, 1980; Suppe ed. 1974, 1989; Thompson 1989) is that the models used in factual science are models in the model-theoretic or logical sense of the word. Witness Suppes (1961, p. 165): "The meaning of the concept of model is the same in mathematics and the empirical sciences". As will be recalled from Section 3.5, logical models_1 are interpretations, within mathematics, of abstract formulas or theories. Since the interpretations (i.e., the semantics) involved in model theory are thus an intramathematical affair, that branch of logic is irrelevant to scientific theories, for these are supposed to refer to extramathematical entities and, consequently, must include construct-fact bridges, i.e., factual interpretations. (Recall also from Sect. 3.5 that what we call 'factual interpretations' have nothing to do with the correspondence rules of the logical empiricists.)

The proposal of equating models in model theory and models in factual science might have three possible joint sources. One is, of course, the unfortunate occurrence of the word 'model' in both cases. (Hence our indexing of the word 'model'.) A second source is the bankrupt verifiability theory of meaning, the neopositivist doctrine according to which a sentence is meaningful just in case it is verifiable. (See also Sect. 2.3.) Indeed, van Fraassen (1972, p. 305) asserts that an interpretation of the syntax [mathematical formalism] of a theory "should issue specifically in a truth-definition": one should be able to decide whether the given abstract formula is satisfied under the given interpretation. This is, indeed, so in the case of abstract mathematics. But in the case of scientific hypotheses and theories, meaning precedes testability because a nonsensical expression cannot be checked for truth. And only actual tests allow us to gain knowledge about the degree of truth of a hypothesis: believing that a hypothesis is true without checking is dogmatism, not science. A third source of the confusion is the belief (van Fraassen l.c.) that "the axiomatic ideal insists on a purely syntactic definition [sic] of the theorems". But this too is false. Even Hilbert, the arch-formalist and champion of axiomatics, warned at the very beginning of his monumental treatise (Hilbert and Bernays 1934, p. 2) that one must distinguish between "inhaltlicher und formaler Axiomatik", and that only the latter abstracts from all content. For example,

Peano's system of postulates for natural numbers is not formal (abstract), for all the constructs occurring in it are interpreted even if they are undefined or primitive. (For example, the first axiom reads "0 is a natural number".) This holds, *a fortiori*, for all the theories in factual science and technology. When properly axiomatized, every one of these theories splits into two parts: a mathematical formalism and a set of semantic assumptions.

To conclude, in the "semantic" conception of theories "[...] 'semantic' is used [...] in the sense of formal semantics or model theory in mathematical logic" (Suppe 1989, p. 4). Thus, the semantics of model theory consists in the mapping of mathematical structures (also called 'systems') into mathematical structures (systems). By contrast, the semantic assumptions in factual science correlate definite mathematical structures with real systems. (Evidently, real systems, though describable in mathematical terms, are not mathematical objects.) Therefore, model theory cannot take care of the semantics of factual science. As a matter of fact, the semantics of factual science takes off where model theory leaves. Thus, for the factual scientist and the realist philosopher of science, the so-called *semantic* view is actually a *nonsemantic* view of scientific theories. (Hence, the double quotation marks around the term 'semantic' when referring to the semantic view in this book.) Since the referents of a scientific model are concrete, and those of a model in pure mathematics are not, model theory and the nonsemantic view of scientific theories are of no use in the philosophy of science and technology.

Sometimes the "semantic" view of theories is also called the 'state space approach to scientific theories', because some authors make use of the notion of a state space (e.g., van Fraassen 1972; Suppe 1989). Lest the impression is evoked that the state space approach is essential to the "semantic" conception of theories and, moreover, an alternative to the conception of a theory as a hypothetico-deductive system, we hasten to point out that such an impression would be wrong. For any hypothetico-deductive system containing state variables, such as position and momentum, or gene frequency and adaptedness, involves a state space, since this is just the space spanned by the state variables. Thus, if the theory contains n such variables, every one of which represents a property of the things of the kind under consideration, then the corresponding state space will be an n-dimensional Cartesian space, every axis of which will be the range of the corresponding state variable (recall Sect. 1.4) In sum, state spaces are not peculiar to the "semantic" conception of theories.

However, what is peculiar to the state space approach within the "semantic" view is the idea that "laws do not describe the behavior of objects in the world; they describe the nature and behavior of an abstract system" (Thompson 1989, p. 72). Thus, there are, for instance, "laws of coexistence" that "serve to select the physically possible set of states in the state space", and there are "laws of succession" that "select the physically possible trajectories in the state space" (l.c., p. 81). At first sight, this sounds very similar to what we have said about nomological state and event spaces in Section 1.4. Yet the differences are fundamental. In our view, a restriction (i.e., law$_2$) on the logically possible set of states of an object within a state space is a law_2 only if it represents a law_1, i.e., the behavior of a real object. By contrast, in the "semantic" view, any arbitrary restriction seems to function as a law. This comes as no surprise, since on the "semantic" view theories are devoid of factual content, i.e., they are not concerned with representing reality but with defining abstract (mathematical) systems. "Laws" thus conceived do indeed not describe the behavior of objects in the world. In other words, in this view there is no distinction between objective regularities ($laws_1$) and $laws_2$ as representations of objective patterns of being and becoming. Yet scientists are not after exercises in abstract mathematics: they are centrally interested in statements ($laws_2$) that describe the lawful behavior of real objects. (More on laws in the "semantic" view in Ereshefsky 1991.)

To be sure, even the "semanticists" assure us that their abstract models have an "implicit" or "intended" application or interpretation. However, since they do not use the notion of factual interpretation of statements, they come up with a curious trick: they say that the highly abstract systems "defined" by a theory "relate to phenomena through a complex hierarchy of other theories" (Thompson 1989, p. 92; similarly p. 82). However, if theories *are* definitions of abstract systems, then *all* of them are abstract and thus *none* of them can have any relation whatsoever to facts ("phenomena"). If consistent, they must construe those "theories of experimental design, goodness of fit, analysis and standardization of data, and so forth" (p. 82) that are supposed to do the relating as ideal or abstract systems as well. How such a hierarchy of ideal systems, then, can eventually relate to something concrete remains the semantic mystery of the "semanticists".

Moreover, since the "semantic" view rejects the construal of scientific theories as sets of logically related statements, it has no use for the notion of factual truth. (The above-mentioned concept of satisfaction in a model is a notion of formal truth.) Therefore, the concepts of representation and truth are substituted for by the notion of isomorphism. That is, instead of saying that a theory or model is true, it is said that the abstract (often misleadingly called 'physical') system defined by a scientific theory would be "isomorphic to a particular empirical system" (Thompson 1989, p. 72); in other words, the two would have "the same causal structure" (p. 82). We have two objections to this idea.

First, the attribution of a causal structure to constructs is clearly a metaphysically ill-conceived notion. Second, the mathematical relation of isomorphism is a purely mathematical relation holding only between sets. Since models (or ideal or abstract systems or structures) are conceptual objects, and "empirical" [read: factual] systems are material objects, the mathematically well-defined relation of isomorphism does not apply. Indeed, in a footnote, Lloyd (1988, p. 168) admits that "in practice, the relationship between theoretical and empirical model [?] is typically weaker than isomorphism, usually a homomorphism, or sometimes an even weaker type of morphism". But neither a homomorphism nor any other intramathematical morphism or mapping can be invoked to render the vague idea more precise that a model somehow "mirrors" its factual referent(s). We must, therefore, suspect that the metaphorical usage of terms such as 'isomorphism' might be a residue of the (wrong) *reflection theory of knowledge* held by dialectical materialists (Lenin 1908), the young Wittgenstein (1922), and some contemporary philosophers (van Fraassen 1980). However, although models and theories may *represent* their referents to some approximation, such representations are not pictures or copies. For many scientific theories contain constructs without real counterparts, which serve at most as computational devices; and every piece of reality ends up by showing traits that had not been foreseen by any theory. Therefore, at best there can be a global correspondence between theory and facts, not a point by point one as is presupposed by the isomorphism thesis. In fact, theories and models are *symbolic constructions* bearing no resemblance whatsoever to the objects they represent (see Postulate 3.5).

Assuming for the sake of argument that it would make sense to use the notion of isomorphism, that is, that one could say that models (or abstract systems) are isomorphic to concrete systems if they have the same structure, we would face another serious problem. This problem is the fact that, by comparing structures, one can at best *describe* concrete systems but *not explain* them. Worse, it is impossible to make predictions from a mathematical structure. Explanations and predictions are logical arguments, that is, they presuppose statements from which further statements can be deduced, thereby either explaining the referents of the propositions or predicting them. How all this should be possible in a nonstatement view of theories is unclear. In short, neither subsumption nor explanation nor prediction is possible if theories are not hypothetico-deductive systems. Scientists should not be prepared to pay such a high price for the fancies of some philosophers.

9.3.2.3 The "Semantic" (Including Structuralist) View of Theories and Biology

Apparently unaware of the failure of the nonstatement views in physics, Beatty (1981, 1987), Lloyd (1988), and Thompson (1983, 1987, 1989) have argued for the application of the "semantic" view to biological theories. Since Thompson (1989) has made the most extensive case in favor of the "semantic" view in biology, we shall focus on his account. Following Beatty's (1981) set-theoretical approach to population genetics, Thompson, in analogy with Suppes's axiomatic definition of a system of particle mechanics, proposes an axiomatic definition of a "Mendelian breeding system" (Thompson 1989, p. 88), which reads as follows:

T: A system $\beta = \langle P, A, f, g \rangle$ is a Mendelian breeding system if and only if the following axioms are satisfied:

Axiom 1. The sets P and A are finite and nonempty.

Axiom 2. For any $a \in P$ and $l, m \in A$, $f(a, l)$ & $f(a, m)$ iff $l = m$.

Axiom 3. For any $a, b \in P$ and $l \in A$, $g(a, l)$ & $g(b, l)$ iff $a = b$.

Axiom 4. For any $a, b \in P$ and $l \in L$ such that $f(a, l)$ and $f(b, l)$, $g(a, l)$ is independent of $g(b, l)$.

Axiom 5. For any $a, b \in P$ and $l, m \in L$ such that $f(a, l)$ and $f(b, m)$, $g(a, l)$ is independent of $g(b, m)$.

All the defects that plagued Suppes's "reconstruction" of particle mechanics re--emerge in the Beatty-Thompson "reconstruction" of Mendelian genetics. First, as with Suppes's expression 'system of particle mechanics' criticized above, the expression 'Mendelian breeding system' in this axiomatic definition is equally ambiguous. It may mean either a reproductive community, i.e., a concrete system, or a theoretical model, i.e., an ideal system, representing a concrete system. Now, $\langle P, A, f, g \rangle$ is, if anything, a *mathematical* system, hence a conceptual object, which cannot be identified with a material one. The only interpretation, therefore, that makes sense (in a normal scientific framework) is that $\langle P, A, f, g \rangle$ *represents*, i.e., is a model or an idealization of, a Mendelian breeding system. But if so, we face the problem that this axiomatic definition does not contain any factual reference: it is a pure formalism not representing anything. Worse, even if we endowed the axioms with a factual interpretation, this interpretation could no longer be said to "define" the systems to which the formalism applies, because one can only define signs and concepts, not things. The latter can only be described by the statements of a theory.

Such general problems apart, there are also several fundamental logical and mathematical problems with this particular model[1]. First of all, a mathematical formalism ought to make (mathematical) sense in itself, i.e., without resort to extratheoretical explications. This is not the case here. For example, the last four axioms make no mathematical sense because the key symbols f and g occurring in them are undefined. Moreover, since the formula $f(a, l)$ and its likes do not correspond to any standard notation, they could mean just anything.

If we, nevertheless, show some good will to make mathematical sense of the axioms, we could try to read the formula $f(a, l)$ and its likes as relational notation. Then a more conventional notation for '$f(a, l)$' would be 'Ral' or 'aRl'. Correspondingly, we could rewrite (the relevant part of) Axiom 2 in the form "Ral & Ram iff $l = m$". Then this axiom would tell us that the relation R, or for that matter f, is a function from P to A. However, if this were the mathematical meaning of this axiom, we wonder why one would want to use relational notation to designate a function. Axiom 3 on the other hand (rewritten as "Ral & Rbl iff $a = b$") would not define the relation g as a function: it tells us only that g is *not* a many-to-one relation. In other words, it could be either a one-to-many or a one-to-one relation.

Showing even more good will, we eventually try to make sense of the formalism by taking Thompson's extratheoretical remarks into account. They are given immediately after the above-quoted formalism and read thus:

> Where P and A are sets and f and g are functions. P is the set of all alleles in the populations [sic], A is the set of all loci in the population. If $a \in P$ and $l \in A$, then $f(a, l)$ is an assignment, in a diploid phase of a cell, of a to l (i.e., f is a function that assigns a as an alternative allele at locus l). If $a \in P$, and $l \in A$, then $g(a, l)$ is the gamete formed, by meiosis, with a being at l in the gamete (the haploid phase of the cell). (p. 88).

---

[1] We owe several of the following arguments to Michael Kary.

These remarks help clarify the matter somewhat, but they do not save the formalism from being ill-formed. (We take the letter L instead of A in the last two axioms to be a misprint.) First, if Axiom 2 were needed to define the relation f as a function, then why state twice in the extratheoretical remarks that f is a function? It would be obvious that f is a function, if Thompson had chosen the standard (and correct) functional notation such as '$f(a) = l$'. Then, however, the axiom would be superfluous for defining f as a function. Second, if Axiom 2 were necessary to define f as a function, then why is there no axiom defining g as a function as well? (Recall that Axiom 3 only states that g is *not* a many-to-one relation.) However, if g is indeed a function, though it is nowhere defined as such, then Axiom 3 says that g is an injective (i.e., a one-to-one) function. But if this were the (mathematical) meaning of Axiom 3, it would be sufficient to write "If $g(a) = l$ & $g(b) = l$, then $a = b$". The converse, that is, "If $a = b$, then $g(a) = l$ & $g(b) = l$", as indicated by the equivalence "iff" in the axiom, would not be required.

Worse, if we take the extratheoretical remarks into account and interpret the formulas in terms of alleles, loci, and gametes, the equivalence "iff" in Axiom 3 says that, in gametes, *every* allele is assigned to the locus l. More precisely, let a designate an arbitrary allele in the set P, and let l designate an arbitrary locus in the set A. Now define b as follows: $b =_{df} a$. Therefore, since $a = b$, by this axiom, $g(a) = l$. Now let c designate some other arbitrary allele in P, different from a. Define d as follows: $d =_{df} c$. Therefore $g(c) = l$ as well. But since $g(a) = l$ and $g(c) = l$, then $a = c$, which contradicts the original supposition that they were different.

A similar argument can be made for Axiom 2. Let a designate an arbitrary allele in P, and let l designate an arbitrary locus in the set A. Define m as follows: $m =_{df} l$. Therefore, by this axiom, $f(a) = l$. Now let n designate some other locus, different from l. Define p as follows: $p =_{df} n$. Therefore $f(a) = n$. And of course, since $f(a) = l$ and $f(a) = n$, then $l = n$, contradicting the original supposition that they were different. So all alleles, whether in normal cells or in gametes, are assigned only to the single locus l.

Another problem with the axiomatic definition and the extratheoretical remarks is that it is unclear whether what are at issue are particular alleles and loci or else allele *kinds* and loci *kinds*: the extratheoretical remarks speak only of alleles and loci in populations. So there is nothing in the formalism that prohibits the assignment of the individual allele a in cell c of organism m in population p to the individual locus l in cell d of organism n in population p.

Furthermore, if f and g do designate functions, then the formulas "$f(a, l)$ & $f(a, m)$" and "$g(a, l)$ & $g(b, l)$" are not well-formed, because the concept "&" is a logical relation between propositions, and there is no indication that the values of these functions are that. This holds in particular for g, where a concrete thing—a gamete—is identified with a symbol. That is, if the symbols '$g(a, l)$' and '$g(b, l)$' *are* gametes, they are neither true nor false, and they do not designate propositions, as required by the presence of the '&' in the axioms. Even if we could say that g *represents* a gamete, we must wonder why a concrete thing should be represented by a function rather than, typically, as an element of some set—particularly since alleles and loci, though objects, are not "represented" by functions but as elements of the sets P and A. (Usually, *properties* of objects are represented by functions that assign to the object some denominate number. For instance, body temperature may be represented by the function T(human body, centigrade scale) = 37; recall Sect. 1.3.3.) As it seems, in structuralism gametes are treated as disembodied properties or, more precisely, as allele-locus relations.

Finally, let us turn to the Axioms 4 and 5, in which the word 'independence' is undefined. Such definition is required because 'independence' designates different concepts in different contexts. For example, since the formalism is uninterpreted, we could read it as some logical or mathematical notion of independence. Thus, the two functions f and g occur in a logical conjunction in Axioms 2 and 3, which only makes sense if the functions represent propositions such as "a is related to l", not concrete things such as gametes. But then the truth values of the two propositions are *not* independent, i.e., both must be true if the conjunction is supposed to be true. Alternatively, we could read 'independent' in terms of functional independence: two functions are independent iff they do not share the same variables. However, both Axiom 4 and 5 only state that the values of the function g are independent. In sum, the word 'independent' in the last two axioms is meaningless as far as the axiomatic definition itself is concerned.

Only by turning to the extratheoretical remarks can we tell that a factual application is intended, and that this application does not refer to, say, political independence but to the notion of the independent assortment of alleles during gametogenesis, i.e., Mendel's second law. Yet if one wants to define this particular concept, one cannot import it from outside the theory in order to make sense of the formula: it ought to be defined, implicitly or explicitly, within the theory. After all, as said at the beginning, a mathematical formalism must make (mathematical) sense in itself.

To conclude, this "example as it stands illustrates adequately the nature of a set--theoretical approach" (Thompson 1989, p. 88). Indeed, it does. However, we still have an ontological quarrel with Thompson's plea in favor of the "semantic" view in biology. This criticism concerns the equating of constructs, such as mathematical functions, with concrete things, such as gametes. This conflation is not accidental: it belongs to the very essence of the formalist view of theories, in which the construct-fact dichotomy is blurred. A multitude of metaphysically ill-formed statements in Thompson's work attests to this. Let us list a few of them.

First, being constructs, neither state spaces (Thompson 1989, p. 12) nor theories can in any way "interact with each other" (p. 52, p. 96). Second, a category mistake is involved in Thompson's statement that an "empirical model is understood as logically equivalent to the phenomenal system to which the theory applies" (p. 72). In fact, neither phenomenal nor transphenomenal facts have any logical properties. Third, it is impossible that "evolutionary theory is related to human behavior through a causal chain of theoretical frameworks" (p. 111), and there just is no "causal sequence of theories" (p. 17). No construct can in any way be causally related to concrete things and their changes. Similarly, the claim that according to the received view "theories are deductively related to phenomena..." (Thompson 1987, p. 29) makes no sense to us because, if anything, only further statements (about facts and phenomena) can be deduced from the statements of a theory. Furthermore, law statements do not "define" things (Thompson 1989, p. 63), but at best describe them. The same holds for theories: population genetic theory does not define "the systems to which it applies" (p. 91), just because there is nothing in the "semantic" view that could relate an abstract formalism to the real world. Not even the correspondence rules in the neopositivist conception of theories "define an empirical model of the formal system" (p. 72): neither an interpretation nor an operationalization of a theory has anything to do with a definition; and concrete systems are not models of anything.

So much for the "success", philosophical and technical, of the "semantic" view of scientific theories from physics to biology.

9.3.2.4 Conclusion

There is nothing wrong with representing, i.e., describing, the structure of concrete systems by means of mathematical structures. Yet in order to do so, a mathematical structure must *explicitly* refer to a concrete system; that is, it must be factually interpreted. Even so, there is still more to material systems than structure: every material system has also a composition and an environment (except the universe as a whole), and, moreover, a history. All these characteristics, in particular the states and changes of state of the system in question, can be taken into account in a state space model. However, a state space (model) cannot be said to be in any sense isomorphic to a concrete system, for the dimensionality of the latter is the ontic dimensionality of spatiotemporality, whereas the *n*-dimensionality of a state space is the purely conceptual dimensionality of a mathematical space.

Neither is there anything wrong with idealization. As we saw in Section 3.5, theoretical models involve specific assumptions or object sketches, so that models$_3$ are *simplified* and *idealized* representations of their real referents. Still, scientists are not interested in ideal systems *per se*: even an idealization is supposed to approach and approximate at least one aspect of a real system, or to help understand real systems by showing in what respect the latter deviates from the ideal case. (Think of ideal populations.)

In sum, we see no reason for giving up the conception of scientific theories as hypothetico-deductive systems and adopting the "semantic" (or nonstatement) view of theories instead. After all, there is nothing in mathematical structures and state spaces that could not be expressed by propositions. Moreover, every virtue (incorrectly) attributed to the "semantic" conception can be (correctly) attributed to a realist conception of theory structure, which avoids the problems that plagued the neopositivist version. In addition, we can spare ourselves scientifically (not mathematically) useless tools such as model theory, and we can—which is even more important—avoid any contamination with certain philosophical positions often associated with the "semantic" view, such as phenomenalism, empiricism, instrumentalism, and antirealism. (Notable exceptions seem to be Giere 1984, 1985, and Suppe 1989, whose versions are avowedly realist.)

Table 9.1 summarizes and compares the realist, the neopositivist, and the "semantic" views of theory structure.

Table 9.1. The realist, neopositivist, and "semantic" views of theory structure

	Realism	**Neopositivism**	**"Semantic" View**
A theory is	A hypothetico-deductive system	A hypothetico-deductive system	A definition (structuralism) or a set of models ("semantic" view)
Formal presuppositions	Predicate logic and all the requisite branches of mathematics	Same	Same plus model theory
Status of axioms	Far removed from data (consisting usually of transempirical concepts)	Empirical generalizations	Abstract
Empirical content	No	Yes (via correspondence rules)	No
Factual reference	Yes (endowed by semantic assumptions)	No (reference only to observables, i.e., phenomena)	No ("application" to facts may be "intended", but if so, it is external to theory)
Indicators	External to theory and added to it in preparation for tests	Included in correspondence rules	Absent
Empirical testability	Only upon enrichment with indicator hypotheses and data	Direct (because of empirical content)	No

9.3.3 The Actual Structure of Evolutionary Theory

Since in our view all theories are hypothetico-deductive systems, all theories have the same general structure: this is a context closed under deduction. More precisely, a theory can be characterized as a quadruple $\mathcal{T} = \langle P, Q, R, \Rightarrow \rangle$, where P designates a set of propositions, Q the set of all predicates occurring in P, R the set of referents in Q (i.e., its reference class or domain), and $\Rightarrow$ designates the implication relation that glues the members of P into a system. Here, P, Q, and R designate the composition of the theory, while the internal structure proper is given by the implication relation (or, alternatively, by the relation of entailment).

Given this general structure of theories, what we are concerned with when talking of *the structure* of ET, is actually exhibiting the composition *and* the structure of ET. That is, if we know the propositions and the predicates as well as the reference class of the theory, the structure of the theory is given by the logical connections among the propositions in P. In other words, we can exhibit the structure proper of a theory only by determining what the postulates and the logical consequences are. This amounts to an axiomatization of the theory in question.

However, axiomatizations of biological theories are very rare. Besides the early attempts of Woodger (1952), the most famous axiomatization in biology until today is Mary Williams's axiomatization of selection theory (1970), which, though often criticized (e.g., by Ruse 1973; Jongeling 1985), has been endorsed by Rosenberg (1985, 1994). A less rigorous axiomatization of parts of evolutionary theory has been provided by Van Valen (1976b) as well as by Kitcher (1989b), and central statements of the genetics of Eukaryotes have been axiomatized by Rizzotti and Zanardo (1986) and formalized by Zanardo and Rizzotti (1986). Finally, Lewis (1980) has compiled (not axiomatized) several central statements of biological theories, and Vollmer (1987b) has listed several principles of the theory of evolution.

Given this state of affairs of theory axiomatization in biology, most of what passes as analyses of the structure of ET must be regarded as preliminary studies of certain components of ET, in particular its reference class. This is not to say that these studies are in any way useless, but it is to point out that they are still far from really exhibiting the structure of ET. For example, it has become fashionable to say that ET or, for that matter, the theory of selection has a "hierarchical structure". Obviously, this claim does not refer to the structure of ET as constituted by the logical connections among its statements and perhaps subtheories. What may be meant is that the theory does not refer to entities on a single level but to entities belonging to different biolevels. Yet the fact that, say, selection may occur among entities belonging in different levels of organization does not make the *theory* of selection hierarchical.

By contrast, a "hierarchical" view of ET in a different sense has been suggested by Tuomi (1981), who distinguishes the generic or "metatheory" of evolution from specific theories and theoretical models. He calls this a 'multi-level model' of ET, which would consist of different "levels of abstraction". (Similarly Darden and Cain 1989.) However, what he actually refers to are not levels of abstraction but degrees of generality (see Sect. 3.5.3). Although Tuomi's view is similar to ours, we have claimed in Section 9.3.1 that we do not believe that there is a unified (hyper)general theory of evolution. However, there are theories in evolutionary biology that are general (possibly hypergeneral) and thus come in different degrees of generality.

As elucidated in Section 9.2.4, the most conspicuous example of a hypergeneral theory in evolutionary biology is the theory of natural selection. If the reference class of this hypergeneral selection theory is narrowed down from the collection of all interactors to the collection of all organisms, we obtain the general theory of organismal selection T. This general selection theory T yields (potentially) millions of special theories or theoretical models (models$_3$) M_i of natural selection, one for each taxon and every major feature, Recent, extinct, or future. More precisely, the relation between the general theory T and the various models M_i is as follows. Every M_i can, at least in principle, be obtained by enriching T with a set S_i of subsidiary assumptions concerning the specific features of the corresponding species or higher taxon. That is,

$$M_i = Cn(T \cup S_i), \text{ with } 1 \leq i \leq n.$$

Thus, the theory of natural selection T may be regarded either as the family of all its theoretical models or as their union. That is,

$$T = \{M_i \mid 1 \leq i \leq n\}, \quad \text{or} \quad T = \bigcup_{i=1}^{n} M_i.$$

However, due to the lack of a generally accepted axiomatized theory of selection, it is hard to tell in practice whether some proposed model of selection is a bound model, i.e., a specification of the general theory of selection, or just a free model (recall the notions of bound and free models from Sect. 3.5). Anyway, if we want to systematize our knowledge, we have to attempt to turn a collection of free models into a family of bound models by constructing a general theory.

What has been said about the general theory of selection can also be said about population genetic theory. Though referring only to eukaryotic organisms, population genetic theory is still a general theory with (potentially) millions of bound models, one for each species and each genotype, Recent, extinct, or future.

In sum, even though there appears to be no (hyper)general theory of the evolution of all organisms, at least two general theories are available, namely the theory of natural selection and population genetic theory. We contend that this generality and its capacity to yield bound models is one of the reasons that these two theories are so prevalent in evolutionary biology.

However, there should be more to evolutionary theory than these two theories: we need not only a theory about the trans-generational distribution of genotypes, but also a theory about the generation of new phenotypes$_O$, i.e., a theory of organismal speciation. For instance, we need a *theory* of mutation and a *theory* of development. Note that by 'theory' we mean a theory of the general mechanisms of such processes, not just a description of them. It is important to emphasize this point, because too often mere description passes for theory. For example, although we have elaborate descriptions of the various steps involved in protein synthesis, we do not have a *theory* of protein synthesis. (And we maintain that, as long as we accept redescriptions of biochemical processes in empty information-speak, there will be no progress in this matter.) In sum, all these theories are necessary for a complete explanation of biotic evolution. However, even if we had these theories at our disposal, this collection of theories would not automatically amount to a unified (hyper)general theory of the evolution of all organisms. So what is the structure, if any, of this collection of theories usually called 'evolutionary theory'?

The first to deal with this question was apparently Beckner (1959), who contended that evolutionary theory is a "family of related models" (p. 160). This idea, which had been formulated rather casually by Beckner himself, was elaborated by Caplan (1978), who called his version the 'ordered-set conception' of theory structure. Lewis (1980) regarded the theory of evolution as a "system of theories", and Wassermann (1981, p. 419) as a "hypertheory which comprises a set of subordinate theories". Finally, also inspired by Beckner, Lloyd (1988) and Thompson (1989) argued at length—though, as has been shown above, on the basis of a wrongheaded conception of theory structure—that ET is a family of interrelated models. The interrelated theories in these views are basically all at the same level. That is, they are not a family of bound models but rather a collection of either free models or else general theories capable of generating bound models. Thus, they fit in with what Tuomi (1981, 1992) called the 'reticulate model' of ET. Indeed, the theories appear to be referentially and evidentially related, i.e., they form a system of consiliences (Whewell 1847; Ruse 1988). Moreover, they are mutually consistent and thus constrain the construction of theories and models in the various disciplines involved (Caplan 1978). In other words, they provide a unified picture of evolution, but they do not constitute a unified general theory of evolution.

So far, then, evolutionary theory is not a theory proper but a *system of theories*. Though otherwise in broad agreement with Caplan (1978), we prefer Lewis's expression 'system of theories' over Caplan's 'ordered set', because an ordered set is a concept in set theory that has nothing to do with the systemicity at issue here. And since a system of theories is not necessarily a theory itself, i.e., a hypothetico-deductive system, Wassermann's term 'hypertheory' is also inadequate. Another consequence of regarding evolutionary "theory" as a system of theories rather than a theory proper is that its component theories are not subtheories: a subtheory is included in the main theory (see Sect. 3.5). Yet the system of theories ET lacks the logical unity of a theory proper, so that the component theories are not (and cannot be) entailed by any other component.

Although our conclusions as to the structure of ET are at first sight similar to those of the "semanticists" (see also Tuomi 1992), we emphasize again that all this is possible without presupposing any of the underlying philosophical ideas of the "semantic" view.

9.3.4 Unification Through Entropyspeak and Infospeak?

We have just argued that the theory of natural selection, together with population genetic theory, does not constitute a complete theory of evolution. Moreover, we have argued that there is no unified general theory of evolution but at most a system of theories, which may jointly account for the biological processes we subsume under the term 'evolution'. While some—with whom we concur—expect major explanatory contributions to problems of evolution from developmental biology, others trust in the unifying power of thermodynamics, statistical mechanics, or information theory (e.g., Prigogine 1973; Brooks and Wiley 1988; Brooks et al. 1989). We maintain that this latter approach is essentially barren. The reasons for this harsh judgment are as follows.

Precisely because thermodynamics and the statistical mechanics of open systems apply to all of these, they are far too general to say anything of particular interest to biology. Indeed, those theories apply not only to organisms but also to chemical reactors and physical systems such as solutions. (In particular, all irreversible processes—such as those of birth, development, and death, or combustion, heat transfer and diffusion—are entropic. That is, they are accompanied by an increase in entropy, if not in the system of interest, at least in its surroundings.) Thermodynamics only describes and restricts the *possible* biological processes: it tells us which processes are physically possible and which are not. And as a hypergeneral black box theory, i.e., one uncommitted to any mechanism, it is even less helpful than either mechanics or electrodynamics. In sum, physics—in particular thermodynamics—"allows" for life but gives no hope that it can be explained in purely physical terms. Therefore, all general talk about thermodynamics and biology (e.g., Brooks et al. 1989) is irrelevant—unlike the detailed thermodynamic calculation of particular biochemical reactions. (Further criticisms in Berry 1995.)

This is only an instance of the old methodological principle *dictum de omni, dictum de nullo*: what holds for everything says nothing in particular. In other words, an extremely general theory can cover only the features common to all the members of its reference class, hence it will miss all peculiarities. (In semi-technical semantic jargon: sense or content is inversely proportional to extension or truth domain.)

Statistical information theory is even less helpful to biology than thermodynamics, and this for two reasons. First, because it does not contain such biologically relevant variables as energy and temperature. Second, because, unlike thermodynamics, it contains no $laws_2$ of nature, so that it cannot even help decide whether a given conceptually possible process is physically possible. In any event, the "information" involved in genetics is nothing but molecular structure, a property which, unlike the information "flowing" in a communications system, is conceptualized in nonprobabilistic terms. Consequently, all talk of genetic code and genetic information transmission is metaphorical (recall Sect. 8.2.3.1).

In sum, the unification aimed at with the help of hypergeneral and scaffolding theories aims too far: biologists cannot be interested in theories so general that they are applicable to all concrete systems. Biologists can only be interested in a general theory unifying biology, not all the factual sciences. If any such theories are available, information theory, thermodynamics, and statistical mechanics are not among them, as being far too general.

Some approaches are not only too general, but also functionalist and formalist. That is, they assume that they can ignore the stuff of the evolutionary process and focus on its formal or abstract or logical character, if any. Let us take a look at a specimen of such approach.

9.3.5 Is Evolution an Algorithm?

Daniel Dennett (1995) has recently put forward the idea that evolution is an algorithm. This has already met with the approval of some evolutionary biologists (e.g., Maynard Smith 1995). Yet the question is: what, if anything, does this thesis add to our understanding of the process of evolution as well as of evolutionary theory? We submit: nothing at all. The reasons for our skepsis are as follows.

To begin with, the proposition "evolution *is* an algorithmic process" (Dennett 1995, p. 60) or, equivalently, "evolution *is* an algorithm" is a metaphysically ill-formed statement, because an algorithm is a formal rule or recipe, hence a construct. For this reason, all one could say at best is that evolution (or, more precisely, individual evolutionary processes) can be *modeled* or *simulated* by some algorithm. Yet, as we shall see in a moment, even this is a dubious assumption.

For another, the standard notion of an algorithm involves that it is a goal-directed, "mechanical", and fool-proof method: once started, it goes ahead regardless of circumstances, so that it is sure to attain the preset goal. Think of the algorithms of long division, of extracting square roots, or of the calculation of derivatives and integrals. Quite obviously, bioevolution is nothing of the sort: it is nonteleological, "opportunistic", and subject to both randomness and historical (e.g., geological and meteorological) accident—hence anything but predictable in detail. Moreover, when designing or using an algorithm, we certainly know the *kind* of thing we must expect it to deliver, e.g., the 10th decimal figure of $\sqrt{2}$, although we cannot predict the particular final result. In evolution, however, we cannot even predict the *kind* of things the alleged evolutionary algorithm is to come up with, namely the emergence of any new *species* of organisms. (Only an artificially, i.e., biotechnologically, produced species can sometimes be predicted.) In short, the concept of speciation cannot be accommodated within the standard notion of an algorithm

However, there are at least two ways out of this dilemma—both of which are pursued by Dennett. The first is to disregard the concept of speciation altogether and reduce evolution to natural selection. Indeed, like many others, Dennett confuses evolution with (or reduces evolution to) natural selection. The most interesting aspect of evolution, namely speciation, is thus reduced to some unspecified random input into a selection or, rather, sorting algorithm. Since, of course, all things can be sorted according to certain properties, the concomitant functionalist claim that "evolution" is substrate-neutral or generic becomes trivially true.

The second way out of the dilemma is to soften the concept of an algorithm so as to accommodate just any old process. Indeed, Dennett believes that essentially *every process is algorithmic* or at least treatable as such (1995, p. 59). However, this notion of an algorithm is so broad that the expression 'algorithmic process' becomes a pleonasm. Thus, when Dennett says that "evolution is an algorithmic process" (p. 60), he says nothing but "evolution is a process". (And we wonder why he thinks that "putting it this way is still controversial" [p. 60].) Similarly, what Dennett calls 'Darwin's dangerous idea', namely the idea that "all the fruits of evolution can be explained as the products of an algorithmic process" (p. 60), thus boils down to the statement "all the fruits of evolution can be explained as the results of a process". Worse, since the term 'evolution' refers to the class of all (individual) evolutionary processes, which is often referred to as *the* evolutionary process, we can further reformulate his statement to read "all the fruits of the evolutionary process are the results of a process". This idea is dangerous only if someone actually believed that it adds anything to our understanding of evolution.

In a broader perspective, however, Dennett's thesis *is* dangerous: it is another instance of functionalism, according to which the material substrate is irrelevant to the function of the system in question, because all that would matter is the "logical form of the process". Yet if not only evolution but also, as we saw before (Sects. 4.4 and 6.2), life and mind are all allegedly substrate-neutral, then not just the theory of selection but all the corresponding theories can be "lifted out of their home base in biology" (Dennett 1995, p. 58), for they are, after all, concerned with the search for *abstract* truths (p. 59). Such a formalist and immaterialist conception of biology seems attractive to some, as sparing them the task of learning anything about biology (i.e., cytology, physiology, genetics, development, morphology, ecology, neurobiology, and so on) as well as about adjacent fields such as biophysics and biochemistry. Ultimately, it is the idea that biology can be pursued by pure mathematicians, nay, that all the factual sciences are reducible to the formal sciences, which are indeed substrate-neutral, since they deal with conceptual objects, not material ones. *This* is a dangerous idea—not Darwin's but the functionalists'.

9.3.6 The Methodological Status of Evolutionary Theory

In the wake of Popperianism, much has been written about the scientific and methodological status of the theory of evolution. In particular, it has been claimed that ET is tautological and irrefutable, that it does not contain $laws_2$, and that it makes no predictions proper. It is well known that Popper pronounced ET a metaphysical research program (1972, 1974) rather than a scientific theory proper—a view he later recanted (1978). Since the concerns and criticisms of Popper, as well as those of others (e.g., Goudge 1961; Smart 1963), have been amply addressed in the literature (see, e.g., M.B. Williams 1973b, 1981; Ruse 1977, 1988; Caplan 1978; Olding 1978; Brady 1979; van der Steen 1983; Bunge 1985b; Rosenberg 1985; van der Steen and Kamminga 1991), we shall make only a few general remarks here.

Since ET is a system of theories rather than a theory proper, it comes as no surprise that it can be tested only indirectly, namely by virtue of the testing of its component theories. Yet it must be admitted that even the component theories are hard to test. First of all, every theory is composed of infinitely many propositions, so that one can only test a finite subset of propositions. Which are actually tested depends mostly on their being interesting to scientists and their ease of testability. Thus, one usually finds clusters of well-tested hypotheses together with hypotheses that are hardly, if ever, tested. Second, the deeper and more general a theory, the harder it is to test (see Sects. 3.5 and 3.7). For example, none of the quantum mechanical formulas about stationary states is directly testable: only transitions among such states are observable. Not surprisingly, this also holds for the Synthetic Theory, i.e., the theory of natural selection plus population genetic theory, which refers to populations of sexually reproducing eukaryotes. In its general form, it is indeed only confirmable but not refutable. To arrive at strong testability (Definition 3.15), we must build bound models_3 of the general theory. That is, we must enrich the general theory with subsidiary assumptions about the particular species or higher taxon of organisms to which the general theory is to be applied, as well as about the special circumstances. (See also Tuomi 1981.) However, as a matter of fact, only a very small number of the virtually millions of possible models of the Synthetic Theory have been tested.

In sum, neither the system of theories called *Evolutionary Theory* nor its component general theories are testable without further ado. What are testable are the bound models of the component theories. Thus, confirmation for the system of theories ET is obtained by what might be called the 'evidential consilience' of the diverse models of its component theories.

Indeed, there is abundant evidence for the component theories of ET, and thus for the system of theories we call 'Evolutionary Theory'. This evidence may be grouped into three types: circumstantial, direct, and historical. For example, *circumstantial* evidence is delivered by functional morphology and it consists of descriptions of imperfections rather than of aptations. Whereas optimal aptation might be taken as evidence of intelligent design and, thus, as support for teleological approaches to biology (e.g., orthogenesis or, worse, creationism), suboptimal performance, obvious flaws in "design", recapitulations, vestigial organs, and the massive occurrence of "junk" DNA make sense only in a theory or a system of theories operating with concepts such as historical accident, variation, and selection. Well--known examples of "design" flaws that can only be explained in an evolutionary context are the crossing of the food and air passages in our throat, and the open connection between the ovaries and the fallopian tubes, which allows for occasional abdominal pregnancies. Another example of bad "design" is the *descensus testiculorum*, namely the fact that in many, though not all, mammalian embryos the testicles initially develop near the kidneys, only to migrate later into the scrotum, a position that makes the spermatic duct go an (unnecessarily) long way around the pubis and that is neither practical nor safe. (More on "design" flaws as well as their bearing on the creationism debate in Mahner 1986.) Another piece of circumstantial evidence, namely the data on extinction, is produced by paleontology, which is also *the* supplier of the *historical* evidence for ET.

Direct evidence for ET is mainly available for the theory of selection. It is threefold: observational, experimental, and practical. The *observational* evidence ranges from the now classic peppered moths to the short-term selection of different beak sizes in Darwin's finches (see, e.g., Dobzhansky et al. 1977; Boag and Grant 1981; Futuyma 1986). The *experiments* on the "power" of selection are well known: they range from the experiments with RNA molecules ("molecular evolution") through bacteria (e.g., sundry resistance phenomena) to *Drosophila*. In particular, the latter are even part of many introductory courses in genetics and evolution. More recent experiments have investigated the role and weight of adaptation versus chance in evolution (Travisano et al. 1995). Finally, the *practical* evidence consists, of course, in the results of artificial selection and breeding, whether prescientific or technological. It is no accident that Darwin devoted the first chapter of his 1859 classic to this subject matter.

What holds for testability also holds for explanation and prediction. Clearly, a system of theories, not being deductively closed, does not allow for straightforward explanations and predictions, which have the form of logical arguments. Furthermore, its component theories are general, and general theories only allow for general explanations and predictions. For example, population genetic theory in general can predict the distribution of any genotypes$_{O}$ in any population but, unless we specify the theory to some species of interest, nothing specific can be predicted. (More on general predictions of selection theory in M.B. Williams 1973b.) The same holds for explanation.

Although predictions need not be about the future—there are also retrodictions (see Sect. 3.6.2)—the nature of evolution seems to render precise forecasts impossible. Indeed, since evolution involves the emergence of qualitative novelties, which, in turn, involve at least partly random events such as point mutations, as well as historical accidents, there appears to be no way of making any long-range forecasts of biotic evolution.

Still, although we may not be able to predict particular phenotypic transformations, knowledge of developmental and phylogenetic constraints should allow us to predict the *range of possible* transformations of the members of a given taxon. However, before we elaborate on this, we should elucidate the notion of phylogenetic constraint. Obviously, phylogeny does not constrain anything in the literal sense, because the past does not act on the present. What 'phylogenetic constraint' can mean is only that, for every organism x (that did not originate by neobiogenesis), there is a class of properties (constraints and laws) of x that can be accounted for only by reference to the properties of x's ancestors. Since the past does not act on the present, all phylogenetic constraints must be developmental constraints.

But of course, the converse does not hold. (More on constraints in Amundson 1994.)

Now, since all phylogenetic constraints are actually developmental constraints, and since the latter (in the context of evolution) are actually developmental laws (in the context of development), the laws of a given taxon specify the range of possible lawful transformations of its members. Knowing such laws, we can infer the complement class of impossible transformations. (As, in practice, it is often easier to state what is impossible than what is possible, biologists usually speak of constraints and prohibitions rather than of laws and possibilities: see Sect. 1.4.3, as well as Gould 1989.) Thus, by building nomological state space models of the members of a given taxon, we should be able to specify the set of their possible transformations. Still, it seems we cannot predict particular transformations because they depend not only on developmental laws but also on historical accidents. Yet this state of affairs neither invalidates the search for laws in biology, nor undermines the scientific status of ET.

Other components of ET seem not to involve laws and general theories, as selection and population genetic theory do, but to consist in mere historiography or in the reconstruction of phylogeny. It should come as no surprise, therefore, to learn that it is often believed that ET would comprise so-called narrative explanations rather than deductive-nomological ones (recall Sect. 3.6.1.4). Yet without reference to mechanisms, evolutionary scenarios and phylogenetic reconstructions are at most descriptions, not explanations. As such, they belong to natural history (more precisely: historiography), not to the theory of evolution, which, after all, is a theory about the mechanisms of evolution.

However, if explanatory, evolutionary models will involve laws and mechanisms, although they need not be stated in a deductive-nomological form. For example, a model or an evolutionary scenario about the evolution of birds will, unless it is a mere just-so story, make use of whatever biological knowledge is available: comparative and functional morphology (e.g., function and structure of the relevant bird traits), developmental biology (e.g., explanation of wing structure, in particular the reduction and fusion of fingers), paleozoology and systematics (sequence of the evolution of bird characters), and ecology (role and adaptive value of bird characters). In this example there is plenty of room for morphological, physiological, developmental, and ecological laws and mechanisms. On the other hand, the involvement of selection theory and population genetic theory is either nil or trivial, because we have no data on either Jurassic selection regimes or Jurassic gene or genotype distributions. So all that is required is the compatibility of the scenario with the general theory of selection and population genetic theory. However, the latter do not contribute any significant new insights to phylogenetic reconstructions, whether descriptive or explanatory: they only constrain the building of evolutionary hypotheses.

The preceding example shows that the laws involved in the emergence (speciation) and submergence (extinction) of organisms of new species may "only" be genetic, developmental, and ecological: they need not be evolutionary laws proper.

Yet, several candidates to the rank of evolutionary law have been proposed. For instance, Rensch (1971) lists a hundred evolutionary laws, such as "There is more rapid emergence of new races and species among small, isolated populations, and those which derive from relatively few individuals, than among very large ones" (p. 133). However, most of these laws are statistical, and they are not evolutionary proper. Rather, they comprise biophysical, biogeographic, ecological, genetic, and developmental statements. Furthermore, some of those "laws" are actually statements as to the occurrence of trends in evolution, such as Cope's "law": "There is a tendency—most marked in larger flightless land animals—toward a progressive increase in hereditary body size" (p. 136). Clearly, this not a law but at best a trend, if any (Gould 1997), so much so that most of the species to which it applies have become extinct. However, trends (i.e., patterns of uniform sequences of states of, or events in, systems) are not laws, but ought to be explained in terms of laws and circumstances (Popper 1957a; Bunge 1967a, 1983a).

Although from the ontological point of view there can be no doubt that biosystems behave lawfully and that there are generalizations that refer to $laws_1$, the problem for biology and, in particular, evolutionary biology seems to be predominantly epistemological: many candidates for biological laws are not firmly embedded in biological theories. (Recall Definition 3.9; see also van der Steen and Kamminga 1991.) Far more work will be needed to solve this problem.

This problem is analogous to that of social science, in particular human history, with the difference that the latter has been studied for two and a half millennia, not just for one and a half centuries. Indeed, here, too, there are a number of plausible candidates for law statements, such as "Social change is more frequent in heterogeneous societies than in homogeneous ones", and "Rapid population growth $\rightarrow$ overcultivation and deforestation $\rightarrow$ erosion and loss of soil fertility $\rightarrow$ decline in food production $\rightarrow$ food shortage $\rightarrow$ political unrest". However, none of these quasilaws will qualify as a historical law proper as long as no fully-fledged theory of history, if any, is produced. (See Bunge 1996.)

The failure of historiography and evolutionary biology to come up with $laws_2$ should give no comfort to the antinomianist or antilaw party in the philosophy of science, according to which those disciplines are necessarily idiographic (i.e., about particulars) rather than nomothetic (i.e., involving laws). Indeed, even if neither historiography nor evolutionary biology were ever to produce any laws of their own, both consume $laws_2$ generated in other fields. This, rather than law_2 production, is sufficient to secure their status as scientific disciplines.

To conclude, even if Popper's assertion (1957a) were vindicated that there is no single, all-encompassing law of evolution, the term 'evolution' could still refer to a class of lawful individual evolutionary processes, and evolutionary biology would still be a scientific discipline, for it involves laws of genetics, development, selection, and so on. Last, but not least, if the reader bothers to apply all the conditions listed in Section 5.5.1 that a research field must fulfill in order to count as a scientific discipline, he or she should be firmly persuaded that evolutionary biology passes this test with flying colors.

10 Teleology

Teleology is one of the classical topics in the philosophy of biology. (See, e.g., Sommerhoff 1950; Beckner 1959; Bunge 1973a, 1979b, 1985b; Ruse 1973, 1982, 1988; Hull 1974; Mayr 1982, 1988; Rosenberg 1985; Sattler 1986.) The popularity of teleological notions in biology is not surprising, given that the teleological mode of thought, central to anthropomorphism, is at least as old as humankind. Teleological thinking proceeds by analogy with human purposive behavior and is thus easily understood. However, as pointed out in Section 3.6, to understand a fact is not the same as to explain it in scientific terms. Understanding is a psychological category and must not be confused with scientific explanation, which is a methodological category. Accordingly, the methodologist's task is not to study how teleological thinking furthers our intuitive understanding of biological facts, but to investigate whether the use of teleological concepts is scientifically necessary and legitimate in a given context and whether there is such a thing as teleological explanation.

Although teleological expressions can occasionally be found in the physical sciences (e.g., "In order to keep moving against a force, a body must draw energy from an outside source"), it is generally acknowledged that such are instances of careless wording. In the biological sciences, however, the situation is different. In fact, here we meet an almost schizophrenic situation. On the one hand, many authors maintain that teleological concepts are legitimate in biology or are even constitutive of biology's (alleged) autonomy; on the other hand, they take pains to point out that biological teleology is somehow not a genuine teleology, but only an *as if*-teleology, occasionally called 'teleonomy'. A similar contradiction can be found in the assurance that teleological explanations in biology could be translated into nonteleological ones, but eliminating teleology altogether would be impossible because "something would get lost" by doing so. Thus, biologists apparently cannot live *with* teleology but they cannot live *without* it either.

In order to shed light on this problem, we shall examine several notions of teleology, as well as some of the attempts to render biological teleology scientifically respectable. (For a historical summary of the notion of teleology see Lennox 1992a.)

10.1 External and Internal Teleology

Two notions of teleology are often distinguished: internal and external, depending on whether the *telos* is taken to be an immanent property of an object or whether it is attributed to it from outside (see, e.g., Goudge 1961; Ayala 1970; Woodfield 1976; Lennox 1992a).

Internal teleology comes in two versions: cosmic and regional. Things which have some immanent or intrinsic property that makes them either purposive or goal-directed, or purpose- or goal-conceiving, can be said to be internally teleological. If internal teleology or finality is a property attributed to *all* things we can speak of cosmic internal teleology, or internal *panteleology*. Well-known examples are animism and the concept of (natural) entelechy occurring in Aristotelian metaphysics.

The regional version of internal teleology holds that finality does not inhere in all things, but only in *some* things, in particular, living beings. Regional internal teleology, or internal *hemiteleology*, can be further analyzed into a naturalistic and an animistic variety (Lennox 1992a). For example, the (alleged) finality of living things may just be an immanent natural property, as is the case with human intentionality, or it may be due to a quasi-intentional immaterial or spirit-like entity, agent, or force inherent in all living beings, as is assumed by psychovitalists.

If purposes and goals are not intrinsic properties of things, but are attributed or assigned to them by a purpose- or goal-conceiving agent, we speak of *external* teleology. Thus, external teleology presupposes at least one entity that is internally teleological and that is able to attribute goals or purposes to other entities.

External teleology, too, comes in cosmic and regional varieties. Cosmic external teleology, or external panteleology, goes back to Plato, who believed that the purposes and goals of things have been conceived and planted by a divine rational agent, i.e., the creator of the world. Cosmic external teleology was easily adopted by Christianity and thus dominated the study of life as long as natural theology was influential in biology. It was also an essential part of Leibnizian metaphysics. However, even after the triumph of evolutionary biology, some sort of cosmic external teleology must be invoked by the religionists who wish to reconcile evolutionary biology with a religious outlook, e.g., by assuming some kind of orthogenetic evolution resulting in the emergence of human beings (Mahner and Bunge 1996). A famous example is the cosmic teleology of Teilhard de Chardin (1964).

If there is no cosmic teleology, there are at least some conscious agents in the universe, like humans, who do have intentions and thus can attribute purposes or goals to things, or create things for a certain purpose. Human artifacts, then, epitomize regional external teleology, or external hemiteleology. It should be noted, however, that since the purposiveness of artifacts is an external one, they have a "function" or purpose only as long as they are used by somebody, that is, as long as there is a purpose-attributing agent. In other words, the purpose or "function" of an artifact is not an intrinsic property of the latter, but a property of the designer-

-artifact or user-artifact system. The artifact has no purpose whatever when nobody uses, or thinks of, it. For example, a computer on a dumping-ground has no purpose or "function" whatsoever, although somebody who retrieves it from there and who is able to rethink the intentions of its designer, e.g., by examining its structure, may (re)attribute some purpose to it, perhaps even its former purpose. Note that, in so doing, an artifact's structure may provide *evidence* of design and purposiveness, but it must not be equated with purposiveness itself. Thus, an artifact may *serve* a purpose, but it does not possess one.

According to whether everything or something is believed to be teleological, whether internally or externally, we have distinguished *panteleology* from *hemiteleology*. The third option, of course, is *ateleology*, that is, the thesis that nothing at all is teleological. (See also Bunge 1985b; Woodfield 1976). Ateleology is obviously false because at least some behavior of higher vertebrates is goal-directed or purposive, such as the authors' writing this book.

Panteleology is also rejected by any scientific metaphysics. The latter has no room for supernatural entities that could conceive any purposes or impose any cosmic purpose. Neither do physics and chemistry attribute any purposiveness to their referents, although teleological sentences can occasionally be found in the physical and chemical literature. But these have to be regarded as careless lapses or elliptical talk because they can easily be eliminated without loss of meaning. For example, the sentence '*E* is the energy *needed to* complete reaction *R*' can, nay, ought to, be formulated as '*E* is the energy *consumed* by the system in the completion of reaction *R*'. Therefore, internal panteleology is also wrong.

We are thus left with regional or hemiteleology, whether external or internal. Since the psychovitalistic form of internal teleology is also at odds with a scientific ontology, we are finally left with those versions of internal and external hemiteleology that are consistent with scientific naturalism. Furthermore, since our subject here is the philosophy of biology, and since the external teleology of artifacts depends upon the internal teleology of some organisms, we can disregard external teleology and focus on the (alleged) internal teleology of living beings.

10.2 Teleonomy

In order to avoid confusion with cosmic or vitalistic teleology, it has been suggested to call a scientifically respectable notion of biological teleology, if any, *teleonomy* (Pittendrigh 1958). Although this proposal has gained acceptance among some biologists (see, in particular, Mayr 1982, 1988; Monod 1971; Hassenstein 1981), most philosophers, though accepting the naturalistic version of hemiteleology, do not follow that terminological distinction. We shall adopt the term 'teleonomy' for the sake of the subsequent examination, but shall return to 'teleology', after having identified the context where we think the use of the concept is legitimate.

Having dismissed ateleology, hence simultaneously ateleonomy, we can, from a logical point of view, distinguish two versions of biological teleonomy. Either *all* living beings can be said to be teleonomic systems, or only *some* living beings are purposive or goal-seeking. These views may be termed *panteleonomy* and *hemiteleonomy*, respectively.

10.2.1 Hemiteleonomy

If some living beings have goals, plans, or intentions, human beings (or, rather, persons) will epitomize this kind of teleonomy. Most likely also some other animals are capable of goal-seeking behavior. Starting from this example of hemiteleonomy, we shall examine whether the notion of teleonomy can be legitimately extended to cover all living beings.

Acknowledging purposive behavior in humans and some other animals is not a mark of obscurantism provided one does not add that forming purposes and devising means to attain them are attributes of the immaterial mind or soul. Purposive behavior can be explained, at least in principle, in terms of processes occurring in highly evolved central nervous systems (see, e.g., Hebb 1966; Bindra 1976; Bunge 1980). One way of explaining purpose is in terms of learning and expectation, where each of these is conceived of as a particular activity of certain neuronal systems. As we have dealt with learning in Chapter 3, we only need to define the notions of expectation and purposive behavior here. Thus, we start with:

DEFINITION 10.1. Animal b *expects* a future event e of kind E when sensing an (external or internal) stimulus r while in state s if, and only if, b has learned to pair r and s with e.

We are now ready for our definition of the concept of purposive behavior:

DEFINITION 10.2. An action x of an animal b has the *purpose* or *goal* y if, and only if,
(i) b may choose *not* to do x;
(ii) b has *learned* that x brings about or enhances the chances of attaining y;
(iii) b *expects* the possible occurrence of y upon doing x, and
(iv) b *values* y (not necessarily consciously).

This definition does not restrict purposive behavior to humans. Yet, the task of comparative neurobiology and ethology is to find out which nonhuman animals do possess the required capabilities such as learning and expectation. On the other hand, this version of hemiteleonomy clearly restricts purposive behavior to a rather small subset of Animalia whose members possess plastic nervous systems of a certain complexity.

Note that, according to Definition 10.2, it is incorrect to impute goal-seeking behavior to systems of animals, in particular to social systems such as schools or firms. Only individuals, when sufficiently evolved, have the ability to form goals and contrive means to attain them. Conceiving of a goal is a process occurring in individual brains, not in groups: the latter have no brains. To be sure, two or more members of an animal group may combine to attain a common goal; in the case of humans, they may discuss both the goal and the optimal means to attain it, and may eventually reach a compromise that fits none of the original aims of the individuals concerned. Still, the compromise goal is nowhere but in the brains of the group members. Social groups, be they wolf packs or schools, have no goals above and beyond those of their individual components. A collective goal is nothing but what is common to the individual goals.

Our version of hemiteleonomy belongs to the family of doctrines that has been called the *intentional view* (Nagel 1977), because it is usually formulated in terms of intentions, desires, wants, beliefs, etc. (Woodfield 1976). It is obvious from the intentional view that teleological concepts cannot be legitimately used beyond the domain stated in Definition 10.2, which can be summarized in the formula: *no plastic brain, no teleology*. However, the attempt has often been made to extend a concept of hemiteleonomy to a form of panteleonomy. So much so, that some philosophers have claimed that "if anything is to count as a teleological system, living organisms must" (Hull 1974, p.104).

10.2.2 Panteleonomy

We shall examine here only two such attempts (Nagel 1977): the *system property view*, which tries to exploit concepts of systems theory and cybernetics, and the *program view*, which tries to conceive of goal-directed processes in terms of information theory. (See also Hull 1974; Engels 1982; Mayr 1988.)

10.2.2.1 Cybernetic Systems and Teleonomy

The system property view is not confined to biology. In fact, it originated in engineering and has become the subject of cybernetics, which studies systems in general, i.e., regardless of the nature of their components. (The classical locus is Rosenblueth et al. 1943.) However, since all biosystems are self-regulating systems (Postulate 4.1), teleonomic concepts can easily be applied to biosystems if all homeostatic systems are said to be teleonomic.

Cybernetics deals with systems in general whose activities are controlled by negative feedback loops. One of the central concepts is the concept of *goal state* or *Sollwert* (ought value) of a system, which refers to the fact that a self-regulating system is structured in such a way that deviations from a certain state are compensated for by negative feedback. It is, of course, tempting to regard this end state of the system as its 'goal', and thus the processes in the system as 'goal-directed'. This temptation is furthered by the old behaviorist and operationalist mistake of regarding as goal-directed any process that exhibits a tendency to reach a certain end state despite varying starting-points and circumstances. Yet this observable plasticity can only be an *indicator* that the process in question might be goal-directed: it cannot define goalhood itself (Woodfield 1976).

If there were any goal in a cybernetic system, it could only be found by studying the system's internal structure, not by merely observing its overt behavior. We submit that there is no such thing as a goal or purpose in a self-regulating system because there is nothing, beyond its internal structure and beyond the operationist tradition in those who study such systems, that could justify goal talk in the case of natural systems. The structure of a system is not a purpose, and its end states are not goals. They only can be made such by definition. Talk of goals and purposes in cybernetics is nothing but a metaphorical extension of human intentionality. In other words, the end states in control theory are goals only by proxy.

Whereas artifacts may exhibit external teleonomy, biosystems can only be regarded, if at all, as instances of internal teleonomy—at least from a naturalistic point of view. Yet some formulations in the literature seem to presuppose an external teleonomy of biosystems. For example, it is often said that an organ or a feature has been "designed by natural selection for a certain function". In doing so, natural selection is reified as an agent which externally endows a biosystem with a goal or purpose. (Obviously, the habit of thinking in the spirit of natural theology dies hard. The fact that God has been replaced by Natural Selection after 1859 did not make any difference with the alleged teleology of biosystems [Ruse 1982].) However, if cornered, most biologists may admit that selection is not an agent and thus their talk of design is a metaphor borrowed from the model of human artifacts. Worse, if there is no external teleonomy in biology, the metaphor is not just an unlucky analogy but it is plainly wrong. This is one of the reasons that we take any attempt at coming up with a concept of teleology that holds for both artifacts and biosystems (as pursued, e.g., by Griffiths 1993) to be wrongheaded.

What about the internal teleonomy of biosystems? Suggesting that processes have goals, while systems have purposes, Ernst Mayr (1988) has proposed to regard as teleonomic only goal-directed processes such as development, not quasi-static systems such as the skeleton, however useful or "purposive". However, the concept of *telos* comprises both goals and purposes. Concerning purposes, biologists still talk of the functionality or purposiveness of biosystems, as if the notion of teleonomy could be legitimately extended from artifacts to biosystems. If biological teleonomy is an internal one, only the internal structure of the system and its resulting range of activities could be equated with "purpose" or "function". Of course, in most cases, the internal structure of a biosystem will be due to adaptation, but the concept of adaptation is not cointensive with the notions of purpose, or "proper function", or role (see Sect. 4.6).

Interestingly, Mayr (1988) admits that the notion of adaptation has replaced the teleonomic view of biosystems. This might be one instance of successful deteleologization of biological concepts if biologists followed him and admitted that functional talk is then metaphorical. Yet most biophilosophers continue to defend teleological explanations in biology in terms of "proper functions" or purposes of biosystems (e.g., Millikan 1989; Neander 1991; Griffiths 1993); or they hold that adaptation explanations would be teleological in the sense that they answer *what for*-questions (Brandon 1990). However, a what for-question that is not answered by reference to a genuine purpose is such only by analogy, hence methodologically illegitimate, although it may provide a starting-point of some heuristic value (recall Sect. 4.8).

Admitting that purpose talk is illegitimate in the case of biosystems, Mayr insists that there *is* a genuine teleonomy in biology, namely the one exhibited by goal-directed processes like development. Let us examine this view more closely.

10.2.2.2 Programs as Goal-Conferring Entities?

There can be no doubt that organisms undergo processes leading to some final state or, more precisely, final stage. A prime example is the development of a zygote into an adult organism. Since there is an initial state of the system as well as a final or end state, which is usually reached under varying, though on the whole favorable, circumstances, there is, thus far, no difference with the case of cybernetic systems in general. Thus, the operationalist temptation returns to conclude that reaching a certain state is the goal of the process. However, as noted previously, the fact that a process reaches a certain final state is not sufficient to regard this final state as a goal. For example, a river usually reaches the sea despite many possible geomorphological obstacles in its way. Nobody, except the panteleologist, will attribute a goal to the river. In order to distinguish such apparently goal-directed processes from (allegedly) genuine goal-directed ones, Mayr (1988) has proposed to call the former *teleomatic* processes.

How, then, are supposedly genuine goal-directed or *teleonomic* processes to be characterized? Biosystems are clearly distinguished from nonliving things by their possession of nucleic acid molecules (see Postulate 4.1), which are used as template molecules in the synthesis of proteins, hence enzymes, and thus influence metabolic processes and codetermine development. It is thus concluded that the goal-directedness of biotic processes is due to the action of a genetic "program". Although we rejected the notion of a genetic program in Section 8.2.3, it will be useful to examine briefly the views of two well-known defenders of the notion of a genetic program, Jacques Monod (1971) and Ernst Mayr (1988).

Monod defines goal-directedness in terms of a so-called "teleonomic project": "All the structures, all the processes, and all the activities that contribute to the success of the essential project will [...] be said to be 'teleonomic' " (p. 27). The problem with this characterization is that it renders the idea of teleonomy inconsistent, i.e., Monod's definition of 'teleonomic' leads to contradiction (Bunge 1979b). In fact, without a favorable environment no organism would be able to carry out its teleonomic project. So, just as to the organism, teleonomy should be assigned to the environment (or habitat) as well, whenever it "contributes to the success of the project". But since the environment contains nonliving things, the latter would also be teleonomic. Thus, no difference between living and nonliving things would remain—which contradicts the thesis that teleonomy is peculiar to biosystems. In short, Monod's attempt to justify panteleonomy fails. The same holds, incidentally, for Wimsatt's (1972) definition of "purpose" as any "contribution to the fitness of an evolutionary unit" (p. 8).

Mayr avoids the contradictory consequences of Monod's definition. He defines a teleonomic process thus (Mayr 1988, p. 45):

"A teleonomic process or behavior is one which owes its goal-directedness to the operation of a program." (1)

Actually, this is two definitions in one. Therefore, we reconstruct (1) as consisting of

A process x is teleonomic $=_{df}$ x is goal-directed; (1a)
x is goal-directed $=_{df}$ x is operated by a program. (1b)

In this definition the nature of the program is not specified. It could be a program in the genetic material or a behavioral program in an animal's brain. By referring to a program, Mayr criticizes definitions of "teleonomy" in terms of the cybernetic view, because the goal-directing part (e.g., a thermostat) in such cases would not *produce* or *cause* the process or behavior in question, but only *regulate* or *control* it. However, the same is true of the genetic program. If we take the genetic program of an organism to be identical to the structure of its DNA molecule(s), it becomes obvious that such a structure does not do anything: only the structured thing is able to do something. As for DNA, which is a rather inert molecule, not even this is the case: DNA *is acted upon* by other molecules. Hence, DNA does not *cause* anything whatsoever (recall Sect. 8.2.3). Of course, DNA contributes to determining the biosystem's metabolism and development, if any, but it does not cause them. Therefore, the genetic program is in the same boat as the thermostat: both only regulate, i.e., codetermine, the processes in question, but do not cause them. However, this is only a minor objection, with which Mayr might agree. There is a much more serious problem with Mayr's proposition if we take a closer look at his definition of "program", which reads thus (Mayr 1988, p. 49):

"[...] *program* might be defined as coded or prearranged information that controls a process (or behavior) leading it toward a given end." (2)

If we now substitute the term 'program' in the definiens of (1b) for the definiens of (2) we obtain something like:

A process x is goal-directed (or teleonomic) $=_{df}$ x is operated by a coded or prearranged information that controls a process leading it toward a given end.

Whether or not "goal-directed" and "end-directed" are taken to be cointensive, this definition is flawed. If "goal-directed" is cointensive with "end-directed", the definition is circular. If not, the definition is operationalist, hence it is a pseudodefinition: end-directedness is, at best, an *indicator* of goal-directedness; it cannot define it. If we try to remedy this situation by eliminating the conjunct containing the indicator hypothesis, it is obvious [see (1b)] that the problematic notion of *telos* is only shifted onto the remaining term 'information'—Mayr's assertion to the contrary notwithstanding.

We have examined several senses of 'information' in Sect. 8.2.3.2, where we concluded that the notion of genetic information does not make use of the essential concepts of information theory, so that the only legitimate use of the term 'information' in molecular biology is in the sense of 'composition and structure of the genetic material'. Only thus can the alleged information content of the genome be said to be permanently present like the "information" locked in a book. However, if this is so, then the teleonomic expression 'genetic program' can be replaced by the nonteleonomic expression 'genetic material'. To say, then, that a system contains genetic material codetermining some of the processes in the system, merely amounts to naming one of the properties of a living system (see Postulate 4.1). Since "structure of the genetic material" and "goal" are hardly cointensive, there is no reason to regard processes in biosystems as in any way goal-directed or teleonomic. Developmental processes are lawful and thus reach a certain end state given certain circumstances. Yet "lawful" is not cointensive with "teleonomic".

The notion of program or information only makes sense in the teleonomic context if it is understood in the sense of instruction. Indeed, many biologists openly use the word 'instruction' interchangeably with 'program' or 'information'. "Instruction", however, is not cointensive with "structure of the genetic material" either: "instruction", too, is an intentional concept (see again Sect. 8.2.3.2). It is therefore unsuitable to make a case for panteleonomy. The attempt to render the concept of goal scientifically respectable by replacing it with the apparently innocent concept of information fails because the latter is used in the sense of 'program' or 'instruction', which are themselves intentional concepts. The hope to find salvation in a retreat to information theory thus turns out to be an illusion.

Furthermore, the program metaphor fails not only in the case of genetic programs, but also in the case of behavioral programs. According to Definition 10.2, goal-seeking behavior presupposes learning. If a behavior is hard-wired or pre-programmed, it lacks the plasticity required for genuine goal-seeking behavior. We therefore submit that, whereas a chimpanzee most likely does climb a tree *in order to* pick a fruit, a turtle most likely does *not* come ashore *in order to* lay eggs. Though striking, this is just an analogy.

Finally, the program view, if correct, would have an ironic consequence. (This is just to mention it, not an *argumentum ad consequentiam*.) Mayr (1988) rightly ridicules the idea that, since all organisms die sooner or later, death would be the goal of life—a well-known theological, psychoanalytic, and existentialist thesis. However, some biologists have conjectured that aging and death are genetically "programmed" because they have selective value by eliminating those individuals from the population that are no longer capable of reproduction (Grant 1978). For example, in some cells the number of mitoses a cell can undergo seems to be genetically limited. If this turned out to be true, then death would obviously be "programmed", thus being the goal of life. ("Programmed" cell death, or *apoptosis*, is also well known in embryogenesis: see Grant 1978; Gilbert 1994.) The defender of the program view, then, vindicates theologians and existentialists, who can hardly be said to embrace a scientific outlook. By contrast, the developmental constructionist, not needing any notion of teleology, does not keep such dubious company.

10.2.2.3 Conclusion

Since both the cybernetic and the program view fail to provide arguments in favor of panteleonomy, we conclude that goal and purpose talk in these cases is only a metaphorical extension of concepts borrowed from the intentional view. It is thus a mere reasoning from analogy, seductive only because of the familiarity of the intentional concepts involved. For the very same reason, it is devoid of explanatory power, although we admit that teleological thinking may play an important *heuristic* role in biology (Ruse 1982, 1988). (Incidentally, already Kant 1790 declared it a *regulative* principle, devoid of explanatory power.) However, two reservations with regard to the heuristic fruitfulness of teleological thinking are in order. First, teleological thinking may lead to adaptationism, i.e., the temptation to see "purposes", i.e., biological roles and adaptations, everywhere and to concoct adaptational just-so stories, which may seem plausible, but are nevertheless unsupported by theory or empirical evidence (Gould and Lewontin 1979). Second, from the fact that the question 'What is this organ for?' is heuristically fruitful it cannot be concluded that teleological thinking, beyond the area of hemiteleonomy outlined in Section 10.2.1, is legitimate in biological theory. As a matter of fact, even the heuristic *what for*-question with regard to an organ x of an organism y can be analyzed as a shortcut for a set of nonteleological questions: What is the specific function of x? What is the role of x in y and in the given environment? What is the biological value of x's function and role in y? Does x confer any selective value upon y? Is x an adaptation? Why was x evolutionarily successful? In sum, in biology there are neither teleological questions proper nor teleological answers.

Consequently, there is no such thing as teleological or functional explanation in biology, for it reduces to answering the preceding list of nonteleological questions (recall Sect. 4.8.2), Yet just as some people, though *de facto* atheists, cannot give up the notion of religion due to early childhood indoctrination, many biologists and biophilosophers cling to the notion of teleology and try to redefine and reredefine it, so that eventually two different concepts—the intentional notion of purpose and the notion of adaptation—are treated under the same label, namely 'teleology'. However, we believe that this leads only to permanent misunderstanding and barren controversy. For the sake of clarity, we suggest dropping both inadequate concepts and misleading terms. Moreover, getting rid of obsolete concepts and misleading terms allows us to take a fresh look at things.

11 Concluding Remarks

We have sought to achieve the following things in this book.

By introducing some ontological, semantical, and epistemological fundamentals we hope to be of service to those biologists who, though interested in philosophical questions, have been justifiably bewildered by the multiplicity of conflicting philosophical views and who therefore may have missed a unified science-oriented philosophical outlook.

By grounding our biophilosophy firmly on such a general philosophical outlook we want to make the case that biophilosophical problems ought to be placed against a broad philosophical foil including not only elementary logic but also ontology, epistemology, methodology, semantics, and occasionally ethics as well. For example, it is confusing to discuss biological processes without the benefit of a general and clear concept of a process. And it is unprofitable to discuss matters of biological theory, explanation, prediction, and testability separately from the general methodology of science.

Likewise, by proposing a systematic (though far from comprehensive) philosophy of biology and, particularly, by adopting a very moderate axiomatic format, we hope to have shown that biophilosophical problems should be approached systemically rather than in a piecemeal fashion; in other words, that biophilosophical questions must be related to one another rather than be tackled in isolation. For example, our systematic approach has enabled us to proceed from the general notion of a process to that of a developmental process and thence to that of an evolutionary or speciation process, thereby providing the philosophical foundations for the long-sought-for unification of developmental and evolutionary biology.

We have shown that the foundations for this unification cannot be supplied by those biophilosophies that rely too heavily on selectionism (often subsumed under the labels of either 'Darwinism' or 'Neodarwinism', thereby doing Darwin an injustice) and on what has been called 'population thinking'. The latter has also been called the *variational* principle of evolution and contrasted with the *transformational* view of evolution, where "transformational" subsumes either pre-Darwinian (e.g., Lamarckian) or, more generally, non-Darwinian ideas of evolutionary change (Lewontin 1983a). Obviously, in the wake of population thinking, there has been no room for development as a mode of evolutionary change, since it is the paragon of a transformational process—a process that, moreover, occurs in organisms, not "species". So if we wish to bring developmental and evolutionary biology together, we must try and combine the variational (or populational or selectionist) model of evolution with an up-dated transformational one, as we have attempted to do in this book. Although some suggestions as to such integration do exist in theoretical biology (see, e.g., Johnston and Gottlieb 1990), the philosophical basis for this integration has not only been lacking so far, but current neonominalist philosophy of biology, being antiessentialist and thus antitransformationist, has also made any progress toward such integration impossible.

This example illustrates another point we wished to make: whereas some philosophies can help solve philosophico-scientific problems, such as 'What is a species?' or 'What is evolution?', and may thus contribute to the advancement of science, or at least to removing some obstacles to it, others fail to do so or even happen to be themselves obstacles to scientific progress. It also illustrates that the stand biologists take in a philosophical controversy is likely to affect their thinking on basic scientific matters. For example, reductionists are more likely to support megabuck projects, such as the Human Genome Project and the (strong) Artificial Life project, whereas biosystemists know that the former will not solve the mystery (or rather problem) of human nature, and that the latter is bound to fail, for it attempts to construct living things "in the dry", i.e., by bypassing chemistry.

Given that philosophy can be of service to biology, and vice versa, it is advisable to cultivate both disciplines together rather than in mutual isolation. In particular, biologists should gather once in a while to discuss biophilosophical problems, and biology students should be encouraged to take courses in the philosophy of biology. Likewise, more philosophers and philosophy students should be encouraged to take an interest in contemporary biology, so as to become able to help rather than hinder biological research. To be sure, such mutual interest and cooperation do exist, but too many biologists are still trained as technicians rather than educated as scholars, and too many philosophers still believe that science, hence biology, is just either one more or one other way of looking at the world, or even totally irrelevant to philosophy.

In sum, we hope to have broken a lance for the cause of an intimate relationship between the life sciences and a rationalist, realist and emergentist-materialist philosophy, which requires taking a fresh look at some ideas currently in fashion in the philosophy of biology.

References

Agassi J, Cohen RS (eds) (1982) Scientific Philosophy Today. Essays in Honor of Mario Bunge. Reidel, Dordrecht

Alberch P (1982) Developmental Constraints in Evolutionary Processes. In: Bonner JT (ed) pp 313-332

Ampère AM (1834/1843) Essai sur la philosophic des sciences, 2 vols. Bachelier, Paris

Amundson R (1994) Two Concepts of Constraint: Adaptationism and the Challenge from Developmental Biology. Philosophy of Science 61: 556-578

Amundson R, Lauder GV (1994) Function Without Purpose: The Uses of Causal Role Function in Evolutionary Biology. Biology and Philosophy 9: 443-469

Andersen NM (1995) Phylogeny and Classification of Aquatic Bugs (Heteroptera, Nepomorpha). An Essay Review of Mahner's 'Systema Cryptoceratorum Phylogeneticum'. Entomologica scandinavica 26: 159-166

Apter MJ, Wolpert L (1965) Cybernetics and Development. I. Information Theory. Journal of theoretical Biology 8: 244-257

Arnold AJ, Fristrup K (1982) The Theory of Evolution by Natural Selection: A Hierarchical Expansion. In: Brandon RN, Burian RM (eds) (1984) pp 292-319

Arnold, SJ (1983) Morphology, Performance and Fitness. American Zoologist 23: 347-361

Aronson LR, Tobach E, Lehrman DS, Rosenblatt JS (eds) (1970) Development and Evolution of Behavior. Freeman, San Francisco

Asquith PD, Giere RN (eds) (1981) PSA 1980, vol 2. Philosophy of Science Association, East Lansing

Atkinson JW (1992) Conceptual Issues in the Reunion of Development and Evolution. Synthese 91: 93-110

Ax P (1984) Das Phylogenetische System. Gustav Fischer, Stuttgart (Engl transl 1987 The Phylogenetic System. Wiley, Chichester)

Ax P (1985) Stem Species and the Stem Lineage Concept. Cladistics 1: 279-287

Ax P (1988) Systematik in der Biologie. Gustav Fischer, Suttgart

Ax P (1995) Das System der Metazoa I. Gustav Fischer, Stuttgart

Ayala FA (1968) Biology as an Autonomous Science. American Scientist 56: 207-221

Ayala FA (1970) Teleological Explanations in Evolutionary Biology. Philosophy of Science 37: 1-15

Ayala FJ (1983) Microevolution and Macroevolution. In: Bendall DS (ed) pp 387-402

Ayala F, Dobzhansky T (eds) (1974 Studies in the Philosophy of Biology. University of California Press, Berkeley

Bartlett MS (1975) Probability, Statistics and Time. Chapman and Hall, London
Beatty J (1981) What's Wrong with the Received View of Evolutionary Theory? In: Asquith PD, Giere RN (eds) pp 397-426
Beatty J (1982) Classes and Cladists. Systematic Zoology 31: 25-34
Beatty J (1987) On Behalf of the Semantic View. Biology and Philosophy 2: 17-23
Beatty J, Finsen S (1989) Rethinking the Propensity Interpretation: A Peek Inside Pandora's Box. In: Ruse M (ed) pp 17-30
Beaumont JG, Kenealy PM, Rogers MJC (1996) The Blackwell Dictionary of Neuropsychology. Blackwell, Cambridge, MA
Bechtel W (ed) (1986) Integrating Scientific Disciplines. Martinus Nijhoff, Dordrecht
Beckner M (1959) The Biological Way of Thought. Columbia University Press, New York
Beckner M (1964) Metaphysical Presuppositions and the Description of Biological Systems. In: Gregg JR, Harris FTC (eds) pp 15-29
Beckner M (1974) Reduction, Hierarchies and Organicism. In: Ayala FJ, Dobzhansky T (eds) pp 163-176
Bell G (1992) Five Properties of Environments. In: Grant PR, Horn HS (eds) Molds, Molecules, and Metazoa. Princeton University Press, Princeton, pp 33-56
Bendall DS (ed) (1983) Evolution from Molecules to Men. Cambridge University Press, Cambridge
Berlin B (1992) Ethnobiological Classification. Principles of Categorization of Plants and Animals in Traditional Societies. Princeton University Press, Princeton
Bernier R (1984) The Species as an Individual. Facing Essentialism. Systematic Zoology 33: 460-469
Bernier R, Pirlot P (1977) Organe et fonction. Essai de biophilosophie. Maloine-Doin--Edisem, Paris, St Hyacinthe
Berry S (1995) Entropy, Irreversibility and Evolution. Journal of theoretical Biology 175: 197-202
Beyerstein BL (1987) Neuroscience and Psi-ence. Behavioral and Brain Sciences 10: 571-572
Bhaskar R (1978) A Realist Theory of Science. Harvester Press, Hassocks
Bigelow J, Pargetter R (1987) Functions. Journal of Philosophy 84: 181-196
Bindra D (1976) A Theory of Intelligent Behavior. Wiley, New York
Birkhoff G, Bartee TC (1970) Modern Applied Algebra. McGraw-Hill, New York
Blackmore S (1986) Cellular Ontogeny. Cladistics 2: 358-362
Blitz D (1992) Emergent Evolution. Qualitative Novelty and the Levels of Reality. Kluwer, Dordrecht
Block N (1980a) What Is Functionalism? In: Block N (ed) pp 171-184
Block N (1980b) Troubles with Functionalism. In: Block N (ed) pp 268-305
Block N (ed) (1980) Readings in Philosophy of Psychology, vol 1. Harvard University Press, Cambridge
Boag PT, Grant PR (1981) Intense Natural Selection in a Population of Darwin's Finches (Geospizinae) in the Galapagos. Science 214: 82-85
Bock WJ (1974) Philosophical Foundations of Classical Evolutionary Classification. Systematic Zoology 22: 375-392
Bock WJ (1980) The Definition and Recognition of Biological Adaptation. American Zoologist 20: 217-227
Bock WJ (1981) Functional-Adaptive Analysis in Evolutionary Classification. American Zoologist 21: 5-20

Bock WJ (1986) Species Concepts, Speciation, and Macroevolution. In: Iwatsuki K, Raven PH, Bock WJ (eds) Modern Aspects of Species. University of Tokyo Press, Tokyo, pp 31-57

Bock WJ, von Wahlert G (1965) Adaptation and the Form-Function Complex. Evolution 19: 269-299

Bonner JT (ed) (1982) Evolution and Development. Springer, Berlin Heidelberg New York

Bonnet C (1762) Considérations sur les Corps Organisés. Marc-Michel Rey, Amsterdam

Boorse C (1977) Health as a Theoretical Concept. Philosophy of Science 44: 542-573

Boster J (1987) Agreement Between Biological Classification Systems Is not Dependent on Cultural Transmission. American Anthropologist 89: 914-920

Boyd R (1984) The Current Status of Scientific Realism. In: Leplin J (ed) pp 41-82

Bradie M (1991) The Evolution of Scientific Lineages. In: Fine A, Forbes M, Wessels L (eds) PSA 1990, vol 2. Philosophy of Science Association, East Lansing, pp 245-254

Bradie M (1994a) Epistemology from an Evolutionary Point of View. In: Sober E (ed) pp 453-475.

Bradie M (1994b) The Secret Chain. Evolution and Ethics. SUNY Press, Albany

Bradie M, Gromko M (1981) The Status of the Principle of Natural Selection. Nature and System 3: 3-12

Brady RH (1979) Natural Selection and the Criteria by Which a Theory Is Judged. Systematic Zoology 28: 600-621

Brady RH (1982) Theoretical Issues and "Pattern Cladistics". Systematic Zoology 31: 286-291

Brady RH (1985) On the Independence of Systematics. Cladistics 1: 113-126

Brandon RN (1981) Biological Teleology: Questions and Explanations. Studies in History and Philosophy of Science 12: 91-105

Brandon RN (1990) Adaptation and Environment. Princeton University Press, Princeton

Brandon RN (1992) Environment. In: Keller EF, Lloyd EA (eds) pp 81-86

Brandon RN, Burian RM (eds) (1984) Genes, Organisms, Populations: Controversies over the Units of Selection. MIT Press, Cambridge

Brandon RN, Antonovics J, Burian RM, Carson S, Cooper G, Davies PS, Horvath C, Mishler BD, Richardson RC, Smith K, Thrall P (1994) Sober on Brandon on Screening-Off and the Levels of Selection. Philosophy of Science 61: 475-486

Bridgman PW (1927) The Logic of Modern Physics. Macmillan, New York

Brière C, Goodwin BC (1988) Geometry and Dynamics of Tip Morphogenesis in *Acetabularia*. Journal of theoretical Biology 131: 461-475

Brooks DR, Wiley EO (1988) Evolution as Entropy: Toward a Unified Theory of Biology. University of Chicago Press, Chicago

Brooks DR, Collier J, Maurer B, Smith JDH, Wiley EO (1989) Entropy and Information in Evolving Biological Systems. Biology and Philosophy 4: 407-432

Buck RC, Hull DL (1966) The Logical Structure of the Linnean Hierarchy. Systematic Zoology 15: 97-111

Bunge M (1959a) Causality. The Place of the Causal Principle in Modern Science. Harvard University Press, Cambridge [also: 3rd edn. (1979) Causality in Modern Science. Dover Publications, New York]

Bunge M (1959b) Metascientific Queries. Charles C Thomas, Evanston
Bunge M (1962) Intuition and Science. Prentice-Hall, Englewood Cliffs
Bunge M (1963) The Myth of Simplicity. Prentice-Hall, Englewood Cliffs
Bunge M (1964) Phenomenological Theories. In: Bunge M (ed) The Critical Approach. Free Press, New York, pp 234-254
Bunge M (1967a) Scientific Research I: The Search for System. Springer, Berlin Heidelberg New York
Bunge M (1967b) Scientific Research II: The Search for Truth. Springer, Berlin Heidelberg New York
Bunge M (1967c) Foundations of Physics. Springer, Berlin Heidelberg New York
Bunge M (1969) Models in Theoretical Science. Proceedings of the XIVth International Congress of Philosophy Vienna 1968 3: 208-217
Bunge M (1971) A Mathematical Theory of the Dimensions and Units of Physical Quantities. In: Bunge M (ed) Problems in the Foundations of Physics. Springer, Berlin Heidelberg New York, pp 1-16
Bunge M (1973a) Method, Model and Matter. Reidel, Dordrecht
Bunge M (1973b) Philosophy of Physics. Reidel, Dordrecht
Bunge M (1973c) [Review of] P Suppes (1970) A Probabilistic Theory of Causality. British Journal for the Philosophy of Science 24: 409-410
Bunge M (ed) (1973) The Methodological Unity of Science. Reidel, Dordrecht
Bunge M (1974a) Semantics I: Sense and Reference. Reidel, Dordrecht
Bunge M (1974b) Semantics II: Interpretation and Truth. Reidel, Dordrecht
Bunge M (1974c) The Relations of Logic and Semantics to Ontology. Journal of Philosophical Logic 3: 195-210
Bunge M (1976) Review of W. Stegmüller's The Structure and Dynamics of Theories. Mathematical Reviews 55: 333 (#2480)
Bunge M (1977a) Ontology I: The Furniture of the World. Reidel, Dordrecht
Bunge M (1977b) Levels and Reduction. American Journal of Physiology 233: R75-R82
Bunge M (1977c) The GST Challenge to the Classical Philosophies of Science. International Journal of General Systems 4: 29-37
Bunge M (1979a) Ontology II: A World of Systems. Reidel, Dordrecht
Bunge M (1979b) Some Topical Problems in Biophilosophy. Journal of Social and Biological Structures 2: 155-172
Bunge M (1980) The Mind-Body Problem. Pergamon Press, Oxford
Bunge M (1981a) Scientific Materialism. Reidel, Dordrecht
Bunge M (1981b) Four Concepts of Probability. Applied Mathematical Modelling 5: 306-312
Bunge M (1981c) Biopopulations, Not Biospecies, Are Individuals and Evolve. Behavioral and Brain Sciences 4: 284-285
Bunge M (1982) Is Chemistry a Branch of Physics? Zeitschrift für allgemeine Wissenschaftstheorie 13: 209-223
Bunge M (1983a) Epistemology & Methodology I: Exploring the World. Reidel, Dordrecht
Bunge M (1983b) Epistemology & Methodology II: Understanding the World. Reidel, Dordrecht
Bunge M (1983c) Epistemologie. Aktuelle Fragen der Wissenschaftstheorie. Bibliographisches Institut, Mannheim

Bunge M (1983d) Speculation: Wild and Sound. New Ideas in Psychology 1: 3-6
Bunge M (1985a) Philosophy of Science and Technology. Part I: Formal and Physical Sciences. Reidel, Dordrecht
Bunge M (1985b) Philosophy of Science and Technology. Part II: Life Science, Social Science and Technology. Reidel, Dordrecht
Bunge M (1987a) Two Controversies in Evolutionary Biology: Saltationism and Cladism. In: Rescher N (ed) Scientific Inquiry in Philosophical Perspective. University Press of America, Washington, pp 129-146
Bunge M (1987b) Why Parapsychology Cannot Become a Science. Behavioral and Brain Sciences 10: 576-577
Bunge M (1988) Two Faces and Three Masks of Probability. In: Agazzi E (ed) Probability in the Sciences. Kluwer, Dordrecht, pp 27-49
Bunge M (1989) Ethics – The Good and the Right. Reidel, Dordrecht
Bunge M (1991a) The Power and Limits od Reduction. In: Agazzi E (ed) The Problem of Reductionism in Science. Kluwer, Dordrecht, pp 31-49
Bunge M (1991b) Five Bridges Between Scientific Disciplines. In: Geyer F (ed) The Cybernetics of Complex Systems: Self-Organization, Evolution, and Social Change. Intersystems Publications, Salinas, pp 1-10
Bunge M (1992) System Boundary. International Journal of General Systems 20: 215-219
Bunge M (1993) Explaining Creativity. In: Brzezinski J et al. (eds) Creativity and Consciousness: Philosophical and Psychological Dimensions. Rodopi, Amsterdam, pp 299-304
Bunge M (1996) Finding Philosophy in Social Science. Yale University Press, New Haven
Bunge M (1997) Moderate Mathematical Fictionism. In: Agazzi E, Darvas G (eds) Philosophy of Mathematics Today. Kluwer, Dordrecht, pp 51-71
Bunge M, Ardila R (1987) Philosophy of Psychology. Springer, Berlin Heidelberg New York
Burian RM (1983) Adaptation. In: Grene M (ed) pp 287-314
Burian RM (1986) On Integrating the Study of Evolution and of Development. In: Bechtel W (ed) pp 209-228
Burian RM (1992) Adaptation – Historical Perspectives. In: Keller EF, Lloyd EA (eds) pp 7-12
Burns TP (1992) Adaptedness, Evolution and a Hierarchical Concept of Fitness. Journal of theoretical Biology 154: 219-237
Byerly H (1986) Fitness As a Function. In: Fine A, Machamer P (eds) PSA 1986, vol 1, pp 494-501
Byerly HC, Michod RE (1991) Fitness and Evolutionary Explanation. Biology and Philosophy 6: 1-22

Callicott JB (1980) Animal Liberation: A Triangular Affair. In: Hargrove EC (ed) (1992) pp 37-69
Callicott JB (1988) Animal Liberation and Environmental Ethics: Back Together Again. In: Hargrove EC (ed) (1992) pp 249-261
Campbell DT (1974) 'Downward Causation' in Hierarchically Organised Biological Systems. In: Ayala FJ, Dobzhansky T (eds) pp 179-186
Canfield J (1964) Teleological Explanation in Biology. British Journal for the Philosophy of Science 14: 285-295

Caplan AL (1978) Testability, Disreputability, and the Structure of the Modern Synthetic Theory of Evolution. Erkenntnis 13: 261-278

Caplan AL (1981) Back to Class: A Note on the Ontology of Species. Philosophy of Science 48: 130-140

Caplan AL, Bock WJ (1988) Haunt Me No Longer. Biology and Philosophy 3: 443-454

Carlson EA (1966) The Gene: A Critical History. Saunders, Philadelphia

Carnap R (1936-37) Testability and Meaning. Philosophy of Science 3: 419-471; 4: 1-40

Carnap R (1939) Foundations of Logic and Mathematics. International Encyclopedia of Unified Science, vol 1, no 3. University of Chicago Press, Chicago

Carpenter SR, Chisholm SW, Krebs CJ, Schindler DW, Wright RF (1995) Ecosystem Experiments. Science 269: 324-327

Causey RL (1977) Unity of Science. Reidel, Dordrecht

Cavalieri LF, Koçak H (1995) Intermittent Transition Between Order and Chaos in an Insect Pest Population. Journal of theoretical Biology 175: 231-234

Chang CC, Keisler HJ (1973) Model Theory. North-Holland Publishing, Amsterdam

Charig AJ (1982) Systematics in Biology: A Fundamental Comparison of Some Major Schools of Thought. In: Joysey KA, Friday AE (eds) pp 363-440

Churchill FB (1974) William Johannsen and the Genotype Concept. Journal of the History of Biology 7: 5-30

Churchland P, Hooker CA (eds) (1985) Images of Science. University of Chicago Press, Chicago

Colless DH (1985) On 'Character' and Related Terms. Systematic Zoology 34: 229-233

Collier J (1988) Supervenience and Reduction in Biological Hierarchies. In: Matthen M, Linsky B (eds) pp 209-234

Colwell RK (1992) Niche: A Bifurcation in the Conceptual Lineage of the Term. In: Keller EF, Lloyd EA (eds) pp 241-248

Copi IM (1968) Introduction to Logic, 3rd edn. Collier-Macmillan, London

Costantino RF, Cushing JM, Dennis B, Desharnais RA (1995) Experimentally Induced Transitions in the Dynamic Behaviour of Insect Populations. Nature 375: 227-230

Cracraft J (1981) The Use of Functional and Adaptive Criteria in Phylogenetic Systematics. American Zoologist 21: 21-36

Cracraft J (1987) Species Concepts and the Ontology of Evolution. Biology and Philosophy 2: 329-346

Cracraft J (1989) Species as Entities of Biological Theory. In: Ruse M (ed) pp 31-52

Cracraft J (1990) The Origin of Evolutionary Novelties: Pattern and Process at Different Hierarchical Levels. In: Nitecki M (ed) pp 21-44

Cracraft J, Eldredge N (eds) (1979) Phylogenetic Analysis and Paleontology. Columbia University Press, New York

Craw R (1992) Margins of Cladistics: Identity, Difference and Place in the Emergence of Phylogenetic Systematics, 1864-1975. In: Griffiths PE (ed) pp 65-107

Culp S, Kitcher P(1989) Theory Structure and Theory Change in Contemporary Molecular Biology. British Journal for the Philosophy of Science 40: 459-483

Cummins R (1975) Functional Analysis. Journal of Philosophy 72: 741-764

Damuth J (1985) Selection Among "Species": A Formulation in Terms of Natural Functional Units. Evolution 39: 1132-1146

Damuth J, Heisler IL (1988) Alternative Formulations of Multilevel Selection. Biology and Philosophy 3: 407-430

Dancoff SM, Quastler H (1953) The Information Content and Error Rate of Living Things. In: Quastler H (ed) Information Theory in Biology. University of Illinois Press, Urbana, pp 263-273

Darden L, Cain JA (1989) Selection Type Theories. Philosophy of Science 56: 106-129

Darden L, Maull N (1977) Interfield Theories. Philosophy of Science 44: 43-64

Darwin C (1859) On the Origin of Species. Harvard University Press, Cambridge (1964)

Dawkins R (1976) The Selfish Gene. Oxford University Press, New York

Dawkins R (1982) The Extended Phenotype. The Gene as the Unit of Selection. Freeman, Oxford

de Jong G (1994) The Fitness of Fitness Concepts and the Description of Natural Selection. Quarterly Review of Biology 69: 3-29

Dennett DC (1978) Brainstorms: Philosophical Essays on Mind and Psychology. Bradford Books, Montgomery

Dennett DC (1995) Darwin's Dangerous Idea. Simon & Schuster, New York

de Queiroz K (1988) Systematics and the Darwinian Revolution. Philosophy of Science 55: 238-259

de Queiroz K (1992) Phylogenetic Definitions and Taxonomic Philosophy. Biology and Philosophy 7: 295-313

de Queiroz K (1994) Replacement of an Essentialistic Perspective on Taxonomic Definitions as Exemplified by the Definition of "Mammalia". Systematic Biology 43: 497-510

de Queiroz K, Donoghue MJ (1988) Phylogenetic Systematics and the Species Problem. Cladistics 4: 317-338

de Queiroz K, Gauthier J (1990) Phylogeny as a Central Principle in Taxonomy: Phylogenetic Definitions of Taxon Names. Systematic Zoology 39: 307-322

de Queiroz K, Gauthier J (1992) Phylogenetic Taxonomy. Annual Review of Ecology and Systematics 23: 449-480

Dobzhansky T, Ayala FJ, Stebbins GL, Valentine JW (1977) Evolution. Freeman, San Francisco

Dretske F (1977) Laws of Nature. Philosophy of Science 44: 248-268

Dretske F (1978) The Role of the Percept in Visual Cognition. In: Savage CW (ed) pp 107-127

Duhem P (1908) ΣΩΖΕΙΝ ΤΑ ΦΑΙΝΟΜΕΝΑ. Essai sur la notion de Théorie physique de Platon à Galilée. A Hermann et Fils, Paris

Dunbar MJ (1972) The Ecosystem as Unit of Natural Selection. In: Deevey ES (ed) Growth by Intussusception. Essays in Honor of G. Evelyn Hutchinson. Connecticut Academy of Arts and Sciences, New Haven, pp 113-130

Dupré J (1981) Natural Kinds and Biological Taxa. Philosophical Review 90: 66-90

Dyson F (1985) Origins of Life. Cambridge University Press, Cambridge

Earman J, Salmon WC (1992) The Confirmation of Scientific Hypotheses. In: Salmon MH et al., pp 42-103

Eigen M, Gardiner W, Schuster P, Winckler-Oswatitsch R (1981) The Origin of Genetic Information. Scientific American 244(4): 88-118

Eldredge N (1979) Cladism and Common Sense. In: Cracraft J, Eldredge N (eds) pp 165-198
Eldredge N (1985a) Unfinished Synthesis. Biological Hierarchies and Modern Evolutionary Thought. Oxford University Press, New York
Eldredge N (1985b) The Ontology of Species. In: Vrba ES (ed) pp 17-20
Emmeche C (1992) Life as an Abstract Phenomenon: Is Artificial Life Possible? In: Varela FJ, Bourgine P (eds) Toward a Practice of Autonomous Systems. Proceedings of the First European Conference on Artificial Life. MIT-Press, Cambridge, pp 466-474
Endler JA (1986) Natural Selection in the Wild. Princeton University Press, Princeton
Engelberg J, Boyarsky LL (1979) The Noncybernetic Nature of Ecosystems. American Naturalist 114: 317-324
Engels E-M (1982) Die Teleologie des Lebendigen. Duncker & Humblot, Berlin
Ereshefsky M (1988) Indivduality and Macroevolutionary Theory. In: Fine A, Leplin J (eds) PSA 1988, vol 1. Philosophy of Science Association, East Lansing, pp 216-222
Ereshefsky M (1991) The Semantic Approach to Evolutionary Theory. Biology and Philosophy 6: 59-80
Ereshefsky M (ed) (1992) The Units of Evolution. Essays on the Nature of Species. MIT-Press, Cambridge
Ereshefsky M (1994) Some Problems with the Linnean Hierarchy. Philosophy of Science 61: 186-205

Fernholm B, Bremer K, Jörnvall H (eds) (1989) The Hierarchy of Life. Molecules and Morphology in Phylogenetic Analysis. Elsevier, Amsterdam
Feyerabend PK (1981) Philosophical Papers, 2 vol. Cambridge University Press, Cambridge
Fodor JA (1981) The Mind-Body Problem. Scientific American 244(1): 114-123
Fox SW (1984) Proteinoid Experiments and Evolutionary Theory. In: Ho M-W, Saunders PT (eds) pp 15-60
Fristrup K (1992) Character: Current Usages. In: Keller EF, Lloyd EA (eds) pp 45-51
Futuyma DJ (1986) Evolutionary Biology. Sinauer, Sunderland

Gaffney ES (1979) An Introduction to the Logic of Phylogeny Reconstruction. In: Cracraft J, Eldredge E (eds) pp 79-111
Gardiner BG (1993) Haematothermia: Warm-Blooded Amniotes. Cladistics 9: 369-395
Gardner M (1989) Gaiaism. Skeptical Inquirer 13: 252-256
Gasper P (1992) Reduction and Instrumentalism in Genetics. Philosophy of Science 59: 655-670
Gayon J (1996) The Individuality of the Species: A Darwinian Theory? — from Buffon to Ghiselin, and back to Darwin. Biology and Philosophy 11: 215-244
Ghiselin MT (1966) On Psychologism in the Logic of Taxonomic Controversies. Systematic Zoology 15: 207-215
Ghiselin MT (1974) A Radical Solution to the Species Problem. Systematic Zoology 23: 536-544
Ghiselin MT (1981) Categories, Life, and Thinking. Behavioral and Brain Sciences 4: 269-283
Ghiselin MT (1984) "Definition", "Character" and Other Equivocal Terms. Systematic Zoology 33: 104-110

Giere RN (1984) Understanding Scientific Reasoning. Holt, Rinehart and Winston, New York
Giere RN (1985) Constructive Realism. In: Churchland P, Hooker CA (eds) pp 75-98
Gifford F (1990) Genetic Traits. Biology and Philosophy 5: 327-347
Gilbert SF (1988) Developmental Biology. Sinauer, Sunderland [also: 4th edn. (1994)]
Gliddon CJ, Gouyon P-H (1989) The Units of Selection. Trends in Ecology and Evolution 4: 204-208
Godfrey-Smith P, Lewontin RC (1993) The Dimensions of Selection. Philosophy of Science 60: 373-395
Goodman M (1989) Emerging Alliance of Phylogenetic Systematics and Molecular Biology: A New Age of Exploration. In: Fernholm B et al. (eds) pp 43-61
Goodwin BC (1982a) Genetic Epistemology and Constructionist Biology. Revue internationale de philosophie 36: 527-548
Goodwin BC (1982b) Development and Evolution. Journal of theoretical Biology 97: 43-55
Goodwin BC (1984) A Relational or Field Theory of Reproduction and Its Evolutionary Implications. In: Ho M-W, Saunders PT (eds) pp 219-241
Goodwin BC (1990) Structuralism in Biology. Science Progress 74: 227-243
Goodwin BC (1994) How the Leopard Changed Its Spots. The Evolution of Complexity. Charles Scribner's Sons, New York
Goodwin BC, Trainor LEH (1980) A Field Description of the Cleavage Process in Embryogenesis. Journal of theoretical Biology 86: 757-770
Goodwin BC, Trainor LEH (1985) Tip and Whorl Morphogenesis in *Acetabularia* by Calcium Regulated Strain Fields. Journal of theoretical Biology 117: 79-106
Goodwin BC, Holder N, Wylie CC (eds) (1983) Development and Evolution. Cambridge University Press, Cambridge
Gottlieb G (1970) Conceptions of Prenatal Behavior. In: Aronson LR et al. (eds) pp 111-137
Gottlieb G (1991) Experiential Canalization of Behavioral Development: Theory. Developmental Psychology 27: 4-13
Goudge TA (1961) The Ascent of Life. University of Toronto Press, Toronto
Gould SJ (1977) Ontogeny and Phylogeny. Belknap Press, Cambridge
Gould SJ (1980) Is a New and General Theory of Evolution Emerging? Paleobiology 6: 119-130
Gould SJ (1989) A Developmental Constraint in *Cerion*, with Comments on the Definition and Interpretation of Constraint in Evolution. Evolution 43: 516-539
Gould SJ (1997) Cope's Rule as a Psychological Artefact. Nature 385: 199-200
Gould SJ, Lewontin RC (1979) The Spandrels of San Marco and the Panglossian Paradigm: A Critique of the Adaptationist Programme. In: Sober E (ed) (1994) pp 73-90
Gould SJ, Vrba ES (1982) Exaptation – A Missing Term in the Science of Form. Paleobiology 8: 4-15
Grant P (1978) Biology of Developing Systems. Holt, Rinehart and Winston, New York
Grant V (1994) Evolution of the Species Concept. Biologisches Zentralblatt 113: 401-415
Gray R (1992) Death of the Gene: Developmental Systems Strike Back. In: Griffiths PE (ed) pp 165-209
Gregg JR (1950) Taxonomy, Language, and Reality. American Naturalist 84: 419-435

Gregg JR (1954) The Language of Taxonomy. Columbia University Press, New York
Gregg JR (1968) Buck and Hull: A Critical Rejoinder. Systematic Zoology 17: 342-344
Gregg JR, Harris FTC (eds) (1964) Form and Strategy in Science. Reidel, Dordrecht
Grene M (ed) (1983) Dimensions of Darwinism. Themes and Counterthemes in Twentieth-Century Evolutionary Biology. Cambridge University Press, Cambridge
Grene M (1987) Hierarchies in Biology. American Scientist 75: 504-510
Griesemer JR (1992) Niche: Historical Perspectives. In: Keller EF, Lloyd EA (eds) pp 231-240
Griffiths GCD (1974) On the Foundations of Biological Systematics. Acta Biotheoretica 23: 85-131
Griffiths PE (ed) (1992) Trees of Life: Essays in the Philosophy of Biology. Kluwer, Dordrecht
Griffiths PE (1993) Functional Analysis and Proper Functions. British Journal for the Philosophy of Science 44: 409-422
Griffiths PE, Gray RD (1994) Developmental Systems and Evolutionary Explanation. Journal of Philosophy 91: 277-304
Grobstein C (1962) Levels and Ontogeny. American Scientist 50: 46-58
Günther K (1950) Ökologische und funktionelle Anmerkungen zur Frage des Nahrungserwerbs bei Tiefseefischen mit einem Exkurs über die ökologischen Zonen und Nischen. In: Grüneberg H, Ulrich W (eds) Moderne Biologie. Festschrift zum 60. Geburtstag von Hans Nachtsheim. FW Peters, Berlin, pp 55-93
Gurdon JB, Harger P, Mitchell A, Lemaire P (1994) Activin Signalling and Response to a Morphogen Gradient. Nature 371: 487-492
Gutmann WF (1995) Evolution von lebendigen Konstruktionen. Warum Erkenntnis unerträglich sein kann. Ethik und Sozialwissenschaften 6: 303-315
Guyot K (1986) Specious Individuals. Philosophica 37: 101-126

Hacking I (1985) Do We See Through a Microscope? In: Churchland P, Hooker CA (eds) pp 132-152
Hacking I (1991) A Tradition of Natural Kinds. Philosophical Studies 61: 109-126
Haeckel E (1866) Generelle Morphologie der Organismen. Walter de Gruyter, Berlin (1988)
Hagen JB (1989) Research Perspectives and the Anomalous State of Modern Ecology. Biology and Philosophy 4: 433-455
Haldane JBS (1929) Heredity. In: Encyclopaedia Britannica, 14th edn, vol 11. Encyclopaedia Britannica Company, London, pp 484-496
Hall BK (1992) Evolutionary Developmental Biology. Chapman & Hall, London
Hargrove EC (ed) (1992) The Animal Rights/Environmental Ethics Debate. SUNY Press, Albany
Harré R (1986) Varieties of Realism. Basil Blackwell, Oxford
Hartmann M (1965) Einführung in die allgemeine Biologie und ihre philosophischen Grund- und Grenzfragen. Walter de Gruyter, Berlin
Hassenstein B (1981) Biologische Teleonomie. Neue Hefte für Philosophie 20: 60-71
Hebb DO (1949) The Organization of Behavior. Wiley, New York
Hebb DO (1966) A Textbook of Psychology. Saunders, Philadelphia
Hebb DO (1980) Essay on Mind. Erlbaum, Hillsdale
Heise H (1981) Universals, Particulars, and Paradigms. Behavioral and Brain Sciences 4: 289-290

Hempel CG (1965) Aspects of Scientific Explanation. The Free Press, New York; Collier-Macmillan, London
Hempel CG, Oppenheim P (1948) Studies in the Logic of Explanation. Philosophy of Science 15: 135-175
Hennig W (1966) Phylogenetic Systematics. University of Illinois Press, Urbana
Hess B, Mikhailov A (1994) Self-Organization in Living Cells. Science 264: 223-224
Hilbert D (1918) Axiomatisches Denken. Mathematische Annalen 78: 405-415
Hilbert D, Bernays P (1934) Grundlagen der Mathematik, 2 vols. Springer, Berlin Heidelberg New York (1968)
Hill CR, Crane PR (1982) Evolutionary Cladistics and Angiosperms. In: Joysey KA, Friday AE (eds) pp 269-362
Hillis DM (1987) Molecular versus Morphological Approaches to Systematics. Annual Review of Ecology and Systematics 18: 23-42
Hillis DM, Moritz C (1990) An Overview of Applications of Molecular Systematics. In: Hillis DM, Moritz C (eds) pp 502-515
Hillis DM, Moritz C (eds) (1990) Molecular Systematics. Sinauer, Sunderland
Ho MW, Fox SW (1988) Processes and Metaphors in Evolution. In: Ho MW, Fox SW (eds) pp 1-16
Ho MW, Fox SW (eds) (1988) Evolutionary Processes and Metaphors. Wiley, Chichester
Ho MW, Saunders PT (1979) Beyond Neo-Darwinism – An Epigenetic Approach to Evolution. Journal of Theoretical Biology 78: 573-591
Ho MW, Saunders PT (1993) Rational Taxonomy and the Natural System with Particular Reference to Segmentation. Acta Biotheoretica 41: 289-304
Ho MW, Saunders PT (eds) (1984) Beyond Neo-Darwinism. Academic Press, London
Hooker CA (1978) An Evolutionary Naturalist Realist Doctrine of Perception and Secondary Qualities. In: Savage CW (ed) pp 405-440
Hooker CA (1987) A Naturalist Realism. Revue internationale de philosophie 41: 5-28
Horan BL (1994) The Statistical Character of Evolutionary Theory. Philosophy of Science 61: 76-95
Horder TJ (1983) Embryological Bases of Evolution. In: Goodwin BC et al. (eds) pp 315-352
Horder TJ (1989) Syllabus for an Embryological Synthesis. In: Wake DB, Roth G (eds) Complex Organismal Functions: Integration and Evolution in Vertebrates. Wiley, Chichester, pp 315-348
Hoyningen-Huene P, Wuketits FM (eds) (1989) Reductionism and Systems Theory in the Life Sciences. Kluwer, Dordrecht
Hubbard R, Wald E (1993) Exploding the Gene Myth. Beacon Press, Boston
Hull DL (1965) The Effect of Essentialism on Taxonomy – Two Thousand Years of Stasis. British Journal for the Philosophy of Science 15: 314-326; 16: 1-18
Hull DL (1968) The Operationalist Imperative. Sense and Non-Sense in Operationalism. Systematic Zoology 17: 432-459
Hull DL (1970) Contemporary Systematic Philosophies. Annual Review of Ecology and Systematics 1: 19-54
Hull DL (1974) Philosophy of Biological Science. Prentice Hall, Englewood Cliffs
Hull DL (1976) Are Species Really Individuals? Systematic Zoology 25: 174-191
Hull DL (1978) A Matter of Individuality. In: Sober E (ed) (1994) pp 193-215
Hull DL (1979) The Limits of Cladism. Systematic Zoology 28: 416-440

Hull DL (1980) Individuality and Selection. Annual Review of Ecology and Systematics 11: 311-332
Hull DL (1987) Genealogical Actors in Ecological Roles. Biology and Philosophy 2: 168-184
Hull DL (1988) Science as a Process. University of Chicago Press, Chicago
Hull DL (1989) The Metaphysics of Evolution. SUNY Press, Albany
Hume D (1739-1740) A Treatise of Human Nature. In: Selby-Bigge LA (ed) Clarendon Press, Oxford (1888)
Hurlbert SH (1981) A Gentle Depilation of the Niche. Dicean Resource Sets in Resource Hyperspace. Evolutionary Theory 5: 177-184
Hutchinson GE (1957) Concluding Remarks. Cold Spring Harbor Symposia on Quantitative Biology 22: 415-427

Jahn I (1990) Grundzüge der Biologiegeschichte. Gustav Fischer, Jena
Janzen DH (1977) What are Dandelions and Aphids? American Naturalist 111: 586-589
Jeuken M (1975) The Biological and Philosophical Definitions of Life. Acta Biotheoretica 24: 14-21
Johannsen WL (1913) Elemente der exakten Erblichkeitslehre. Gustav Fischer, Jena [also: 3rd edn. (1926)]
Johnston T, Gottlieb G (1990) Neophenogenesis: A Developmental Theory of Phenotypic Evolution. Journal of theoretical Biology 147: 471-495
Jonckers LHM (1973) The Concept of Population in Biology. Acta Biotheoretica 22: 78-108
Jongeling TB (1985) On an Axiomatization of Evolutionary Theory. Journal of theoretical Biology 117: 529-543
Jordan CF (1981) Do Ecosystems Exist? American Naturalist 118: 284-287
Joysey KA, Friday AE (eds) (1982) Problems of Phylogenetic Reconstruction. Academic Press, London

Kanitscheider B (1989) Realism from a Biological Point of View. Dialectica 43: 141-156
Kant I (1787) Kritik der reinen Vernunft. Felix Meiner, Hamburg (1952)
Kant I (1790) Kritik der Urteilskraft. In: Königlich Preußische Akademie der Wissenschaften (ed) (1908) Kant's gesammelte Schriften, vol 5. Georg Reimer, Berlin
Kary M (1990) Information Theory and the Treatise : Towards a New Understanding. In: Weingartner P, Dorn GJW (eds) pp 263-280
Katz MJ (1982) Ontogenetic Mechanisms: The Middle Ground of Evolution. In: Bonner JT (ed) pp 207-212
Kauffman SA (1985) New Questions in Genetics and Evolution. Cladistics 1: 247-265
Kauffman SA (1993) The Origins of Order. Self-Organization and Selection in Evolution. Oxford University Press, New York
Keller EF, Lloyd EA (eds) (1992) Keywords in Evolutionary Biology. Harvard University Press, Cambridge
Keosian J (1974) Life's Beginnings – Origin or Evolution? Origins of Life 5: 285-293
Kim J (1978) Supervenience and Nomological Incommensurables. American Philosophical Quarterly 15: 149-156
Kitcher P (1984a) Species. Philosophy of Science 51: 308-333
Kitcher P (1984b) 1953 and All That: A Tale of Two Sciences. In: Sober E (ed) (1994) pp 379-399

Kitcher P (1987) Ghostly Whispers: Mayr, Ghiselin, and the "Philosophers" on the Ontological Status of Species. Biology and Philosophy 2: 184-192

Kitcher P (1989a) Some Puzzles About Species. In: Ruse M (ed) pp 183-208

Kitcher P (1989b) Explanatory Unification and the Causal Structure of the World. In: Kitcher P, Salmon WC (eds) pp 410-506

Kitcher P (1992) Gene: Current Usages. In: Keller EF, Lloyd EA (eds) pp128-131

Kitcher P, Salmon WC (eds) (1989) Minnesota Studies in the Philosophy of Science, vol 13: Scientific Explanation. University of Minnesota Press, Minneapolis

Kitts DB, Kitts DJ (1979) Biological Species as Natural Kinds. Philosophy of Science 46: 613-622

Kosslyn SM, Koenig O (1995) Wet Mind: The New Cognitive Neuroscience. Free Press, New York

Kovac D, Maschwitz U (1989) Secretion-Grooming in the Water Bug *Plea minutissima*: A Chemical Defence Against Microorganisms Interfering with the Hydrofuge Properties of the Respiratory Region. Ecological Entomology 14: 403-411

Kump LR (1996) The Physiology of the Planet. Nature 381: 111-112

Küppers B-O (1979) Towards an Experimental Analysis of Molecular Self-Organization and Precellular Darwinian Evolution. Naturwissenschaften 66: 228-243

Lander ES, Schork NJ (1994) Genetic Dissection of Complex Traits. Science 265: 2037-2048

Lang HS (1983) Aristotle and Darwin. The Problem of Species. International Philosophical Quarterly 23: 141-153

Lange M (1995) Are There Natural Laws Concerning Particular Biological Species? Journal of Philosophy 92: 430-451

Langton CG (1989) Artificial Life. In: Langton CG (ed) pp 1-47

Langton CG (ed) (1989) Artificial Life. Addison-Wesley, Redwood City

Langton CG (1991) Introduction. In: Langton CG et al. (eds) pp 3-23

Langton CG, Taylor C, Farmer JD, Rasmussen S (eds) (1991) Artificial Life II. Addison-Wesley, Redwood City

Lauder GV (1982) Historical Biology and the Problem of Design. Journal of theoretical Biology 97: 57-67

Lawton JH (1995) Ecological Experiments with Model Systems. Science 269: 328-331

Lazarsfeld PF, Menzel H (1961) On the Relation Between Individual and Collective Properties. In: Etzioni A (ed) Complex Organizations: A Sociological Reader. Holt, Rinehart & Winston, New York, pp 422-440

Lehrman DS (1970) Semantic and Conceptual Issues in the Nature-Nurture Problem. In: Aronson LR et al. (eds) pp 17-52

Leibniz GW (1704) Nouveaux Essais sur l'entendement humain [New Essays on Human Understanding]. Cambridge University Press, Cambridge (1981)

Lenartowicz P (1975) Phenotype-Genotype Dichotomy. An Essay in Theoretical Biology. Typis Pontificiae Universitatis Gregorianae, Rome

Lenin VI (1908) Materialism and Empirio-Criticism. Lawrence and Wisehart, London (1950)

Lennox JG (1991) Commentary on Byerly and Michod. Biology and Philosophy 6: 33-37

Lennox JG (1992a) Teleology. In: Keller EF, Lloyd EA (eds) pp 324-333

Lennox JG (1992b) Philosophy of Biology. In: MH Salmon et al. pp 269-309

Lenoir T (1982) The Strategy of Life. Teleology and Mechanics in Nineteenth Century German Biology. Reidel, Dordrecht

Leplin J (ed) (1984) Scientific Realism. University of California Press, Berkeley

Leroux N (1993) What are Biological Species? The Impact of the Debate in Taxonomy on the Species Problem. MA Thesis, Department of Philosophy, McGill University, Montreal

Levine L (1993) GAIA: Goddess and Idea. BioSystems 31: 85-92

Levins R, Lewontin RC (1982) Dialectics and Reductionism in Ecology. In: Saarinen E (ed) pp 107-138.

Levins R, Lewontin RC (1985) The Dialectical Biologist. Harvard University Press, Cambridge

Lewes GH (1879) Problems of Life and Mind. Trubner, London

Lewis RW (1980) Evolution: A System of Theories. Perspectives in Biology and Medicine 23: 551-572

Lewontin RC (1970) The Units of Selection. Annual Review of Ecology and Systematics 1: 1-18

Lewontin RC (1974) The Genetic Basis of Evolutionary Change. Columbia University Press, New York

Lewontin RC (1978) Adaptation. Scientific American 239(3): 212-232

Lewontin RC (1983a) The Organism as the Subject and Object of Evolution. Scientia 118: 65-82

Lewontin RC (1983b) Gene, Organism and Environment. In: Bendall DS (ed) pp 273-285

Lewontin RC (1984) Laws of Biology and Laws in Social Science. In: Keyfitz N (ed) Population and Biology. Ordina Editions, Liège, pp 19-28

Lewontin RC (1991) Biology as Ideology. The Doctrine of DNA. Anansi, Concord

Lewontin RC (1992a) The Dream of the Human Genome. New York Review of Books 39(10): 31-40

Lewontin RC (1992b) Genotype and Phenotype. In: Keller EF, Lloyd EA (eds) pp 137-144

Linsky B, Zalta EN (1995) Naturalized Platonism versus Platonized Naturalism. Journal of Philosophy 92: 525-555

Lloyd EA (1988) The Structure and Confirmation of Evolutionary Theory. Greenwood Press, New York

Lloyd EA (1992) Unit of Selection. In: Keller EF, Lloyd EA (eds) pp 334-340

Loehle C (1988) Philosophical Tools: Potential Contributions to Ecology. Oikos 51: 97-104

Löther R (1972) Die Beherrschung der Mannigfaltigkeit. Philosophische Grundlagen der Taxonomie. Gustav Fischer, Jena

Løvtrup S (1973) Classification, Convention, and Logic. Zoologica Scripta 2: 49-61

Løvtrup S (1974) Epigenetics. A Treatise on Theoretical Biology. Wiley, London

Løvtrup S (1987) On Species and Other Taxa. Cladistics 3: 157-177

Lowe EJ (1995) The Metaphysics of Abstract Objects. Journal of Philosophy 92: 509-524

Lwoff A (1962) Biological Order. MIT-Press, Cambrige

MacMahon JA, Phillips DL, Robinson JV, Schimpf DJ (1978) Levels of Biological Organization: An Organism-Centered Approach. Bioscience 28: 700-704

MacMahon JA, Schimpf DJ, Andersen DC, Smith KG, Bayn RL (1981) An Organism-Centered Approach to Some Community and Ecosystem Concepts. Journal of theoretical Biology 88: 287-307

Mahner M (1986) Kreationismus – Inhalt und Struktur antievolutionistischer Argumentation. Pädagogisches Zentrum, Berlin

Mahner M (1989) Warum eine Schöpfungstheorie nicht wissenschaftlich sein kann. Praxis der Naturwissenschaften – Biologic 38(8): 33-36

Mahner M (1993a) What Is a Species? A Contribution to the Never Ending Species Debate in Biology. Journal for General Philosophy of Science 24: 103-126

Mahner M (1993b) Systema Cryptoceratorum Phylogeneticum (Insecta, Heteroptera). E Schweizerbart'sche Verlagsbuchhandlung, Stuttgart (Zoologica vol 143)

Mahner M (1994a) Phänomenalistische Erblast in der Biologie. Biologisches Zentralblatt 113: 435-448 [modified Engl. version forthcoming: Operationalist Fallacies in Biology. Science & Education]

Mahner M (1994b) Anmerkungen zu Ernst Mayrs "Evolution — Grundfragen und Mißverständnisse". Ethik und Sozialwissenschaften 5: 234-237

Mahner M (1995) Hydraulischer *Dies irae* in Frankfurt. Ethik und Sozialwissenschaften 6: 336-339

Mahner M, Bunge M (1996) Is Religious Education Compatible with Science Education? Science & Education 5: 101-123, 189-199

Mahner M, Kary M (1997) What Exactly Are Genomes, Genotypes and Phenotypes? And What About Phenomes? Journal of theoretical Biology (in press)

Marquis JP (1990) Partial Truths About Partial Truth. In: Weingartner P, Dorn GJW (eds) pp 61-78

Maschwitz U (1971) Wasserstoffperoxid als Antispetikum bei einer Wasserwanze. Naturwissenschaften 58: 572

Matthen M, Linsky B (eds) (1988) Philosophy & Biology. University of Calgary Press, Calgary

May RM (1974) Biological Populations with Nonoverlapping Generations: Stable Points, Stable Cycles, and Chaos. Science 186: 645-647

Maynard Smith J (1983) Evolution and Development. In: Goodwin BC et al. (eds) pp 33-45

Maynard Smith J (1984) Group Selection. In: Brandon RN, Burian RM (eds) pp 238-249

Maynard Smith J (1986) The Problems of Biology. Oxford University Press, Oxford

Maynard Smith J (1995) Genes, Memes, and Minds. New York Review of Books 42(19): 46-48

Maynard Smith J, Burian RM, Kauffman S, Alberch P, Campbell J, Goodwin BC, Lande R, Raup D, Wolpert L (1985) Developmental Constraints and Evolution. Quarterly Review of Biology 60: 265-287

Mayo DG, Gilinsky NL (1987) Models of Group Selection. Philosophy of Science 54: 515-538

Mayr E (1960) The Emergence of Evolutionary Novelties. In: Tax S (ed) Evolution After Darwin, vol 1, The Evolution of Life. University of Chicago Press, Chicago, pp 349-380

Mayr E (1963) Animal Species and Evolution. Harvard University Press, Cambridge

Mayr E (1974) Cladistic Analysis or Cladistic Classification? Zeitschrift für zoologische Systematik und Evolutionsforschung 12: 94-128

Mayr E (1982) The Growth of Biological Thought. Harvard University Press, Cambridge
Mayr E (1988) Towards a New Philosophy of Biology. Harvard Unversity Press, Cambridge
Mayr E (1989) Attaching Names to Objects. In: Ruse M (ed) pp 235-243
Mayr E (1994) Evolution – Grundfragen und Mißverständnisse. Ethik und Sozialwissenschaften 5: 203-209
Mayr E (1995) Systems of Ordering Data. Biology and Philosophy 10: 419-434
Mayr E (1996) The Autonomy of Biology: The Position of Biology Among the Sciences. Quarterly Review of Biology 71: 97-106
Mayr E, Ashlock PD (1991) Principles of Systematic Zoology. McGraw Hill, New York
McClamrock R (1995) Screening-Off and the Levels of Selection. Erkenntnis 42: 107-112
McIntosh RP (1982) The Background and Some Current Problems of Theoretical Ecology. In: Saarinen E (ed) pp 1-61
McIntosh RP (1995) H.A. Gleason's 'Individualistic Concept' and Theory of Animal Communities: A Continuing Controversy. Biological Reviews 70: 317-357
McKinsey JCC, Sugar AC, Suppes P (1953) Axiomatic Foundations of Classical Particle Mechanics. Journal of Rational Mechanics and Analysis 2: 253-289
Merton RK (1973) The Sociology of Science. Theoretical and Empirical Investigations. University of Chicago Press, Chicago
Mertz DB, McCauley DE (1982) The Domain of Laboratory Ecology. In: Saarinen E (ed) pp 229-244
Mill JS (1875) A System of Logic, 8th edn. Longmans, Green, London (1952)
Millikan R (1989) In Defense of Proper Functions. Philosophy of Science 56: 288-302
Mills SK, J Beatty J (1979) The Propensity Interpretation of Fitness. Philosophy of Science 46: 263-286
Minshall GW, Petersen RC, Nimz CF (1985) Species Richness in Streams of Different Size From the Same Drainage Basin. American Naturalist 125: 16-38
Mitcham C (1994) Thinking Through Technology. University of Chicago Press, Chicago
Mohr H (1981) Biologische Erkenntnis. Teubner, Stuttgart
Monod J (1971) Chance and Necessity. Knopf, New York
Montagu WP (1925) The Ways of Knowing. Allen & Unwin, London
Moritz C, Hillis DM (1990) Molecular Systematics: Context and Controversies. In: Hillis DM, Moritz C (eds) pp 1-10
Moss L (1992) A Kernel of Truth? On the Reality of the Genetic Program. In: Hull DL, Forbes M, Okruhlik K (eds) PSA 1992, vol 1, pp 335-348
Müller GB (1990) Developmental Mechanisms at the Origin of Morphological Novelty: A Side-Effect Hypothesis. In: Nitecki M (ed) pp 99-130
Müller GB, Wagner GP (1991) Novelty in Evolution: Restructuring the Concept. Annual Review of Ecology and Systematics 22: 229-256
Munson R (1971) Biological Adaptation. Philosophy of Science 38: 200-215

Nagel E (1956) Logic Without Metaphysics. Free Press, Glencoe
Nagel E (1961) The Structure of Science. Harcourt, Brace and World, New York
Nagel E (1977) Teleology Revisited. Journal of Philosophy 74: 261-301
National Institutes of Health (1990) The US Human Genome Project – The First Five Years: Fiscal Years 1991-1995. NIH Publication No 90-1590

Naylor GJP (1992) The Phylogenetic Relationships Among Requiem and Hammerhead Sharks: Inferring Phylogeny when Thousands of Equally Most Parsimonious Trees Result. Cladistics 8: 295-318

Neander K (1991) Functions as Selected Effects: The Conceptual Analysist's Defense. Philosophy of Science 58: 168-184

Nelson GJ (1970) Outline of a Theory of Comparative Biology. Systematic Zoology 19: 373-384

Nelson GJ (1983) Reticulation in Cladograms. In: Platnick NI, Funk VA (eds) Advances in Cladistics, vol 2. Columbia University Press, New York, pp 105-111

Nelson GJ (1989) Cladistics and Evolutionary Models. Cladistics 5: 275-289

Nelson GJ, Platnick NI (1981) Systematics and Biogeography. Cladistics and Vicariance. Columbia University Press, New York

Nitecki M (ed) (1990) Evolutionary Innovations. University of Chicago Press, Chicago

Novikoff AB (1945) The Concept of Integrative Levels and Biology. Science 101: 209-215

Odell GM, Oster G, Alberch P, Burnside B (1981) The Mechanical Basis of Morphogenesis. I. Epithelial Folding and Invagination. Developmental Biology 85: 446-462

Odum EP (1971) Fundamentals of Ecology. Saunders, Philadelphia

Olding A (1978) A Defence of Evolutionary Laws. British Journal for the Philosophy of Science 29: 131-143

Osche G (1972) Evolution. Herder, Freiburg

Oyama S (1985) The Ontogeny of Information. Cambridge University Press, Cambridge

Oyama S (1988) Stasis, Development and Heredity. In: Ho M-W, Fox SW (eds) pp 255-274

Patel NH (1994) Developmental Evolution: Insights from Studies of Insect Segmentation. Science 266: 581-590

Paterson HEH (1985) The Recognition Concept of Species. In: Vrba ES (ed) pp 21-29

Pattee HH (1989) Simulations, Realizations, and Theories of Life. In: Langton CG (ed) pp 63-77

Patterson C (1982) Morphological Characters and Homology. In: Joysey KA, Friday AE (eds) pp 21-74

Patterson C (ed) (1987) Molecules and Morphology in Evolution: Conflict or Compromise? Cambridge University Press, Cambridge

Peano G (1921) La definizioni in matematica. Periodico di matematiche 1: 175-189

Peirce CS (1892-93) Scientific Metaphysics, vol VI. In: Hartshorne C, Weiss P (eds) Collected Papers. Harvard University Press, Cambridge (1935)

Peters RH (1991) A Critique for Ecology. Cambridge University Press, Cambridge

Petersen AF (1983) On Downward Causation in Biological and Behavioural Systems. History and Philosophy of the Life Sciences 5: 69-86

Pirlot P, Bernier R (1973) Preliminary Remarks on the Organ-Function Relation. In: Bunge M (ed) (1973) pp 71-83

Pittendrigh CS (1958) Adaptation, Natural Selection, and Behavior. In: Roe A, Simpson GG (eds) Behavior and Evolution. Yale University Press, New Haven, pp 390-416

Platnick NI (1977) Cladograms, Phylogenetic Trees, and Hypothesis Testing. Systematic Zoology 26: 438-442
Platnick NI (1979) Philosophy and the Transformation of Cladistics. Systematic Zoology 28: 537-546
Platnick NI (1985) Philosophy and the Transformation of Cladistics Revisited. Cladistics 1: 87-94
Platnick NI (1989) Cladistics and Phylogenetic Analysis Today. In: Fernholm B et al. (eds) pp 17-24
Pomiankowski A, Hurst LD (1993) Siberian Mice Upset Mendel. Nature 363: 396-397
Popper KR (1957a) The Poverty of Historicism. Routledge & Kegan Paul, London
Popper KR (1957b) Propensities, Probabilities, and the Quantum Theory. In: Miller D (ed) (1985) Popper Selections. Princeton University Press, Princeton, pp 199-206
Popper KR (1959) The Logic of Scientific Discovery. Hutchinson, London
Popper KR (1962) Conjectures and Refutations. Basic Books, New York
Popper KR (1972) Objective Knowledge. Clarendon Press, Oxford
Popper KR (1974) Darwinism as a Metaphysical Research Programme. In: Schilpp PA (ed) The Philosophy of Karl Popper, vol 1. Open Court, LaSalle, pp 133-143
Popper KR (1978) Natural Selection and the Emergence of Mind. Dialectica 32: 339-355
Popper KR, Eccles JC (1977) The Self and Its Brain. Springer, Berlin Heidelberg New York
Portin P (1993) The Concept of the Gene: Short History and Present Status. Quarterly Review of Biology 68: 173-223
Priestley J (1776) Disquisitions Relating to Matter and Spirit. Arno Press, New York (1975)
Prigogine I (1973) In Round Table with Ilya Prigogine: Can Thermodynamics Explain Biological Order? Impact of Science on Society 23: 159-179
Prior EW (1985) What Is Wrong with Etiological Accounts of Biological Function? Pacific Philosophical Quarterly 66: 310-328
Putnam H (1975) Philosophical Papers, 2 vols. Cambridge University Press, Cambridge
Putnam H (1983) Realism and Reason (Philosophical Papers, vol 3). Cambridge University Press, Cambridge
Putnam H (1994) Sense, Nonsense, and the Senses: An Inquiry into the Powers of the Human Mind. Journal of Philosophy 91: 445-517
Pylyshyn ZW (1984) Computation and Cognition. MIT Press, Cambridge

Raff RA, Parr BA, Parks AL, Wray GA (1990) Heterochrony and Other Mechanisms of Radical Evolutionary Change in Early Development. In: Nitecki M (ed) pp 71-98
Rasmussen S (1991) Aspects of Information, Life, Reality, and Physics. In: Langton CG et al. (eds) pp 767-773
Reig OA (1982) The Reality of Biological Species: A Conceptualistic and Systemic Approach. In: Cohen LJ et al. (eds) Logic, Methodology, and Philosophy of Science VI. North Holland Publishing, Amsterdam, pp 479-499
Rensch B (1971) Biophilosophy. Columbia University Press, New York
Rescher N (1987) Scientific Realism. A Critical Reappraisal. Reidel, Dordrecht
Resnik D (1994) The Rebirth of Rational Morphology: Process Structuralism's Philosophy of Biology. Acta Biotheoretica 42: 1-14

Richardson RC, Burian RM (1992) A Defense of Propensity Interpretations of Fitness. In: Hull DL, Forbes M, Okruhlik K (eds) PSA 1992, vol 1, pp 349-362
Ricklefs RE (1990) Ecology. Freeman, New York
Ridley M (1989) The Cladistic Solution to the Species Problem. Biology and Philosophy 4: 1-16
Riedl R (1980) Biologie der Erkenntnis. Paul Parey, Hamburg
Riedl R, Wuketits FM (eds) (1987) Die Evolutionäre Erkenntnistheorie. Bedingungen-Lösungen-Kontroversen. Paul Parey, Berlin
Rieger R, Michaelis A, Green MM (1991) Glossary of Genetics and Cytogenetics. Springer, Berlin Heidelberg New York
Rieppel O (1988) Fundamentals of Comparative Biology. Birkhäuser, Basel
Rieppel O (1990) Structuralism, Functionalism, and the Four Aristotelian Causes. Journal of the History of Biology 23: 291-320
Rizzotti M, Zanardo A (1986) Axiomatization of Genetics 1. Biological Meaning. Journal of theoretical Biology 118: 61-71
Robinson A (1965) Introduction to Model Theory and to the Metamathematics of Algebra. North-Holland Publishing, Amsterdam
Rosenberg A (1978) The Supervenience of Biological Concepts. Philosophy of Science 45: 368-386
Rosenberg A (1985) The Structure of Biological Science. Cambridge University Press, Cambridge
Rosenberg A (1987) Why Does the Nature of Species Matter? Comments on Ghiselin and Mayr. Biology and Philosophy 2: 192-197
Rosenberg A (1989) From Reductionism to Instrumentalism? In: Ruse M (ed) pp 245-262
Rosenberg A (1994) Instrumental Biology or the Disunity of Science. University of Chicago Press, Chicago.
Rosenblueth A, Wiener N, Bigelow J (1943) Behavior, Purpose, and Teleology. Philosophy of Science 10: 18-24
Roush W (1995) When Rigor Meets Reality. Science 269: 313-315
Ruse M (1969) Definitions of Species in Biology. British Journal for the Philosophy of Science 20: 97-119.
Ruse M (1971) Narrative Explanation and the Theory of Evolution. Canadian Journal of Philosophy 1: 59-74
Ruse M (1973) The Philosophy of Biology. Hutchinson University Library, London
Ruse M (1977) Karl Popper's Philosophy of Biology. Philosophy of Science 44: 638-661
Ruse M (1981) Species as Individuals: Logical, Biological, and Philosophical Problems. Behavioral and Brain Sciences 4: 299-300
Ruse M (1982) Teleology Redux. In: Agassi J, Cohen RS (eds) pp 299-309
Ruse M (1986) Taking Darwin Seriously. Basil Blackwell, Oxford
Ruse M (1987) Biological Species: Natural Kinds, Individuals, or What? British Journal for the Philosophy of Science 38: 225-242
Ruse M (1988) Philosophy of Biology Today. SUNY Press, Albany
Ruse M (ed) (1989) What the Philosophy of Biology Is. Kluwer, Dordrecht
Russell B (1918) The Philosophy of Logical Atomism. In: Marsh RC (ed) (1956) Logic and Knowledge. Allen & Unwin, London, pp 175-281

Saarinen E (ed) (1982) Conceptual Issues in Ecology. Reidel, Dordrecht

Salmon MH, Earman J, Glymour C, Lennox JG, Machamer P, McGuire JE, Norton JD, Salmon WC, Schaffner KF (1992) Introduction to the Philosophy of Science. Prentice Hall, Englewood Cliffs

Salmon WC (1989) Four Decades of Scientific Explanation. In: Kitcher P, Salmon WC (eds) pp 3-219

Salt GW (1979) A Comment on the Use of the Term Emergent Properties. American Naturalist 113: 145-148

Salthe SN (1985) Evolving Hierarchical Systems. Columbia University Press, New York

Sattler R (1986) Biophilosophy. Springer, Berlin Heidelberg New York

Savage CW (ed) (1978) Minnesota Studies in the Philosophy of Science, vol 9: Perception and Cognition. Issues in the Foundations of Psychology. University of Minnesota Press, Minneapolis

Schaffner KF (1969) The Watson-Crick Model and Reductionism. British Journal for the Philosophy of Science 20: 325-348

Schejter A, Agassi J (1982) Molecular Phylogenetics: Biological Parsimony and Methodological Extravagance. In: Agassi J, Cohen RS (eds) pp 333-356

Schmitt M (1987) 'Ecological Niche' sensu Günther and 'Ecological Licence' sensu Osche – Two Valuable but Poorly Appreciated Explanatory Concepts. Zoologische Beiträge NF 31: 49-60

Schoener TW (1986) Mechanistic Approaches to Community Ecology: A New Reductionism? American Zoologist 26: 81-106

Schoener TW (1989) The Ecological Niche. In: Cherrett JM (ed) Ecological Concepts. Blackwell, Oxford, pp 79-113

Schrödinger E (1944) What Is Life? Cambridge University Press, Cambridge

Schuh RT, Slater JA (1995) True Bugs of the World (Hemiptera: Heteroptera). Cornell University Press, Ithaca, NY

Schwartz SS (1981) Natural Kinds. Behavioral and Brain Sciences 4: 301-302

Sellars W (1963) Science, Perception and Reality. Routledge & Kegan Paul, London

Shrader-Frechette KS, McCoy ED (1993) Method in Ecology: Strategies for Conservation. Cambridge University Press, Cambridge

Siegel H (1987) Relativism Refuted: A Critique of Contemporary Epistemological Relativism. Reidel, Dordrecht

Simberloff D (1982) A Succession of Paradigms in Ecology: Essentialism to Materialism to Probabilism. In: Saarinen E (ed) pp 63-99

Simon MA (1971) The Matter of Life. Philosophical Problems of Biology. Yale University Press, New Haven

Simpson GG (1953) The Major Features of Evolution. Columbia University Press, New York

Simpson GG (1961) Principles of Animal Taxonomy. Columbia University Press, New York

Sites RW, Willig MR (1994a) Efficacy of Mensural Characters in Discriminating Among Species of Naucoridae (Insecta: Hemiptera): Multivariate Approaches and Ontogenetic Perspectives. Annals of the Entomological Society of America 87: 803-814

Sites RW, Willig MR (1994b) Interspecific Morphometric Affinities in *Ambrysus* (Hemiptera: Naucoridae). Proceedings of the Entomological Society of Washington 96: 527-532

Sklar A (1964) On Category Overlapping in Taxonomy. In: Gregg JR, Harris FTC (eds) pp 395-401
Slack JMW, Holland PWH, Graham CF (1993) The Zootype and the Phylotypic Stage. Nature 361: 490-492
Sloep PB (1986) Null Hypotheses in Ecology: Towards the Dissolution of a Controversy. In: Fine A, Machamer P (eds) PSA 1986, vol 1, pp 307-313
Sloep PB, van der Steen WJ (1987) The Nature of Evolutionary Theory: The Semantic Challenge. Biology and Philosophy 2: 1-15
Smart JJC (1963) Philosophy and Scientific Realism. Humanities Press, New York
Smith KC (1992a) The New Problem of Genetics: A Response to Gifford. Biology and Philosophy 7: 331-348
Smith KC (1992b) Neo-rationalism and Neo-Darwinism: Integrating Development and Evolution. Biology and Philosophy 7: 431-452
Sneath P, Sokal RR (1973) Numerical Taxonomy. Freeman, San Fancisco
Sober, E (1980) Evolution, Population Thinking, and Essentialism. In: Sober E (ed) (1994) pp 161-189
Sober E (1981) Evolutionary Theory and the Ontological Status of Properties. Philosophical Studies 40: 147-176
Sober E (1982) Why Logically Equivalent Predicates May Pick Out Different Properties. American Philosophical Quarterly 19: 183-189
Sober E (1984) The Nature of Selection. Evolutionary Theory in Philosophical Focus. MIT Press, Cambridge
Sober E (1991) Learning from Functionalism – Prospects for Strong Artificial Life. In: Langton CG et al. (eds) pp 749-765
Sober E (1992) Screening-Off and the Units of Selection. Philosophy of Science 59: 142-152
Sober E (1993) Philosophy of Biology. Westview Press, Boulder
Sober E (ed) (1994) Conceptual Issues in Evolutionary Biology. MIT-Press, Cambridge
Sober E, Lewontin RC (1982) Artifact, Cause and Genic Selection. Philosophy of Science 49: 157-180
Sober E, Wilson DS (1994) A Critical Review of Philosophical Work on the Units of Selection Problem. Philosophy of Science 61: 534-555
Sokal RR, Sneath P (1963) Principles of Numerical Taxonomy. Freeman, San Francisco
Solbrig OT, Solbrig DJ (1979) Introduction to Population Biology and Evolution. Addison-Wesley, Reading
Sommerhoff G (1950) Analytical Biology. Oxford University Press, London
Splitter LJ (1988) Species and Identity. Philosophy of Science 55: 323-348
Stent GS (1981) Strength and Weakness of the Genetic Approach to the Development of the Nervous System. In: Cowan WM (ed) Studies in Developmental Neurobiology. Oxford University Press, New York, pp 288-321
Sterelny K (1995) Understanding Life: Recent Work in Philosophy of Biology. British Journal for the Philosophy of Science 46: 155-183
Sterelny K, Kitcher P (1988) The Return of the Gene. Journal of Philosophy 85: 339-361
Sterelny K, Smith KC, Dickinson M (1996) The Extended Replicator. Biology and Philosophy 11: 377-403

Stidd BM, Wade DL (1995) Is Species Selection Dependent upon Emergent Characters? Biology and Philosophy 10: 55-76

Strong DR (1982) Null Hypotheses in Ecology. In: Saarinen E (ed) pp 245-259

Sudhaus W, Rehfeld K (1992) Einführung in die Phylogenetik und Systematik. Gustav Fischer, Stuttgart

Suppe F (1972) What's Wrong with the Received View on the Structure of Scientific Theories. Philosophy of Science 39: 1-19

Suppe F (ed) (1974) The Structure of Scientific Theories. University of Illinois Press, Urbana, IL

Suppe F (1989) The Semantic Conception of Theories and Scientific Realism. University of Illinois Press, Urbana

Suppes P (1957) Introduction to Logic. Van Nostrand Reinhold, New York

Suppes P (1961) A Comparison of the Meaning and Uses of Models in Mathematics and the Empirical Sciences. In: Freudenthal H (ed) The Concept and the Role of the Model in Mathematics and Natural and Social Sciences. Reidel, Dordrecht, pp 163-177

Tabony J (1994) Morphological Bifurcations Involving Reaction-Diffusion Processes During Microtubule Formation. Science 264: 245-248

Tarski A (1953) A General Method in Proofs of Undecidability. In: Tarski A, Mostowski A, Robinson RM, Undecidable Theories. North-Holland Publishing, Amsterdam, pp 1-36

Tarski A (1954-55) Contributions to the Theory of Models. Proceedings of the Academy of Sciences of the Netherlands, ser A, 57: 572-588; 58: 56-64

Taylor P (1992) Community. In: Keller EF, Lloyd EA (eds) pp 52-60

Teilhard de Chardin P (1964) The Phenomenon of Man. Harper, New York

Theiss J, Prager J, Streng R (1983) Underwater Stridulation by Corixids. Stridulatory Signals and Sound-Producing Mechanism in *Corixa dentipes* and *Corixa punctata*. Journal of Insect Physiology 29: 761-771

Thom R (1972) Stabilité structurelle et morphogenèse. WA Benjamin, Reading

Thom R (1983) Mathematical Models of Morphogenesis. Ellis Horwood, Chichester

Thompson D'AW (1917) On Growth and Form. Cambridge University Press, Cambridge

Thompson P (1983) The Structure of Evolutionary Theory: A Semantic Approach. Studies in History and Philosophy of Science 14: 215-229

Thompson P (1987) A Defense of the Semantic Conception of Evolutionary Theory. Biology and Philosophy 2: 26-32

Thompson P (1989) The Structure of Biological Theories. SUNY Press, Albany

Thomson KS (1988) Morphogenesis and Evolution. Oxford University Press, New York

Thorpe NO (1984) Cell Biology. Wiley, New York

Travisano M, Mongold JA, Bennett AF, Lenski RE (1995) Experimental Tests of the Roles of Adaptation, Chance, and History in Evolution. Science 267: 87-90

Truesdell C (1982) Our Debt to the French Tradition: "Catastrophes" and Our Search for Structure Today. Scientia 76: 63-77

Truesdell C (1984) An Idiot's Fugitive Essays on Science. Springer, Berlin Heidelberg New York

Tuomi J (1981) Structure and Dynamics of Darwinian Evolutionary Theory. Systematic Zoology 30: 22-31

Tuomi J (1992) Evolutionary Synthesis: A Search for the Strategy. Philosophy of Science 59: 429-438

Tuomivaara T (1994) On Idealization in Ecology. In: Kuokkanen M (ed) Idealization VII: Structuralism, Idealization and Approximation. Rodopi, Amsterdam, pp 217-241

van Brakel J (1992) Natural Kinds and Manifest Forms of Life. Dialectica 46: 243-261
van der Steen WJ (1983) Methodological Problems in Evolutionary Biology. I. Testability and Tautologies. Acta Biotheoretica 32: 207-215
van der Steen WJ (1991) Natural Selection as Natural History. Biology and Philosophy 6: 41-44
van der Steen WJ (1993) A Practical Philosophy for the Life Sciences. SUNY Press, Albany
van der Steen WJ (1994) New Ways to Look at Fitness. History and Philosophy of the Life Sciences 16: 479-492
van der Steen WJ (1996) Discussion: Screening-Off and Natural Selection. Philosophy of Science 63: 115-121
van der Steen WJ, Kamminga H (1991) Laws and Natural History in Biology. British Journal for the Philosophy of Science 42: 445-467
van der Steen WJ, Voorzanger B (1986) Methodological Problems in Evolutionary Biology. VII. The Species Plague. Acta Biotheoretica 35: 205-221
van der Weele C (1993) Metaphors and the Privileging of Causes. The Place of Environmental Influences in Explanations of Development. Acta Biotheoretica 41: 315-327
van Fraassen BC (1972) A Formal Approach to the Philosophy of Science. In: Colodny RG (ed) Paradigms & Paradoxes. University of Pittsburgh Press, Pittsburgh, pp 303-366
van Fraassen BC (1980) The Scientific Image. Clarendon Press, Oxford
Van Valen L (1976a) Individualistic Classes. Philosophy of Science 43: 539-541
Van Valen L (1976b) Domains, Deduction, the Predictive Method, and Darwin. Evolutionary Theory 1: 231-245
Van Valen L (1976c) Ecological Species, Multispecies, and Oaks. Taxon 25: 233-239
Varela FJ, Maturana HR, Uribe R (1974) Autopoiesis: The Organization of Living Systems, Its Characterization and a Model. BioSystems 5: 187-196
Vollmer G (1975) Evolutionäre Erkenntnistheorie. Hirzel, Stuttgart
Vollmer G (1983) Mesocosm and Objective Knowledge – On Problems Solved by Evolutionary Epistemology. In: Wuketits FM (ed) Concepts and Approaches in Evolutionary Epistemology. Reidel, Dordrecht, pp 69-121
Vollmer G (1985) Was können wir wissen? 2 vols. Hirzel, Stuttgart
Vollmer G (1987a) What Evolutionary Epistemology Is Not. In: Callebaut W, Pinxten R (eds) Evolutionary Epistemology: A Multiparadigm Program. Reidel, Dordrecht, pp 203-221
Vollmer G (1987b) The Status of the Theory of Evolution in the Philosophy of Science. In: Andersen S, Peacocke A (eds) Evolution and Creation. Århus University Press, Århus, pp 70-77
Vollmer G (1989) The Concept of Evolution as a Synthetic Tool in Science. Its Strengths and Limits. In: Koch WA (ed) The Nature of Culture. Brockmeyer, Bochum, pp 500-520
Vollmer G (1990) Against Instrumentalism. In: Weingartner P, Dorn GJW (eds) pp 245-259
Vollmer G (1995) Biophilosophie. Reclam, Stuttgart

von Bertalanffy L (1952) Problems of Life. Harper, New York
von Bertalanffy L (1968) General Systems Theory. Braziller, New York
von Helmholtz H (1873) Popular Lectures on Scientific Subjects. Longmans, Green, London
Vrba ES (1984) What Is Species Selection? Systematic Zoology 33: 318-328
Vrba ES (ed) (1985) Species and Speciation. Transvaal Museum, Pretoria (Transvaal Museum Monograph no 4)
Vrba ES, Eldredge N (1984) Individuals, Hierarchies and Processes: Towards a More Complete Evolutionary Theory. Paleobiology 10: 146-171
Vrba ES, Gould SJ (1986) The Hierarchical Expansion of Sorting and Selection: Sorting and Selection Cannot Be Equated. Paleobiology 12: 217-228

Wachbroit R (1994) Normality as a Biological Concept. Philosophy of Science 61: 579-591
Waddington CH (1970) Concepts and Theories of Growth, Development, Differentiation and Morphogenesis. In: Waddington CH (ed) Towards a Theoretical Biology, vol 3. Aldine, Chicago, pp 177-197
Wägele JW (1994) Review of Methodological Problems of 'Computer Cladistics' Exemplified with a Case Study on Isopod Phylogeny (Crustacea: Isopoda). Zeitschrift für zoologische Systematik und Evolutionsforschung 32: 81-107
Wägele JW, Wetzel R (1994) Nucleic Acid Sequence Data Are Not *per se* Reliable for Inference of Phylogenies. Journal of Natural History 28: 749-761
Walker B (1985) Can Ecosystems Evolve or Are They Merely Epiphenomena? In: Vrba ES (ed) pp 173-176
Walton D (1991) The Units of Selection and the Bases of Selection. Philosophy of Science 58: 417-435
Warren MA (1983) The Rights of the Nonhuman World. In: Hargrove EC (ed) (1992 pp 185-210
Wassermann GD (1981) On the Nature of the Theory of Evolution. Philosophy of Science 48: 416-437
Waters CK (1994) Genes Made Molecular. Philosophy of Science 61: 163-185
Webster G (1984) The Relations of Natural Forms. In: Ho M-W, Saunders PT (eds) pp 193-217
Webster G (1993) Causes, Kinds and Forms. Acta Biotheoretica 41: 275-287
Webster G, Goodwin BC (1982) The Origin of Species: A Structuralist Approach. Journal of Social and Biological Structures 5: 15-47
Weingartner P (1990) The Non-Statement View. A Dialogue between Socrates and Theaetetus. In: Weingartner P, Dorn GJW (eds) pp 455-465
Weingartner P, Dorn GJW (eds) (1990) Studies on Mario Bunge's *Treatise*. Rodopi, Amsterdam
Weiss PA (1970) Whither Life Science? American Scientist 58: 156-163
Weiss PA (1973) The Science of Life: The Living System - A System for Living. Futura Publishing, Mount Kisco
Wessells NK (1982) A Catalogue of Processes Responsible for Metazoan Morphogenesis. In: Bonner JT (ed) pp 115-154
West-Eberhard MJ (1992) Adaptation – Current Usages. In: Keller EF, Lloyd EA (eds) pp 13-18
Weston T (1992) Approximate Truth and Scientific Realism. Philosophy of Science 59: 53-74

Whewell W (1847) Philosophy of the Inductive Sciences, 2 vols. Johnson Reprint, New York (1967)

Whitehead AN (1929) Process and Reality. Macmillan, New York (1969)

Wiley EO (1975) Karl. R. Popper, Systematics, and Classification: A Reply to Walter Bock and Other Evolutionary Taxonomists. Systematic Zoology 24: 233-243

Wiley EO (1978) The Evolutionary Species Concept Reconsidered. Systematic Zoology 27: 17-26

Wiley EO (1980) Is the Evolutionary Species Fiction? – A Consideration of Classes, Individuals, and Historical Entities. Systematic Zoology 29: 76-80

Wiley EO (1981) Phylogenetics. The Theory and Practice of Phylogenetic Systematics. Wiley, New York

Wiley EO (1989) Kinds, Individuals, and Theories. In: Ruse M (ed) (1989) pp 289-300

Wilkerson TE (1993) Species, Essences and the Names of Natural Kinds. Philosophical Quarterly 43(170): 1-19

Williams GC (1966) Adaptation and Natural Selection. Princeton University Press, Princeton

Williams GC (1992a) Natural Selection. Domains, Levels, Challenges. Oxford University Press, New York

Williams GC (1992b) GAIA, Nature Worship and Biocentric Fallacies. Quarterly Review of Biology 67: 479-486

Williams MB (1970) Deducing the Consequences of Evolution. Journal of theoretical Biology 29: 343-385

Williams MB (1973a) The Logical Status of the Theory of Natural Selection and Other Evolutionary Controversies. In: Bunge M (ed) pp 84-102

Williams MB (1973b) Falsifiable Predictions of Evolutionary Theory. Philosophy of Science 40: 518-537

Williams MB (1981) Similarities and Differences Between Evolutionary Theory and the Theories of Physics. In: Asquith PD, Giere RN (eds) pp 385-396

Williams MB (1985) Species Are Individuals: Theoretical Foundations for the Claim. Philosophy of Science 52: 578-590

Williams MB (1986) The Logical Skeleton of Darwin's Historical Methodology. In: Fine A, Machamer P (eds) PSA 1986, vol 1, pp 514-521

Williams PA (1992) Confusion in Cladism. Synthese 91: 135-152

Willmann R (1985) Die Art in Raum und Zeit. Paul Parey, Berlin

Wilson BE (1995) A (Not-So-Radical) Solution to the Species Problem. Biology and Philosophy 10: 339-356

Wilson DS (1989) Levels of Selection: An Alternative to Individualism in Biology and the Human Sciences. In: Sober E (ed) (1994) pp 143-154

Wilson DS, Sober E (1989) Reviving the Superorganism. Journal of theoretical Biology 136: 337-356

Wimsatt WC (1972) Teleology and the Logical Structure of Function Statements. Studies in History and Philosophy of Science 3: 1-80

Wimsatt WC (1982a) Reductionistic Research Strategies and Their Biases in the Units of Selection Controversy. In: Saarinen E (ed) pp 155-201

Wimsatt WC (1982b) Randomness and Perceived-Randomness in Evolutionary Biology. In: Saarinen E (ed) pp 279-322

Wimsatt WC (1986) Developmental Constraints, Generative Entrenchment, and the Innate-Acquired Distinction. In: Bechtel W (ed) pp 185-208

Wittgenstein L (1922) Tractatus Logico-Philosophicus. Routledge & Kegan Paul, London (1951)

Wolff C (1740) Philosophia Rationalis Sive Logica, Pars II. In: École J (ed). Georg Olms, Hildesheim (1983)

Wolpert L (1991) The Triumph of the Embryo. Oxford University Press, Oxford

Woodfield A (1976) Teleology. Cambridge University Press, Cambridge

Woodger JH (1929) Biological Principles. Routledge & Kegan Paul, London, Humanities Press, New York (1967)

Woodger JH (1952) Biology and Language. Cambridge University Press, Cambridge

Woodward J (1989) The Causal Mechanical Model of Explanation. In: Kitcher P, Salmon WC (eds) pp 357-383

Wouters A (1995) Viability Explanation. Biology and Philosophy 10: 435-457

Wright L (1973) Functions. Philosophical Review 82: 139-168

Wright L (1976) Teleological Explanations. University of California Press, Berkeley

Wuketits FM (1983) Biologische Erkenntnis: Grundlagen und Probleme. Gustav Fischer, Stuttgart

Wuketits FM (1989) Organisms, Vital Forces, and Machines: Classical Controversies and the Contemporary Discussion 'Reductionism vs. Holism'. In: Hoyningen-Huene P, Wuketits FM (eds) pp 3-28

Young BA (1993) On the Necessity of an Archetypical Concept in Morphology: With Special Reference to the Concepts of "Structure" and "Homology". Biology and Philosophy 8: 225-248

Zanardo A, Rizzotti M (1986) Axiomatization of Genetics 2. Formal Development. Journal of theoretical Biology 118: 145-152

Zandee M, Geesink R (1987) Phylogenetics and Legumes: A Desire for the Impossible? In: Stirton CH (ed) Advances in Legume Systematics, Part 3. Royal Botanic Gardens, Kew, pp 131-167

Zeki S (1993) A Vision of the Brain. Blackwell, Oxford

Zylstra U (1992) Living Things As Hierarchically Organized Structures. Synthese 91: 111-133.

Name Index

Subject Index

Springer-Verlag and the Environment

We at Springer-Verlag firmly believe that an international science publisher has a special obligation to the environment, and our corporate policies consistently reflect this conviction.

We also expect our business partners – paper mills, printers, packaging manufacturers, etc. – to commit themselves to using environmentally friendly materials and production processes.

The paper in this book is made from low- or no-chlorine pulp and is acid free, in conformance with international standards for paper permanency.